al

Notes in
athematics

siam

Philadelphia 1987

Mathematical Modelling:

Classroom Notes in Applied Mathematics

Edited by Murray S. Klamkin

To Irene, and not only for her patience and assistance in the preparation of
this monograph.

Library of Congress Catalog Card Number 86-60090
ISBN 0-89871-204-1

The Doctrine of the Whole Man*

Mathematics has elements that are spatial, kinesthetic, elements that are arithmetic or algebraic, elements that are verbal, programmatic. It has elements that are logical, didactic and elements that are intuitive, or even counter-intuitive. It has elements that are related to the exterior world and elements that seem to be self generated. It has elements that are rational and elements that are irrational or mystical. These may be compared to different modes of consciousness.

To place undue emphasis on one element or group of elements upsets a balance. It results in an impoverishment of the science and represents an unfulfilled potential. The doctrine of the whole man says that we must bring everything we have to bear on our subject. We must not block off arbitrarily any mode of experience or thought. "There is a Nemesis," says Whitehead, "which waits upon those who deliberately avoid avenues of knowledge."

We must realize that the future of the subject depends only in part on the contribution of those who have rigid establishment interest or training in the subject. As regards this training and our own teaching we must

restore geometry,
restore kinesthetics and mechanics,
restore combinatorics,
restore intuitive and experimental mathematics,
deemphasize somewhat the theorem-proof type of lecturing,
give a proper place to computing and programmatics,
make full use of computer graphics,
eliminate the fear of metaphysics, recognizing that in such principles may lie seeds of future growth.

What we want to do is to create as rich and diverse a brew of thought and action as we can. This is the kind of culture which has fostered mathematics in the past and should be our very present hope for the future.

*From P. J. Davis and J. A. Anderson *Nonanalytic aspects of mathematics and their implications for research and education*, SIAM Rev., (1979), pp. 112-125.

Preface

"Classroom Notes in Applied Mathematics" was started in the *SIAM Review* in 1975 to offer classroom instructors and their students brief, essentially self-contained applications of mathematics that could be used to illustrate the relevance of mathematics to current disciplines in a teaching setting. The section has focused on modern applications of mathematics to real world problems; however, classical applications were also welcomed.

It was the plan from the beginning to gather the Notes into a collection to be published by SIAM. The present volume is the result. The only changes that have been made to the original Notes as they appeared in the *SIAM Review* have been to correct some misprints and to add a number of author postscripts. I have also included a small selection of applied problems, almost all with solutions, from the "Problems and Solutions" section of the Review when I felt them to be particularly illustrative of the material under discussion.

The Notes have been classified into two broad sections: (I) Physical and Mathematical Sciences, and (II) Life Sciences. These two sections have been further divided into subsections. At the end of each section or subsection I have added a list of supplementary references. Although not meant to be exhaustive, the listings should provide the instructor and the student with considerable resource and background material in many fields of applications. I must confess to the reader that there is a certain arbitrariness in the classification of the Notes in the table of contents. At times this is because I had a paucity of Notes in certain fields, but often it is simply due to the multifaceted character of the Notes themselves. I regret the fact that we had no Notes submitted in the behavioral sciences and, therefore, for readers who are also inclined to this broad field, I have included an unsolved problem and a list of references as an appendix to the volume. Because there is only one Note pertaining to music, entitled "Optimal temperment," it is included in the section on Optimization. However, it has its own list of supplementary references in the field of music. Similarly, since the three Notes pertaining to sports deal with probabilities, they are included separately under Stochastic Models but have their own list of supplementary references in sports. Similar treatments are accorded to some other sets of Notes.

Many of the references were selected deliberately from a group of journals that are easily accessible or that I consider to be well written. These include papers, both recent and non-recent, from the *American Mathematical Monthly* (AMM), *American Journal of Physics, American Scientist, Bulletin of the Institute of Mathematics and Its Applications* (BIMA), *Mathematical Gazette, Mathematics Magazine* (MM), *Scientific American,* and the *SIAM Review.* A number of the references contain their own large bibliographies; for example, "The Flying Circus of Physics" by J. Walker has 1,630 citations.

Naturally, many of the references reflect my own interests, in particular, in mechanics and applied geometry. I believe that mathematical models from mechanics offer a number of advantages in a classroom setting: they can be easily interpreted in physical terms, their analysis often does not require formidable mathematics, and (for instance, when couched in terms of sports) they have wide appeal. Geometric models often have the same potential advantages, but they are usually bypassed due to the unfortunate neglect of geometry in secondary schools and colleges in the United States and Canada, a situation that contrasts sharply with, for example, the state of geometry education in Hungary and the Soviet Union. In 1971, I had voiced some criticisms (not new) concerning the deplorable state of our geometry education (See *On the ideal role of an industrial mathematician and its educational implications*, Amer. Math. Monthly, (1971), pp. 53–76; reprinted with additional footnotes in Educ. Studies Math., (1971), pp. 244–269). That these criticisms are still being raised is shown by the recent papers of M. Atiyah (*What is geometry?* Math. Gazette, (1982), pp. 179–185) and P. J. Davis and J. A. Anderson (*Nonanalytic aspects of mathematics and their implications for research and education*, SIAM Rev. (1979), pp. 112–125), as well as the earlier work of F. J. Weyl (*The perplexing case of geometry*, SIAM Rev., (1962), pp. 197–201). To encourage renewed attention to geometry, I have included a large list of applied geometry references.

I wish to thank all the contributors and all the respective referees over the past 12 years for their concern and involvement with applied mathematics in the classroom setting and their willingness to develop and present examples in a useful way. I would also like to thank SIAM's editorial and production staff, in particular, Meredith Allen, Alison A. Anderson, and Claire Tanzer, for their help in polishing and putting this volume in final form.

Contents

Part I   Physical and Mathematical Sciences

1. Mechanics

Appendix

1. An Unsolved Problem in the Behavioral Sciences

Introduction

For the beginning of this monograph I feel it is fitting, especially for instructors and students who are new or relatively new to "Mathematical Modelling," to describe the process. I do so by reprinting a short note from the journal, Mathematical Modelling 1(1980)63–69. Instructors using any of the subsequent notes for the classroom should be prepared to augment the material as well as to delineate the six stages of the process as described in this first introductory note.

MATHEMATICAL MODELLING:
DIE CUTTING FOR A FRESNEL LENS*

M. S. KLAMKIN†
Communicated by E. Y. Rodin

Abstract—The various stages in the mathematical modelling of real problems are first discussed briefly and then illustrated stepwise with the problem of cutting a die for the lenses of automobile stoplights. This requires the determination of the shape of a groove cut into a rotating metal blank as a function of the shape of a cross-section of a rotating cutting tool.

1. INTRODUCTION

Mathematical modelling has become more and more visible in course offerings at various departments of mathematics in our universities and more books and papers are appearing in this field (see [1–14]). The reasons are not only economic but also that mathematics is being used in more fields than heretofore. It is thus fitting, in the inaugural issue of this journal, to give a specific example of *mathematical modelling* from the very beginning where a more or less imprecise question is raised to a final answer of the question made more precise. The problem treated here, the cutting of a die for manufacturing Fresnel lenses for automobile stoplights arose when I was with the Ford Motor Company Scientific Research Laboratories and except for some brief mention of the problem at some colloquia [13], it has only appeared as an internal report.

To set the stage for *mathematical modelling* or equivalently "the applied mathematician at work," I now briefly include some aspects of the process which has been presented in more detail elsewhere [6, 13, 15].

Joe Keller once defined *applied mathematics* as that science of which pure mathematics is a branch. Although this is a sort of tongue-in-cheek definition, it does make the point that to do applied mathematics, you have to know things other than mathematics. Related to this, Sydney Goldstein once said that to be a good applied mathematician, you have to be one-third physicist, one-third engineer and one-third mathematician. At present, since mathematics is being applied much more broadly than then, we would have to have different mixes.

A better description of applied mathematics had been given by John L. Synge. He noted that applying mathematics to a real problem involves three stages. The first stage is a dive from the world of reality into the world of mathematics. The second stage is a swim in the world of mathematics. The third stage is a climb back from the world of mathematics into the world of reality and, importantly, with a prediction in your teeth. With respect to this analogy, one of the criticisms of mathematics teaching is that too much time is spent almost exclusively to the middle stage, the swim in the ocean of mathematics. This stage is certainly important but insofar as applications of mathematics in the real world are concerned, the first and third stages are equally important. And it is because these stages have not been stressed as much as the second stage, that there arre difficulties in applying mathematics.

Other mathematicians such as Crank [16], Pollak [15], Sutton [17], Ver Planck and Teare [18] amplified the above three stages into five or more stages. In terms of one word descriptions of Henry Pollak, we have:

*Copyright 1980, Pergamon Press, Ltd. Reprinted from *Mathematical Modelling*, 1, (1980), pp. 63–70 with permission of the publisher. Received by the editors December 1979.
†University of Alberta, Edmonton, Alberta, Canada.

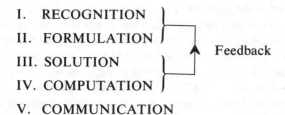

I.   RECOGNITION
II. FORMULATION
III. SOLUTION   Feedback
IV. COMPUTATION
V.   COMMUNICATION

While there are no sharp boundaries between these stages, they do catch the mathematician in different attitudes.

I.   First of all one has to *recognize* that there is a problem and what the problem is.
II. Once the problem is recognized and determined, there must be a mathematical *formulation* of the problem.
III. After the formulation, one has to obtain a mathematical *solution* of the problem which usually will be an approximation.
IV. The mathematical solution will usually involve *computation* and will require the use of computers.

   Coupled with the last two stages, there should be feedback to the first two stages to ensure that the problem actually being solved is indeed the problem which was supposed to be solved and not some other one, albeit interesting, that is related to it.

V.   Finally, there is the last stage of *communication.* If the solution is not communicated well to the persons needing the solution, all the previous work is to no avail, except possibly for publication in some journal. Unfortunately, the "publish or perish" syndrome is still with us.

   In regard to the first two stages of *recognition* and *formulation,* Synge [19], back in 1944, had this to say about the purist mathematicians (this is less valid now):

Nature will throw out mighty problems but they will never reach the mathematician. He will sit in his ivory tower waiting for the enemy with an arsenal of guns, but the enemy will never come to him. Nature does not offer her problems ready formulated. They must be dug up by pick and shovel, and he who will not soil his hands will never see them.

Also, there is a big difference in attitude between an applied and a pure mathematician. For instance, a pure mathematician can work on a given problem for years, and if it turns out to be intractable, he can alter the problem. To paraphrase a line from *Finian's Rainbow,* if a mathematician cannot handle the problem on hand, he'll handle the problem out of hand. The applied mathematician when he is given a problem must handle the particular problem on hand; his job may depend on it. First of all, he'll have to check to see if the problem that's given to him is really the problem that they're trying to solve, because many times the problem they give to you is not really the problem they want. He must come up with an answer, usually an approximation, that is optimal, in a sense, with regard to cost and time. You can't work on the problem for five years unless it's a very important problem to the company, and usually they're not that important. Finally, if he's successful he must communicate his results in such a way that they're understandable to his superiors.

   It is not time for me to discuss the problem of the Fresnel lens in relation to the above five stages.

2. RECOGNITION: DIE CUTTING FOR A FRESNEL LENS

   If you look at the brake lights on automobiles you will see them covered by red lenses. These are Fresnel lenses designed to concentrate the light emitted by the brake lights. A number of years ago, the U.S. government decided that whatever the light intensity was at various points behind the brake lights when lit, they would all have to be increased considerably; they figured it would be safer. One simple response to this new requirement would be to just increase the candle power output of the brake bulbs. However, this is not an efficient solution since car batteries are already loaded with lots of periferal equipment. To do a better job, the optical engineers spend a lot of effort redesigning the Fresnel lenses by means of ray tracings to do a more efficient job of concentrating the light. The problem now was to make a die from which to form the plastic lenses. Since Ford Motor Company does not make everything it needs, they gave the design of the Fresnel lenses to an outside vendor who was supposed to make the corresponding die. Subsequently, the vendor came back to the company and asked for a relaxation on the tolerances for several of the inner most grooves of the lens since otherwise they could not make the die. The optical engineers were not sure why the inner grooves could not be cut to their specified tolerances. One of them, R. M. Ferrell, once heard me give a lecture on solid geometry at a sectional M.A.A. meeting in Minnesota and figuring correctly that this was indeed a solid geometry problem, came to me and simply asked why the inner grooves could not be cut to the specified tolerances. I did not know what the reason was at that time or even after thinking about the problem for several days. Consequently, I called Ferrell back and told him I would like some samples of the brake light lenses as well as any description of the cutting process he could give me. He produced the samples as well as some description of the cutting process. I was then able to carry out the next stage.

3. FORMULATION

The grooves to be cut into the die are to be surfaces of revolution about a common axis. A cross section, containing the axis, of a typical groove is to be bounded by a straight line and a circular arc as indicated in Figure 1 (note that only one of the many grooves are shown in the figure).

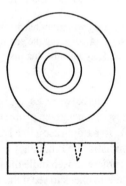

Fig. 1.

The grooves are to be cut out of a rotating solid metal blank by means of a rotating tool which is a body of revolution. Schematically, we have Figure 2 for half the cross-section (by a plane containing the axis of the cutting tool and the axis of the blank) of the cutting tool (the cross-hatched region) and the metal blank. The cutting tool rotates about the axis PP''. The surface of the metal blank corresponds to TOX after the groove has been completely cut out by gradually moving the rotating blank up into the cutting tool.

At this stage, I still did not know why the inner grooves could not be cut to the desired tolerances. I then had a fruitful discussion about the problem with my then immediate supervisor,

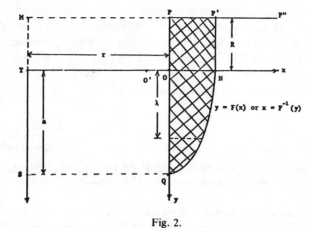

Fig. 2.

A. O. Overhauser, a physicist and head of the Physical Sciences Department. He noted that even if the cross-section of the cutting tool was an infinitesimally thin blade, the segment OQ, a groove would still be cut out and consisting of an inner cylindrical and an outer toroidal surface. Note that as POQ rotates, the distance of point Q from the axis TS of the metal blank increases monotonically from a minimum value when POQ is perpendicular to the blank to a maximum value when Q is on the surface of the blank. A cross-section of this groove is given in Figure 3. Here, $OA=[r^2+(R+a)^2-R^2]^{\frac{1}{2}}-r$ and QA is an arc of a circle centered at M (the intersection of the two axes).

It now finally dawned on me where the difficulty lay. The cross-section of the cutting tool, for any particular groove, was designed to be the same as the cross-section of that groove. However, the groove actually cut out by the cutting tool need not exactly correspond to the cross-section of the cutting tool as indicated above by cutting with a line segment. The mathematical problem here was to determine the shape of the actual groove cut out as a function of the shape of the cutting tool. Then one could determine the distortion, if any, in the groove that was actually cut. So here, finally, was a clear mathematical problem in solid geometry but again I was stymied for a while. Although, I am reasonably competent in solid geometry problems now, I think my early training in geometry was much at fault. I never had any solid geometry until my 8th term in high school and that was of the formal axiomatic type which has been roundly and soundly criticized in the plane geometry case. Nevertheless, I was still fortunate since the vast majority of high school students never got any solid geometry. The late Professor Steenrod had taught elementary calculus courses to see what they involved. He came to the conclusion

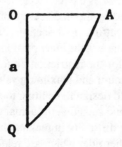

Fig. 3.

that in calculus, there is lots of geometry, but unfortunately the students get little out of it because due to limitations of time and everything else, the instructor essentially does all the geometry and leaves just the analysis (if that) for the student. Then when they get answers, the reinterpretation back into geometric form is again done by the instructor [20]. Also, when I worked for AVCO Research and Advanced Development Division, I found that one of the biggest failings in setting up problems by the engineers and scientists was their rudimentary background in geometry [6].

4. SOLUTION

The shape of the groove actually cut out can be determined mathematically in two steps:

1. We rotate the cutting cross-section PQP' about axis PP'' keeping the flank fixed in its final position. This produces a cut out groove in the blank (which does not go all the way around the axis of the blank).

2. We find the envelope of the groove by rotating the cut out blank about its axis ST.

As with the cutting with a line segment, the width of the groove cut out by the actual cutting tool at depth $y=\lambda$ could be greater than that by the width of the cutting tool at that depth [i.e., $F^{-1}(\lambda)$]. This would be due to some point of the cross-section of the cutting tool below $y=\lambda$ being further from the axis ST after the cross-section has rotated such that the point is also at depth $y=\lambda$. Thus, the cross-section actually cut out is obtained by taking the envelope of circles drawn with M as a center and passing through each point of the curve QN of the cutting tool cross-section. Analytically, the width $w(\lambda)$ of the cross-section of the groove cut out at depth λ is given by

$$w(\lambda)=\max\left\{F^{-1}(\lambda),\ \max\left([x+r]^2+[F(x)+R]^2-[\lambda+R]^2\}^{\frac{1}{2}}-r\right)\right\}$$

$$0\leq x\leq F^{-1}(\lambda).$$

The maximum of the second term in the parenthesis occurs for that point (x,y) on the curve QN, in the given range, which is furthest from M. If the curve QN is such that Q is the furthest point, then the cross-section of the groove cut out is that given by Figure 3. If the distance from M to points of QN increases monotonically from Q to N, then the cross-section cut out will exactly correspond to the cross-section of the cutting tool (OQN). If the distance from M to points of QN first increases monotonically from W to N' (an interior point) and then decreases monotonically from N' to N, the other curve of the cross-section cut out will consist of QN' plus a circular arc from N' to the x-axis with center at M.

In order that the cross-section of the groove cut out corresponds to the cross-section of the cutting tool, the value of R cannot exceed a certain maximum value. There is also a lower limit imposed on R by the practical consideration of clearance for the mounting of the cutting tool. A simple way of determining the maximum R is given in Figure 4. QN is an arc of a circle centered at C. It is assumed that the distance from T to points of QN increases monotonically from Q to N. Then max $R=M'T$ where M' is the intersection of CN with ST. If C falls on the left hand side of ST, then M' is determined by the intersection of line QC with ST.

Actually, the point O of the cutting cross-section (Fig. 2) will fall at O' where OO' is small. However, this only changes the shape of the inner portion of the groove to a conical surface which is desired and not our analysis for the restriction on R.

If it turns out that the minimum and maximum values of R are incompatible (here C falls below the x axis), then one could design the cutting tool to have a cross-section different than the cross-section of the groove and such that the actual groove cut out is the one we want. This would produce a manufacturing difficulty in making the cutting tools since the surface would no longer be cylindrical and spheroidal which are relatively easy to make.

Before we carry out the next stage, *Computation*, we need some actual data. *Data* should

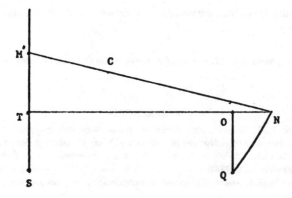

Fig. 4.

have been included as a stage between *Recognition* and *Formulation* as was rightly pointed out by Peter Hilton. This is an important and possibly difficult stage. The data could be experimental, statistical or both. How and when to stop collecting data and their subsequent treatment is a subject in itself. As an indication of possible pitfalls in collecting statistical data, we have the following two classic cases:

Back in the days of the Roosevelt-Landon presidential election, the *Literary Digest*, one of the top circulation magazines of that era, took a very large poll of 2,000,000 people by means of random phone calls across the country. The sampling indicated a landslide for Landon whereas actually there was a landslide for Roosevelt. Subsequently, the magazine went out of business since it was felt that the prediction was a dishonest attempt to influence the election. There was no overt dishonesty. It was a case of a heavily biased sample despite it being large. At that time, people who had telephones were likely to be wealthy and wealthy people were likely to be Republicans who would vote for Landon.

For the second sample, there was a city that bragged about its cultured inhabitants as indicated by the very high attendance at its museum by actual daily count. Soon after a small building was built across the street from the museum, the museum attendance dropped off sharply. What was the small building? (It was a public lavatory!)

5. COMPUTATION

From the actual dimensions given for the grooves, it turned out that

$$\max R = M'T = 0.56 \text{ in.}$$

Since the shaft of the cutting tool has to be reasonably hefty, otherwise there would be chatter during the cutting process, its diameter could be taken as 0.5 in. This leaves a clearance of 0.06 in. which although quite small is still adequate.

The machinists of the outside vendor, from their practical experience, felt that there would not be enough clearance to cut the inner grooves without distortion. And they were very nearly correct.

6. COMMUNICATION

The above analysis was transmitted in my report to the Body Light Section with a conclusion in English stating that there would be enough of a clearance, using a shaft diameter of the cutting tool as 0.5 in., such that the grooves could be cut to the tolerances required. This was then transmitted back to the vendor who went ahead and cut the die to the prescribed tolerances.

The example described was fortunately one of those rather "clean" ones after their formulation. Frequently, this is not the case and one will have to make various approximations before getting a final answer.

Acknowledgement—I am grateful to R. M. Ferrell and A. O. Overhauser for fruitful discussions which led to the final resolution of this problem.

REFERENCES

[1] J. G. ANDREWS AND R. R. McLONE, eds., *Mathematical Modelling*. Butterworths, London (1976).

[2] E. A. BENDER, *An Introduction to Mathematical Modeling*. Wiley-Interscience, New York (1978).

[3] R. BELLMAN AND G. BORG, *Mathematics, systems and society: An informal essay*. Tech. Rep. 70–58, Univ. of Southern California, L.A. (1970).

[4] R. E. GASKELL AND M. S. KLAMKIN, The industrial mathematician views his profession. *Amer. Math. Monthly* **81**, 699–716 (1974).

[5] R. HABERMAN, *Mathematical Models in Mechanical Vibrations, Population Dynamics and Traffic Flow.* Prentice-Hall, Englewood Cliffs, N.J. (1977).

[6] M. S. KLAMKIN, On the ideal role of an industrial mathematician and its educational implications. *Amer. Math. Monthly* **78**, 53–76 (1971); reprinted in *Educational Studies in Mathematics* **3**, 244–269 (1971), with additional footnotes.

[7] P. LANCASTER, *Mathematics, Models of the Real World*. Prentice-Hall, Englewood Cliffs, N.J. (1976).

[8] C. C. LIN AND L. A. SEGAL, *Mathematics Applied to Deterministic Problems in the Natural Sciences*, MacMillan, New York (1974).

[9] E. W. MONTROLL AND W. W. BADGER, *Introduction to Quantitative Aspects of Social Phenomenon*. Gordon and Breach, New York, (1974).

[10] B. NOBLE, *Applications of Undergraduate Mathematics in Engineering*. MacMillan, New York, (1967).

[11] H. POLLARD, *Applied Mathematics*. Addison-Wesley, Reading, Mass. (1972).

[12] F. S. ROBERTS, *Discrete Mathematical Models*. Prentice-Hall, Englewood Cliffs, N.J. (1976).

[13] The Mathematical Training of the Non-Academic Mathematician, *SIAM Review*, **17**, 541–555 (1975).

[14] *The Role of Applications in the Undergraduate Mathematics Curriculum*, National Academy of Sciences, 1979.

[15] H. O. POLLAK, Mathematical research in the communications industry. *Pi Mu Epsilon Jour.* **2**, 494–6 (1959).

[16] J. CRANK, *Mathematics and Industry.* Oxford University Press, London (1962).

[17] G. SUTTON, *Mathematics in Action*. G. Bell, London. (1954).

[18] D. W. VER PLANCK AND B. R. TEARE, *Engineering Analysis*. Chapman and Hall, London (1954).

[19] J. L. SYNGE, Focal properties of optical and electromagnetic systems. *Amer. Math. Monthly* **51**, 185–200 (1944).

[20] N. STEENROD, The geometric content of freshman and sophomore mathematics courses. *CUPM Geometry Conf. Report II*, 1–52 (1967).

General Supplementary References

The following supplementary list of references are to books and papers dealing with Mathematical Modelling in general as well as pedagogical papers on or related to the subject. Quite a few of these books are good sources of a variety of mathematical models, from elementary to advanced.

[1] J. L. AGNEW AND M. S. KEENER, *"A case-study course in applied mathematics using regional industries,"* AMM (1980) pp. 55–59.
[2] E. ALARCON AND C. A. BREBBIA, *Applied Numerical Modelling,* Pentech Press, Plymouth, 1979.
[3] R. S. ANDERSSEN AND F. R. DEHOOG, *The Application of Mathematics in Industry,* Nijholl, 1982.
[4] R. ARIS, *Mathematical Modelling Techniques,* Pitman, London, 1978.
[5] R. ARIS AND M. PENN, *"The mere notion of a model,"* Math. Modelling (1980) pp. 1–12.
[6] M. ATIYAH, *"What is geometry?"* Math. Gaz. (1982) pp. 179–185.
[7] X. J. R. AVULA, ed., *Proceedings of the First International Conference on Mathematical Modeling I, II, III, IV, V,* University of Missouri-Rolla, 1977.
[8] X. J. R. AVULA, R. BELLMAN, Y. L. LUKE AND A. C. RIGLER, eds., *Proceedings of the Second International Conference on Mathematical Modelling I, II,* University of Missouri-Rolla, 1979.
[9] X. J. R. AVULA, R. E. KALMAN, A. I. LIAPIS AND E. Y. RODIN, eds., *Mathematical Modelling in Science and Technology,* Pergamon, N.Y., 1984.
[10] K. S. BANERJEE, *Weighing Designs for Chemistry, Medicine, Economics, Operations Research, Statistics,* Marcel Dekker, N.Y., 1975.
[11] E. H. BAREISS, *"The college preparation for a mathematician in industry,"* AMM (1972) pp. 972–984.
[12] E. F. BECKENBACH, ed., *Modern Mathematics for Engineers,* McGraw-Hill, York, Pa., 1956.
[13] ———, *Modern Mathematics for Engineers,* 2nd series, McGraw-Hill, York, Pa., 1961.
[14] R. BELLMAN, *Some Vistas of Modern Mathematics,* University of Kentucky Press.
[15] R. BELLMAN AND G. BORG, *Mathematics, Systems and Society,* Report No. 2, Comm. on Research, Swedish Natural Science Research Council, Stockholm, 1972.
[16] E. A. BENDER, *"Teaching applicable mathematics,"* AMM (1973) pp. 302–306.
[17] J. BERRY, et al, *Solving Real Problems with Mathematics I, II, Solving Real Problems with C.S.E. Mathematics,* CIT Press, 1981–1983.
[18] M. A. BIOT, *"Applied mathematics, an art and a science,"* Jour. Aero. Science (1956) pp. 406–411, 489.
[19] G. D. BIRKHOFF, *"The mathematical nature of physical theories,"* Amer. Sci. (1943) 281–310.
[20] V. BOFF AND H. NEUNZERT, eds., *Applications of Mathematics in Technology,* Teubner, Stuttgart, 1984.
[21] F. F. BONSALL, *"A down-to-earth view of mathematics,"* AMM (1982) pp. 8–15.
[22] W. E. BOYCE, ed., *Case Studies in Mathematical Modeling,* Pitman, London, 1981.
[23] R. BRADLEY, R. D. GIBSON AND M. CROSS, *Case Studies in Mathematical Modelling,* Halsted Press, Great Britain, 1981.
[24] M. BRAUN, *Differential Equations and their Applications,* Springer-Verlag, Heidelberg, 1983.
[25] C. A. BREBBIA AND J. J. CONNOR, eds., *Applied Mathematical Modeling,* IPC Science and Technology Press, Guilford, England, 1977.

[26] F. E. BROWDER, "The relevance of mathematics," AMM (1976) pp. 249–254.

[27] D. N. BURGHES, "Mathematical modelling: A positive direction for the teaching of applications of mathematics at school," Educ. Studies in Math. (1980) pp. 113–131.

[28] D. N. BURGHES AND M. S. BORRIE, Modelling with Differential Equations, Horwood, Great Britain, 1981.

[29] D. N. BURGHES, I. HUNTLEY AND J. McDONALD, Applying Mathematics: A Course in Mathematical Modelling, Horwood Ltd., Great Britian. 1982.

[30] R. S. BURINGTON, "On the nature of applied mathematics," AMM (1949) pp. 221–242.

[31] ——, "The problem of formulating a problem," Proc. Amer. Philos. Soc. (1960) pp. 429–443.

[32] H. BURKHARDT, "Learning to use mathematics," BIMA (1979) pp. 238–243.

[33] ——, ed., The Real World of Mathematics, Shell Center for Mathematical Education, Nottingham University, Nottingham, 1979.

[34] D. BUSHAW, et al., eds., A Sourcebook of Applications of School Mathematics, NCTM, 1980.

[35] D. N. CHORAFAS, Systems and Simulation, Academic Press, N.Y., 1965.

[36] R. R. CLEMENTS, "The development of methodologies of mathematical modelling," Teach. Math. Applications (1982) pp. 125–131.

[37] ——, "On the role of notation in the formulation of mathematical models," Int. J. Math. Educ. Sci. Tech. (1982) pp. 543–549.

[38] C. V. COFFMAN AND G. J. FIX, eds., Constructive Approaches to Mathematical Models, Academic Press, N.Y., 1979.

[39] H. COHEN, "Mathematical applications, computation, and complexity," Quart. Appl. Math. (1972) pp. 109–121.

[40] O. COULSON, The Spirit of Applied Mathematics, Clarendon Press, Oxford, 1953.

[41] D. DANKOFF AND J. KRUSKAL, Theory and Practice of Sequence Comparison: Time-Warps, String Edits, and Macromolecules, Addison-Wesley, Reading, 1983.

[42] P. J. DAVIS AND J. A. ANDERSON, "Nonanalytic aspects of mathematics and their implications for research and education," SIAM Rev. (1979) pp. 112–125.

[43] R. A. D'INVERNO AND R. R. McLONE, "A modelling approach to traditional applied mathematics," Math. Gaz. (1977) pp. 92–104.

[44] C. L. DYM AND E. S. IVEY, Principles of Mathematical Modeling, Academic Press, N.Y., 1980.

[45] F. DYSON, "Missed opportunities," Bull. AMS (1972) pp. 635–652.

[46] D. P. ELLERMAN, "Arbitrage theory: A mathematical introduction," SIAM Rev. (1984) pp. 241–261.

[47] A. ENGEL, "The relevanced of modern fields of applied mathematics for Mathematical Education," Educ. Studies in Math. (1969) pp. 257–260.

[48] B. Ford and G. C. Hall, "Model building—an educational philosophy for applied mathematics," Int. J. Math. Educ. Sci. Tech. (1970) pp. 77–83.

[49] H. FREUDENTHAL, "Why to teach mathematics so as to be useful," Educ. Studies in Math. (1968) pp. 3–8.

[50] T. C. FRY, "Industrial mathematics," AMM, No. 6, part II, (1941) pp. 1–38.

[51] ——, "Mathematics as a profession today in industry," AMM (1955) pp. 71–80.

[52] M. P. GAFFNEY, L. A. STEEN AND P. J. CAMPBELL, Annotated Bibliography of Expository Writing in the Mathematical Sciences, MAA, 1976.

[53] R. E. GASKELL, "The practice of mathematics," AMM (1957) pp. 557–566.

[54] F. R. GIORDANO AND M. D. WEIR, A First Course in Mathematical Modeling, Brooks/Cole, Monterey, California, 1985.

[55] D. GREENSPAN, Arithmetic Applied Mathematics, Pergamon, Oxford, 1980.

[56] ——, Discrete Models, Addison-Wesley, Reading, 1973.

[57] H. P. GREENSPAN, "Applied mathematics as a science," AMM (1961) pp. 872–880.

[58] C. A. HALL, "Industrial mathematics: A course in realism," MAA (1975) pp. 651–659.

[59] P. R. HALMOS, "Applied mathematics is bad mathematics," in Mathematics Tomorrow, L. A. Steen, ed., Springer-Verlag, N.Y., 1981, pp. 9–20.

[60] R. W. HAMMING, "The unreasonable effectiveness of mathematics," AMM (1980) pp. 81–90.

[61] N. HAWKES, ed., International Seminar on Trends in Mathematical Modeling, Springer-Verlag, Heidelberg, 1973.

[62] M. HAZEWINKEL, "Experimental mathematics," Math. Modelling (1985) pp. 175–211.

[63] N. HELLY, Urban Systems Models, Academic, N.Y., 1975.

[64] F. R. HODSON, D. G. KENDALL AND P. TAUTU, eds., Mathematics in the Archeological and Historical Sciences, Edinburgh University Press, Edinburgh, 1971.

[65] M. HOLT AND D. T. E. MARJORAM, Mathematics in a Changing World, Heinemann, London, 1974.

[66] R. HOOKE AND D. SHAFFER, Math and Aftermath, Walker, N.Y., 1965.

[67] G. HOWSON AND R. McLONE, eds., Maths at Work, Heinemann, London, 1983.

[68] A. JAFFE, "Ordering the universe: The role of mathematics," SIAM Rev. (1984) pp. 473–500.

[69] D. J. G. JAMES, AND J. J. MacDONALD, Case Studies in Mathematics, S. Thornes, 1981.

[70] J. G. KEMENY, J. L. SNELL AND G. L. THOMPSON, Introduction to Finite Mathematics, Prentice-Hall, N.J., 1974.

[71] J. G. KEMENY, "What every college president should know about mathematics," AMM (1973) pp. 889–901.

[72] Y. KHURGIN, Did You Say Math? Mir, Moscow, 1974.

[73] M. S. KLAMKIN, "On the teaching of mathematics so as to be useful," Educ. Studies in Math. (1968) pp. 126–160.

[74] M. KLINE, ed., Mathematics: An Introduction to its Spirit and Use, Freeman, San Francisco, 1979.

[75] M. KLINE, Mathematics and the Physical World, Crowell, N.Y., 1959.

[76] P. J. KNOPP AND G. H. MEYER, eds., Proceedings of a Conference on the Application of Undergraduate Mathematics in Engineering, Life, Managerial and Social Sciences, Georgia Institute of Technology, Atlanta, 1973.

[77] J. LIGHTHILL, "The interaction between mathematics and society," BIMA (1976) pp. 288–299.

[78] ———, ed., Newer Uses of Mathematics, Penguin, Great Britain, 1978.

[79] ———, "The art of teaching the art of applying mathematics," Math. Gaz. (1971) pp. 249–270.

[80] C. C. LIN, "Education in applied mathematics," SIAM Rev. (1978) pp. 838–845.

[81] ———, "On the role of applied mathematics," Advances in Math. (1976) pp. 267–288.

[82] W. F. LUCAS, ed., Modules in Applied Mathematics, Springer-Verlag, N.Y. 1983.
    Vol. 1, M. BRAUN, C. C. COLEMAN, AND D. A. DREW, eds., Differential Equation Models; Vol. 2, S. J. BRAMS, W. F. LUCAS, AND P. D. STRAFFIN, JR., eds., Political and Related Models; Vol. 3, W. F. LUCAS, F. S. ROBERTS, AND R. M. THRALL, eds., Discrete and System Models; Vol. 4, H. MARCUS-ROBERTS AND M. THOMPSON, eds., Life Science Models.

[83] D. P. MAKI AND M. THOMPSON, Mathematical Models and Applications, Prentice Hall, N.J., 1973.

[84] G. I. MARCHUK, ed., Modelling and Optimization of Complex Systems, Springer-Verlag, N.Y., 1979.

[85] N. H. McCLAMROCH, State Models of Dynamic Systems: A Case Study Approach, Springer-Verlag, N.Y., 1980.

[86] Z. A. MELZAK, Mathematical Ideas, Modeling & Applications, Wiley, N.Y., 1976.

[87] W. J. MEYER, Concepts of Mathematical Modeling, McGraw-Hill, N.Y., 1984.

[88] F. J. MURRAY, Applied Mathematics: An Intellectual Orientation, Plenum Press, N.Y., 1978.

[89] ———, "Education for applied mathematics," AMM (1962) pp. 347–357.

[90] Model Building Seminar, Math. Dept., Concordia College, Moorehead, Montana, 1978.

[91] H. NEUNZERT, ed., Mathematics in Industry, Teubner, Stuttgart, 1984.

[92] Y. NIEVERGELT, "Case studies in lower-division mathematics: A bibliography and related teaching issues," UMAP J. (1985) pp. 37–56.

[93] M. OLINICK, "Mathematical models in the social and life sciences: A selected bibliography," Math. Modelling (1981) pp. 237–258.

[94] H. O. POLLAK, "Applications of mathematics," in Mathematical Education, University of Chicago Press, Chicago, 1970.

[95] ———, "How can we teach applications of mathematics," Educ. Studies in Math. (1969) pp. 393–404.

[96] ———, "On some of the problems of teaching applications of mathematics," Educ. Studies Math. (1968–69) pp. 24–30.

[97] S. V. POLLACK, Educational Studies in Computer Sciences, MAA, 1982.

[98] G. POLYA, Mathematical Methods in Science, MAA, 1977.

[99] Proceedings of the Pittsburgh Conferences on Modeling and Simulation, vols. 1–9, Instrument Society of America, 1969–1978.

[100] Proceedings of the Summer Conference for College Teachers on Applied Mathematics, Univ. of Missouri-Rolla, 1971.

[101] G. P. PROTOMASTRO AND C. R. HALLUM, "A mathematical modeling approach to introductory mathematics," AMM (1977) pp. 383–385.

[102] A. L. RABENSTEIN, Introduction to Ordinary Differential Equations with Applications, Academic Press, N.Y., 1972.

[103] E. Y. RODIN, "Modular applied mathematics for beginning students," AMM (1977) pp. 555–560.

[104] R. L. RUBIN, "Model formation using intermediate systems," AMM (1979) pp. 299–303.

[105] T. L. SAATY AND F. J. WEYL, eds., The Spirit and Uses of Mathematical Sciences, McGraw-Hill, N.Y., 1969.

[106] T. L. SAATY AND J. M. ALEXANDER, Thinking with Models, Pergamon, Oxford, 1981.

[107] M. M. SCHIFFER AND L. BOWDEN, The Role of Mathematics in Science, MAA, 1984.

[108] M. R. SCHROEDER, Number Theory in Science and Communication – with Applications in Cryptography, Physics, Biology, Digital Information, and Computers, Springer-Verlag, Berlin, 1984.

[109] T. I. SEIDMAN, "A proposal for a professional program in mathematics," AMM (1975) pp. 162–167.

[110] S. SHARON AND R. REYS, eds., Applications in School Mathematics, NCTM, 1979.

[111] C. L. SMITH, R. W. PIKE AND P. W. MURRILL, Formulation and Optimization of Mathematical Models, Int. Textbook Co., 1970.

[112] D. A. SMITH, "A seminar in mathematical model-building," AMM (1979) pp. 777–783.

[113] J. SPANIER, "The Claremont mathematics clinic," SIAM Rev. (1977) pp. 536–549.

[114] M. R. SPIEGEL, Applied Differential Equations, Prentice-Hall, N.J., 1981.

[115] I. STAKGOLD, ed., Nonlinear Problems in the Physical Sciences and Biology, Springer-Verlag, N.Y., 1973.

[116] M. H. STONE, "Mathematics and the future of science," Bull. AMS (1957) pp. 61–76.

[117] O. G. SUTTON, *Mathematics in Action*, Bell, London, 1954.

[118] *Symposium, "Education in applied mathematics,"* SIAM Rev. (1967) pp. 289–415.

[119] ——, *"The future of applied mathematics,"* Quart. Appl. Math. (1972) pp. 1–125.

[120] ——, *"The mathematical training of the nonacademic mathematician,"* SIAM Rev. (1975) pp. 541–557.

[121] ——, *"What is needed in research and education,"* SIAM Rev. (1962) pp. 297–320.

[122] J. L. SYNGE, *"Postcards on applied mathematics,"* AMM (1939) pp. 152–159.

[123] A. H. TAUB, *Studies in Applied Mathematics*, MAA, 1971.

[124] A. B. TAYLOR, *Mathematical Models in Applied Mathematics*, Cambridge University Press, Cambridge, 1985.

[125] J. W. TUKEY, *"The teaching of concrete mathematics,"* AMM (1958) pp. 1–9.

[126] UMAP *Modules 1977–1979: Tools for Teaching*, Birkhauser, Boston, 1981.

[127] UMAP *Modules 1980: Tools for Teaching*, Birkhauser, Boston, 1981.

[128] UMAP *Modules 1982: Tools for Teaching*, COMAP, Boston, 1983.

[129] J. WALKER, *The Flying Circus of Physics with Answers*, Wiley, N.Y., 1977.

[130] P. C. C. WANG, A. L. SHOENSTADT, B. I. RUSSAK AND C. COMSTOCK, eds., *Information Linkage Between Applied Mathematics and Industry*, Academic Press, N.Y., 1979.

[131] W. C. WATERHOUSE, *"Some errors in applied mathematics,"* AMM (1977) pp. 25–27.

[132] F. J. WEYL, *"The perplexing case of geometry,"* SIAM Rev. (1962) pp. 197–201.

[133] H. WEYL, *Symmetry*, Princeton University Press, Princeton, 1952.

[134] B. P. ZIEGLER, *Theory of Modelling and Simulation*, John Wiley, New York, 1976.

Part I Physical and Mathematical Sciences

1. Mechanics

A DRIVING HAZARD REVISITED*

EDWARD A. BENDER†

John Baylis [1] considered the following problem: When making a right hand turn on British roadways one moves as far to the right as possible on one's side of the roadway and then turns. Unfortunately, the rear of the vehicle moves *leftward* as the right hand turn is begun—toward the unsuspecting driver passing on the left. This can be quite noticeable if the turning vehicle is a long bus.

We assume familiarity with Baylis's paper and notation. Recall that he defines

l = length of wheelbase,

h = length of rear overhang,

ϕ = angle between bus and direction of roadway,

θ = angle of front wheels relative to the bus,

V = speed of the bus.

We add

$2w$ = width of the bus.

(The factor of 2 simplifies some later formulae.) In the process of turning there is some slippage of the front wheels. Consequently the wheels follow different paths and the definition of θ is nonsense. If w is small compared to l, we can ignore this. We would follow Baylis and define θ to be the angle between the path of the outside front wheel and the side of the bus. However, it seems more natural to let θ be angle between the path of R and the center line of the bus. See Fig. 1. (One could argue that θ should be the angle between the path of the inside front wheel and the side of the bus since this reflects how tight the turn is.) This article is adapted from [2, § 8.1].

As Baylis points out, we are interested in the *path* of the bus, not the rate the path is traversed. Hence we may assume that V is constant. (We could even set $V = 1$.) Baylis shows that

$$(1) \qquad l\dot\phi = V \sin \theta$$

and that the rightward displacement of R in Fig. 1 is

$$(2) \qquad x = \int_0^t V \sin (\theta + \phi)\, dt = \int_0^\phi \sin (\theta + \phi) \frac{l\, d\phi}{\sin \theta}$$

by (1). The *leftward* displacement of P is

$$(h + l) \sin \phi - x$$

and so the leftward displacement of N is

$$(3) \qquad f(\phi) = w(\cos \phi - 1) + (h + l) \sin \phi - x.$$

---

* Received by the editors March 15, 1978. Adapted, by permission, from the model in Section 8.1 (pp. 148–152) of AN INTRODUCTION TO MATHEMATICAL MODELING, by E. A. Bender. Copyright © 1978 by John Wiley & Sons, Inc.

† Mathematics Department, University of California, San Diego, La Jolla, California 92093.

BENDER

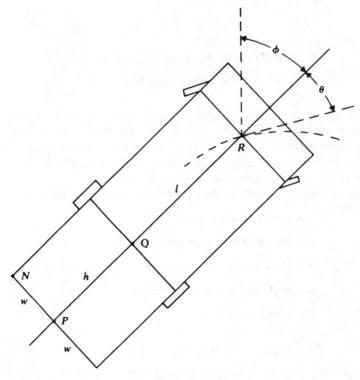

FIG. 1

We wish to maximize f so we set its derivative equal to zero and use (2):

(4)
$$0 = -w \sin \phi + (h+l) \cos \phi - \frac{l \sin (\theta + \phi)}{\sin \theta}.$$

Our basic equations are (2)–(4).

To make further progress we must describe how θ varies throughout the turn. Baylis assumes $\theta = K - \phi$ for some constant K because "this at least has the advantage of analytic solution". The same can be said for the assumption $\theta = C$. We will compare the results of the two assumptions.

First suppose $\theta = C$. Using $\sin (\theta + \phi) = \sin \theta \cos \phi + \cos \theta \sin \phi$ and solving (4) we have

(5)
$$\cot \phi = (w + l \cot C)/h.$$

Integrating (2) we obtain

(6)
$$x = l(\cos C - \cos(\phi + C))/\sin C.$$

Given the values of w, l, h and C we can find the optimum ϕ from (5) and use this in (6) and (3) to determine the maximum leftward displacement of the rear of the vehicle. We have used $l = 16'$, $h = 10'$ and $w = 4'$ for a bus to obtain Table 1. The last row will be explained later.

TABLE 1

C	20°	30°	40°	50°	60°	70°
ϕ	12°	18°	23°	30°	37°	46°
f	1.0'	1.5'	2.1'	2.7'	3.4'	4.2'
$\bar{K}$	26°	39°	52°	65°	79°	93°

Now we consider $\theta + \phi = K$. Substituting $\theta = K - \phi$ in (4), multiplying by $\sin \theta = \sin K \cos \phi - \cos K \sin \phi$, and rearranging, we obtain

$$((h + l) \cos K + w \sin K) \sin (2\phi) - ((h + l) \sin K - w \cos K) \cos (2\phi)$$

$$= (h - l) \sin K + w \cos K =: B.$$

The left side of this equation is $A \sin (2\phi - \delta)$ where

$$A = \sqrt{(h + l)^2 + w^2},$$

$$\sin \delta = ((h + l) \sin K - w \cos K)/A$$

and $-90° < \delta < 90°$. Hence

(7) $\phi = \frac{1}{2}(\text{arc sin } (B/A) + \delta).$

Integrating (2) we obtain

(8) $x = l \sin K \log \dfrac{\tan (K/2)}{\tan ((K - \phi)/2)}.$

Using (7), (8) and (3) as before we obtain Table 2. The last row will be explained shortly.

TABLE 2

K	20°	30°	40°	50°	60°	70°
ϕ	7°	11°	15°	18°	22°	26°
f	.8'	1.1'	1.5'	1.9'	2.3'	2.7'
$\bar{C}$	16°	24°	33°	41°	49°	57°

We would like to somehow compare Tables 1 and 2 to see if they make similar predictions concerning f. Since f depends heavily on C in Table 1 and on K in Table 2, it is not clear how to do this. We adopt the following scheme. Start with a column of Table 1. Let $\bar{K}$ be the average value of $\theta + \phi$ as the bus moves from its initial position to the point at which $f(\phi)$ is a maximum. We will compare Table 1 with the Table 2 entry for K equal to $\bar{K}$. For example, $C = 20°$ gives $\bar{K} = 26°$. Using linear interpolation on $K = 20°$ and $K = 30°$, we find that Table 2 gives $f = 1.0'$, which agrees with Table 1. Since this definition of $\bar{K}$ involves a lot of calculation, we have replaced it by the average of the largest and smallest values of $\theta + \phi$, i.e., $\bar{K} = \theta + \phi/2$. For similar reasons we define $\bar{C} = K - \phi/2$ for Table 2. Now we can compare one table with another. The predictions are fairly close. This suggests that for any reasonable turning scheme the maximum value of f will depend primarily on the average value of θ (or $\theta + \phi$ if you prefer) as the driver bus moves from its initial position to the position at which f is a maximum. In particular, sharper turns produce larger leftward displacements.

REFERENCES

[1] J. BAYLIS, The mathematics of driving hazard, Math. Gaz., 57 (1973), pp. 23–26.
[2] E. BENDER, An Introduction to Mathematical Modeling, Wiley-Interscience, New York, 1978.

A MATHEMATICAL MODEL FOR TRAILER–TRUCK JACKKNIFING*

TIMOTHY V. FOSSUM† AND GILBERT N. LEWIS‡

Abstract. We present a differential equation which is a model for the position of a trailer relative to the cab which is pulling it. The solution is given for two examples, and the results are generalized in a theorem.

Suppose that a cab is pulling a trailer which is d units long. With suitable scaling, d can be taken to be 1. We can represent the positions of the cab and trailer by two vectors. Let $\mathbf{X}$ be a position vector whose terminal point is at the trailer hitch on the cab, and let $\mathbf{Y}$ be a position vector whose terminal point is at the midpoint between the wheels of the trailer. We represent the truck–trailer combination as shown in Fig. 1. $\mathbf{X}$ will be a given, "smooth" function of time t. We would like to be able to predict $\mathbf{Y}$. That is, for a given path $\mathbf{X} = \mathbf{X}(t)$, we want to know if the truck–trailer will jackknife. Alternatively, we want to determine what conditions we must impose on $\mathbf{X}(t)$ to prevent jackknifing.

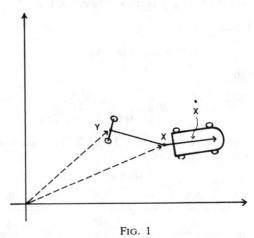

FIG. 1

We model the motion of $\mathbf{Y}$ as follows. First, the trailer length $|\mathbf{X} - \mathbf{Y}| = 1$ is constant, which shows that

$$(1) \qquad (\mathbf{X} - \mathbf{Y}) \cdot (\mathbf{X} - \mathbf{Y}) = 1.$$

where $\cdot$ is the vector dot product. Also, the wheels of the trailer constrain the vector $\mathbf{Y}$ so that its velocity vector is directed along the trailer's lateral axis $\mathbf{X} - \mathbf{Y}$. That is,

$$(2) \qquad \dot{\mathbf{Y}} = \lambda (\mathbf{X} - \mathbf{Y})$$

for some λ, where $\dot{\mathbf{Y}} = d\mathbf{Y}/dt$.
Differentiating (1) yields $2(\dot{\mathbf{X}} - \dot{\mathbf{Y}}) \cdot (\mathbf{X} - \mathbf{Y}) = 0$ and so

$$(3) \qquad \dot{\mathbf{X}} \cdot (\mathbf{X} - \mathbf{Y}) = \dot{\mathbf{Y}} \cdot (\mathbf{X} - \mathbf{Y}).$$

* Received by the editors April 9, 1979, and in revised form August 25, 1979.

† Department of Mathematics, University of Wisconsin–Parkside, Kenosha, Wisconsin 53141.

‡ Department of Mathematical and Computer Sciences, Michigan Technological University, Houghton, Michigan 49931.

Taking the dot product of (2) on both sides by $\mathbf{X}-\mathbf{Y}$ yields $\dot{\mathbf{Y}} \cdot (\mathbf{X}-\mathbf{Y}) = \lambda(\mathbf{X}-\mathbf{Y}) \cdot (\mathbf{X}-\mathbf{Y}) = \lambda$, and so by (3), $\lambda = \dot{\mathbf{X}} \cdot (\mathbf{X}-\mathbf{Y})$. Therefore, (2) becomes

$$(4) \qquad\qquad \dot{\mathbf{Y}} = [\dot{\mathbf{X}} \cdot (\mathbf{X}-\mathbf{Y})](\mathbf{X}-\mathbf{Y}).$$

We introduce Cartesian coordinates to describe $\mathbf{X}$ and $\mathbf{Y}$. Let $\mathbf{X} = (x_1, x_2)$ and $\mathbf{Y} = (y_1, y_2)$. Equation (4) can be written as the system

$$(5) \qquad \begin{aligned} \dot{y}_1 &= x_1^2 \dot{x}_1 - 2x_1 \dot{x}_1 y_1 + x_1 x_2 \dot{x}_2 - x_1 \dot{x}_2 y_2 + \dot{x}_1 y_1^2 - x_2 \dot{x}_2 y_1 + \dot{x}_2 y_2 y_1, \\ \dot{y}_2 &= x_1 \dot{x}_1 x_2 - \dot{x}_1 x_2 y_1 + \dot{x}_2 x_2^2 - 2x_2 \dot{x}_2 y_2 - x_1 \dot{x}_1 y_2 + \dot{x}_1 y_1 y_2 + \dot{x}_2 y_2^2. \end{aligned}$$

For convenience, let $\mathbf{Z} = \mathbf{X} - \mathbf{Y}$, so that (4) can be rewritten as

$$(4') \qquad\qquad \dot{\mathbf{Y}} = (\dot{\mathbf{X}} \cdot \mathbf{Z})\mathbf{Z}.$$

If the cab is moving forward, we say the cab and trailer are *jackknifed* if $\dot{\mathbf{X}} \cdot \mathbf{Z} < 0$; otherwise, we say they are *unjackknifed*. These two configurations are illustrated in Fig. 2. If the cab is backing up, the above situation is reversed. That is, the cab and trailer are *jackknifed* if $\dot{\mathbf{X}} \cdot \mathbf{Z} > 0$ and *unjackknifed* otherwise; see Fig. 3.

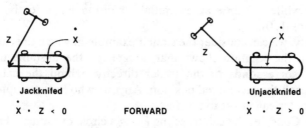

| Jackknifed | FORWARD | Unjackknifed |
| $\dot{\mathbf{X}} \cdot \mathbf{Z} < 0$ | | $\dot{\mathbf{X}} \cdot \mathbf{Z} > 0$ |

FIG. 2

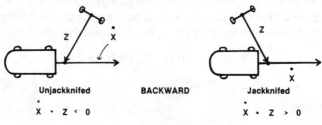

| Unjackknifed | BACKWARD | Jackknifed |
| $\dot{\mathbf{X}} \cdot \mathbf{Z} < 0$ | | $\dot{\mathbf{X}} \cdot \mathbf{Z} > 0$ |

FIG. 3

Example 1. Consider the cab moving forward in a straight line. We may assume the cab travels along the positive x-axis, with $\mathbf{X}(t) = (t, 0)$. Since $|\mathbf{Z}| = 1$, the point $\mathbf{Z}(0) = \mathbf{X}(0) - \mathbf{Y}(0) = -\mathbf{Y}(0)$ must lie on the unit circle, so we may assume $\mathbf{Y}(0) = (\cos \alpha, \sin \alpha)$ for some real α. Then the system (5) becomes

$$\dot{y}_1 = t^2 - 2ty_1 + y_1^2 = (t - y_1)^2, \qquad \dot{y}_2 = -ty_2 + y_1 y_2,$$

whose solution is

$$y_1(t) = t + \frac{e^{-2t} + C_1}{e^{-2t} - C_1},$$

$$y_2(t) = C_2 \exp\left\{ \int_0^t (y_1(s) - s) \, ds \right\}.$$

Applying the initial conditions, $y_1(0) = \cos \alpha$, $y_2(0) = \sin \alpha$, we obtain

$$C_1 = \frac{\cos \alpha - 1}{\cos \alpha + 1}, \qquad C_2 = \sin \alpha.$$

(Note: If $\alpha = 2n\pi$, $C_1 = 0$. If $\alpha = (2n+1)\pi$, $C_1 = \infty$). Thus,

$$y_1(t) = \begin{cases} t+1 & \text{if } \alpha = 2n\pi, \\ t-1 & \text{if } \alpha = (2n+1)\pi, \\ t-1+o(1) & \text{if } \alpha \neq n\pi \end{cases}$$

and

$$y_2(t) = \frac{(C_1 - 1) \sin \alpha}{C_1 e^t - e^{-t}} = \begin{cases} 0 & \text{if } \alpha = n\pi, \\ 0 + o(1) & \text{if } \alpha \neq n\pi. \end{cases}$$

This solution shows that unless the cab and trailer start in the (rather unrealistic) completely jackknifed position with $\alpha = 2n\pi$, the trailer will approach the position following the cab. That is, $\alpha = 2n\pi$ is an unstable initial condition with solution $\mathbf{Y}(t) = (t+1, 0)$, while $\alpha \neq 2n\pi$ as an initial condition leads to the stable limiting solution $\mathbf{Y}(t) = (t-1, 0)$.

Alternatively, we can consider the similar example in which the cab is backing up. In this situation, the solution shows that, *except* for the unstable initial condition $\alpha = 2n\pi$, which corresponds to the trailer directly behind the cab, *all* solutions ultimately approach the jackknifed position. Anyone who has attempted to back up a vehicle with a trailer can attest to this fact.

Example 2. Consider the cab traveling along a circle of radius r. For the moment, assume $r > 1$. We may assume $\mathbf{X}(t) = (r \cos t, r \sin t)$. Again, since $\mathbf{Z}(0) = \mathbf{X}(0) - \mathbf{Y}(0)$ has length 1, we may assume $\mathbf{Y}(0) = \mathbf{X}(0) + (\cos \alpha, \sin \alpha)$. Intuition, aided by computer graphics, tells us that the trailer should approach an asymptotically stable state. It seems reasonable that the cab–trailer combination should approach the configuration shown in Fig. 4. In fact, direct substitution into (4) or (5) shows that the function

$$\mathbf{Y}_1(t) = c(\cos (t - \theta), \sin (t - \theta))$$
(6)
$$= (y_{11}, y_{21})$$

is a solution, where

$$c = \sqrt{r^2 - 1}, \quad \sin \theta = 1/r, \quad \cos \theta = c/r.$$

In the usual terminology of phase plane analysis, $\mathbf{Y}_1(t)$ represents a periodic solution, and the path of $\mathbf{Y}_1(t)$ represents a stable limit cycle. We verify this in the following way.

The differential equations (5) become

(7)
$$\dot{y}_1 = r(y_1 \sin t - y_2 \cos t)(r \cos t - y_1),$$
$$\dot{y}_2 = r(y_1 \sin t - y_2 \cos t)(r \sin t - y_2).$$

We introduce new variables by letting

(8) $u = y_1 \sin t - y_2 \cos t, \qquad v = y_1 \cos t + y_2 \sin t.$

The differential equations for u and v are

(9) $\dot{u} = -ru^2 + v, \qquad \dot{v} = (r^2 - rv - 1)u.$

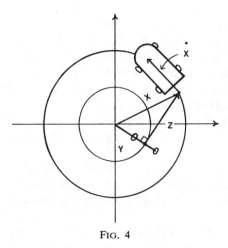

This system has three critical points; $(0, 0)$, $(c/r, c^2/r)$ and $(-c/r, c^2/r)$, all of which correspond to actual solutions.

The first can be eliminated from consideration, since $(u, v) = (0, 0)$ implies $(y_1, y_2) = (0, 0)$. If this were the case, then $1 = |X - Y| = |X| = r > 1$, a contradiction. It should be noted that $(0, 0)$ is a saddle point for (9).

The second critical point corresponds to the solution Y_1. The linearized version of (9) about this solution is

$$\dot{w} = z - 2cw, \qquad \dot{z} = -cz,$$

where $w = u - c/r$, $z = v - c^2/r$. Both characteristic values ($-c$ and $-2c$) of this system are negative. Therefore, by standard theorems of phase plane analysis (see, for example, [1, Chapt. 11]), the solution $(c/r, c^2/r)$ of (9) is stable, and all other solution paths which start close to it will approach it. The point $(c/r, c^2/r)$ is a stable node for (9).

The third critical point corresponds to another solution

$$Y_2(t) = c(\cos(t + \theta), \sin(t + \theta))$$

of (7). Analysis similar to that given above shows that this is an unstable (node) solution. In fact, it represents the same orbit as that for Y_1. In this case, the cab–trailer combination is in a jackknifed position initially and will remain in that position. However, any deviation from that initial position will cause the cab–trailer to wander farther away from the initial configuration.

Now, any solution of (4) must satisfy $r - 1 \le |Y| \le r + 1$ since $|X - Y| = 1$ and $|X| = r$. Hence, any physically meaningful solution of (9) other than $(-c/r, c^2/r)$ and $(0, 0)$, must approach $(c/r, c^2/r)$. Any bounded solution cannot approach $(-c/r, c^2/r)$ or $(0, 0)$, and there cannot be another periodic solution (limit cycle). If there were, there would have to be one enclosing none of the previously mentioned critical points, and thus there would have to be more critical points. Thus, all bounded solutions of (7) approach the stable periodic solution Y_1, with the exception of $(-c/r, c^2/r)$ and $(0, 0)$. In terms of our original model, the trailer approaches the periodic solution of an unjackknifed trailer shown in Fig. 4. We assume, of course, that the trailer is free to rotate 360° around the hitch without the cab getting in the way. In general, as long as $Y(0) \ne Y_2(0)$, then $Y(t) \to Y_1(t)$ as $t \to \infty$.

As before, if the cab is backing up, then the jackknifed solution $Y_2(t)$ is stable, and all others, except $Y_1(t)$, approach it. The solution $Y_1(t)$ is the only solution in which the cab–trailer combination remains unjackknifed, yet it is unstable.

In the above example, we assumed that $r>1$. If $r=1$, then $\mathbf{Y}(t)\to(0,0)$. If $r<1$, then the cab–trailer will enter a jackknifed position, even if it started in an unjackknifed position. This will hold whether the cab is going forward or backing up.

We gain some insight into the general situation from the above example. In particular, if r is too small in Example 2, the cab–trailer jackknifes. From this, we postulate the following theorem.

THEOREM. *Assume* $\mathbf{X}$ *is twice continuously differentiable. If the length of the trailer is 1, if* $1<r(t)$, *where* $r(t)$ *is the radius of curvature of* $\mathbf{X}$, *and if* $\dot{\mathbf{X}}(0)\cdot\mathbf{Z}(0)>0$, *then* $\dot{\mathbf{X}}(t)\cdot\mathbf{Z}(t)>0$ *for all* $t>0$.

In other words, if the cab is moving forward and the cab–trailer combination is not originally jackknifed, then it will remain unjackknifed. On the other hand, if the cab is moving backward and the cab–trailer is originally jackknifed, then it will remain jackknifed.

Proof. Let $f(t)=\dot{\mathbf{X}}\cdot\mathbf{Z}$. Then $f(0)>0$, and f is continuously differentiable. Suppose the conclusion of the theorem is false. Then there exists $t_1>0$ such that $f(t_1)=0$ and $f'(t_1)\leq0$. Assuming that $|\mathbf{Z}|=|\mathbf{X}-\mathbf{Y}|=1$ (trailer has length 1) and $|\dot{\mathbf{X}}|\neq0$ (cab doesn't stop), we have $\dot{\mathbf{X}}(t_1)\cdot\mathbf{Z}(t_1)=0$ so that $\dot{\mathbf{X}}(t_1)\perp\mathbf{Z}(t_1)$. Also,

$$
\begin{aligned}
f'(t) &= \dot{\mathbf{X}}\cdot\dot{\mathbf{Z}}+\ddot{\mathbf{X}}\cdot\mathbf{Z}\\
&= \dot{\mathbf{X}}\cdot(\dot{\mathbf{X}}-\dot{\mathbf{Y}})+\ddot{\mathbf{X}}\cdot\mathbf{Z}\\
&= |\dot{\mathbf{X}}|^2-\dot{\mathbf{X}}\cdot\dot{\mathbf{Y}}+\ddot{\mathbf{X}}\cdot\mathbf{Z}\\
&= |\dot{\mathbf{X}}|^2-(\dot{\mathbf{X}}\cdot\mathbf{Z})\mathbf{Z}\cdot\dot{\mathbf{X}}+\ddot{\mathbf{X}}\cdot\mathbf{Z}\\
&= |\dot{\mathbf{X}}|^2-(\dot{\mathbf{X}}\cdot\mathbf{Z})^2+\ddot{\mathbf{X}}\cdot\mathbf{Z}\\
&= |\dot{\mathbf{X}}|^2-f^2(t)+\ddot{\mathbf{X}}\cdot\mathbf{Z}
\end{aligned}
$$

and

$$
f'(t_1)=|\dot{\mathbf{X}}(t_1)|^2+\ddot{\mathbf{X}}(t_1)\cdot\mathbf{Z}(t_1).
$$

We now use a familiar formula for acceleration (see, for example, [2, p. 423]):

$$
\ddot{\mathbf{X}}=\frac{d|\dot{\mathbf{X}}|}{dt}\mathbf{t}+\kappa|\dot{\mathbf{X}}|^2\mathbf{n}\qquad\left(\kappa=\frac{1}{r}\right),
$$

where $\mathbf{t}$ and $\mathbf{n}$ are unit tangent and normal vectors respectively. Then $\mathbf{t}(t_1)$ is parallel to $\dot{\mathbf{X}}(t_1)$, and $\mathbf{Z}(t_1)$ is parallel to $\mathbf{n}(t_1)$. Therefore, $\ddot{\mathbf{X}}(t_1)\cdot\mathbf{Z}(t_1)=\pm\kappa(t_1)|\dot{\mathbf{X}}(t_1)|^2$, and

$$
\begin{aligned}
f'(t_1) &= |\dot{\mathbf{X}}(t_1)|^2\pm\kappa(t_1)|\dot{\mathbf{X}}(t_1)|^2\\
&= (1\pm\kappa(t_1))|\dot{\mathbf{X}}(t_1)|^2.
\end{aligned}
$$

We conclude that $f'(t_1)>0$, since $\kappa(t_1)=1/r(t_1)<1$. This contradicts our assumption that $f'(t_1)\leq0$, and the theorem is proved.

Acknowledgment. The authors would like to express their thanks to Professor Otto Ruehr for his valuable assistance.

REFERENCES

[1] W. KAPLAN, *Ordinary Differential Equations*, Addison-Wesley, Reading, MA, 1962.
[2] G. B. THOMAS, *Calculus*, Addison-Wesley, Reading, MA, 1969.

DETERMINING THE PATH OF THE REAR WHEELS OF A BUS*

H. I. FREEDMAN† AND S. D. RIEMENSCHNEIDER†

Abstract. Given the locus of the path travelled by the front wheels of a bus, we derive a differential equation describing the path of the rear wheels which we explicitly solve in the case of a straight line and a circle. These solutions can be applied to such specific problems as a bus turning a corner or a bus changing lanes.

Introduction. Suppose that the front wheels of a bus track a given path. What is the path traversed by the rear wheels of the bus? Obviously, some knowledge of the solution to this problem is useful in highway design and in the placement of curbs at intersections. A related problem concerns a driving hazard considered by John Baylis [1] and Edward A. Bender [2]. The hazard is caused by the fact that as a bus turns a corner, the rear of the bus initially moves in the direction opposite the turn, possibly into the path of a passing motorist. Baylis and Bender wanted to know the maximal displacement of the rear of the bus in the direction opposite the turn, a quantity easily determined if the path of the rear wheels is known.

Statement and solution. Let $\mathbf{P}(t) := (p_1(t), p_2(t))$ be the coördinates of the midpoint of the front axle of the bus, which describes a suitably smooth path of motion as a function of t. The rear axle, at a distance L from the front axle, will have the coordinates of its midpoint given by $\mathbf{Q}(t) := (q_1(t), q_2(t))$. Our problem is to find $\mathbf{Q}(t)$ if we are given $\mathbf{P}(t)$ and the initial position of the bus. We could have taken $\mathbf{P}(t)$ and $\mathbf{Q}(t)$ to be the positions of the front and rear wheels on one side of the bus. In any event, once $\mathbf{Q}(t)$ is determined then so is the path followed by any point on the bus. For example, in Fig. 1 the path of the front right wheel is given by

$$\mathbf{OR} = \mathbf{P}(t) + \frac{W}{2L} (\mathbf{P}(t) - \mathbf{Q}(t))^{\perp},$$

and the path of the left rear corner is

$$\mathbf{OT} = \mathbf{Q}(t) - \frac{h}{L} (\mathbf{P}(t) - \mathbf{Q}(t)) - \frac{W}{2L} (\mathbf{P}(t) - \mathbf{Q}(t))^{\perp},$$

where

$$(\mathbf{P}(t) - \mathbf{Q}(t))^{\perp} = (p_2(t) - q_2(t), \; q_1(t) - p_1(t)),$$

Clearly, $\mathbf{Q}(t)$ moves in the direction of $\mathbf{P}(t)$ at any instant. The motion of $\mathbf{Q}(t)$ is described by

$$\frac{d\mathbf{Q}(t)}{dt} = \gamma(t)(\mathbf{P}(t) - \mathbf{Q}(t))$$

(1)   or

$$\frac{dq_i}{dt} = \gamma(t)(p_i(t) - q_i(t)), \qquad q_i(0) = q_{i,0}, \qquad i = 1, 2.$$

---

*Received by the editors July 27, 1982, and in revised form February 24, 1983. Research for this paper was partially supported by the Natural Sciences and Engineering Research Council of Canada, under grant NSERC-A-4823 for the first author and NSERC-A-7687 for the second author.
†Department of Mathematics, University of Alberta, Edmonton, Alberta, Canada T6G 2G1.

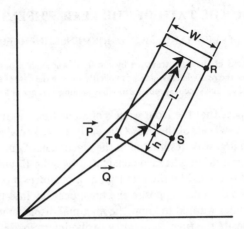

FIG. 1.   *The wheel positions can be computed if the axle positions are known.*

The necessity of the unknown scalar factor $\gamma(t)$ is what makes this problem interesting.
    The solution of (1) is

$$(2) \quad q_i(t) = \exp\left(-\int_0^t \gamma(s)\,ds\right)\left\{q_{i,0} + \int_0^t \exp\left(\int_0^s \gamma(u)\,du\right)\gamma(s)p_i(s)\,ds\right\}, \qquad i = 1, 2.$$

Integrating by parts and setting $p_i(0) = p_{i,0}$, $i = 1, 2$, we obtain

$$(3) \quad p_i(t) - q_i(t) = \exp\left(-\int_0^t \gamma(s)\,ds\right)\left\{p_{i,0} - q_{i,0} + \int_0^t p_i'(s)\exp\left(\int_0^s \gamma(u)\,du\right)ds\right\},$$
$$i = 1, 2.$$

Let $\xi(t) = \exp\left(\int_0^t \gamma(s)\,ds\right)$. Using $|\mathbf{P}(t) - \mathbf{Q}(t)| = L$, from (3) we have

$$(4) \quad L^2\xi^2(t) = \sum_{i=1}^{2}\left\{p_{i,0} - q_{i,0} + \int_0^t \xi(s)p_i'(s)\,ds\right\}^2.$$

Differentiating both sides of (4) and cancelling common factors, we obtain

$$(5) \quad L^2\xi'(t) = \sum_{i=1}^{2}\left\{p_{i,0} - q_{i,0} + \int_0^t \xi(s)p_i'(s)\,ds\right\}p_i'(t).$$

Let $\eta(t)$ be given by $\eta(t) = p_{1,0} - q_{1,0} + \int_0^t \xi(s)p_1'(s)\,ds$, so that $\eta'(t) = \xi(t)p_1'(t)$. Using equation (5) to solve for $p_{2,0} - q_{2,0} + \int_0^t \xi(s)p_2'(s)\,ds$ and the definition of $\eta(t)$, (4) becomes

$$L^2\left(\frac{p_2'(t)\eta'(t)}{p_1'(t)}\right)^2 = \left[L^2\left(\frac{\eta'(t)}{p_1'(t)}\right)' - \eta(t)p_1'(t)\right]^2 + (\eta(t)p_2'(t))^2.$$

Unfortunately, we are unable to explicitly solve this equation.
    We are able to get a simple differential equation for $\xi(t)$ when the motion of the front of the bus is described as a solution of the second order linear differential equation,

$$(6) \qquad\qquad a\mathbf{P}''(t) + b\mathbf{P}'(t) + c\mathbf{P}(t) = \mathbf{d}$$

where a, b, c are constant and $\mathbf{d} = (d_1, d_2)$ is a fixed vector. Differentiating (5) two more times yields the equations

(7) $$L^2 \xi''(t) = \sum_{i=1}^{2} \left\{ \left(p_{i,0} - q_{i,0} + \int_0^t \xi(s) p_i'(s) \, ds \right) p_i''(t) + \xi(t) p_i'(t)^2 \right\},$$

(8)
$$L^2 \xi'''(t) = \sum_{i=1}^{2} \left\{ \left(p_{i,0} - q_{i,0} + \int_0^t \xi(s) p_i'(s) \, ds \right) p_i'''(t) \right.$$
$$\left. + 3\xi(t) p_i'(t) p_i''(t) + \xi'(t) p_i'(t)^2 \right\}.$$

Using (6) we find that

(9)
$$L^2 (a\xi'''(t) + b\xi''(t) + c\xi'(t))$$
$$= \sum_{i=1}^{2} \{ (b p_i'(t)^2 + 3a p_i'(t) p_i''(t)) \xi(t) + a\xi'(t) \, p_i'(t)^2 \},$$

or, using (6) again, $\xi(t)$ satisfies the third order linear homogeneous differential equation

(10)
$$aL^2 \xi'''(t) + bL^2 \xi''(t) + (L^2 c - a(p_1'(t)^2 + p_2'(t)^2)) \xi'(t)$$
$$+ \{ 2b(p_1'(t)^2 + p_2'(t)^2) + 3c(p_1(t) p_1'(t) + p_2(t) p_2'(t))$$
$$- 3(d_1 p_1'(t) + d_2 p_2'(t)) \} \xi(t) = 0.$$

More generally, if the differential operator on the left side of (6) were replaced by an nth order constant coefficient differential operator, or even if (6) where replaced by the $(n+1)$st order homogeneous linear differential equation $\sum_{j=1}^{n+1} a_j(t) \mathbf{P}^{(j)}(t) = 0$, then this method would result in an $(n+1)$st order linear homogeneous differential equation for $\xi(t)$. Equation (10) is suitable for our purposes if we assume that in turning a corner or changing lanes the bus follows a path described by circles and straight lines. In these cases equation (10) will be solved in the sequel.

Straight line motion. As one would expect, if $\mathbf{P}(t)$ follows a straight line, then from any initial direction of $\mathbf{P}(t) - \mathbf{Q}(t)$ the motion of $\mathbf{Q}(t)$ will asymptotically approach the same straight line. In (6) we take $a = 1$, $b = c = 0$, $\mathbf{d} = \mathbf{0}$, and write $p_i(t) = \alpha_i t + \beta_i$, $i = 1, 2$. Then equation (10) reduces to

$$L^2 \xi'''(t) - (\alpha_1^2 + \alpha_2^2) \xi'(t) = 0,$$

which has the solution

$$\xi(t) = k_1 + k_2 \cosh \lambda t + k_3 \sinh \lambda t$$

where

(11) $$\lambda = (\alpha_1^2 + \alpha_2^2)^{1/2}/L = |\mathbf{P}'(0)|/|\mathbf{P}(0) - \mathbf{Q}(0)|.$$

Since $\xi(0) = 1$, we have $k_1 + k_2 = 1$. From (5), (6)

$$\xi'(0) = k_3 \lambda = L^{-2} \{ (\beta_1 - q_{1,0}) \alpha_1 + (\beta_2 - q_{2,0}) \alpha_2 \}$$
$$= [\mathbf{P}'(0) \cdot (\mathbf{P}(0) - \mathbf{Q}(0))]/|\mathbf{P}(0) - \mathbf{Q}(0)|^2,$$

which implies

(12)
$$k_3 = [\lambda L^2]^{-1} \{ (\beta_1 - q_{1,0}) \alpha_1 + (\beta_2 - q_{2,0}) \alpha_2 \}$$
$$= \mathbf{P}'(0) \cdot (\mathbf{P}(0) - \mathbf{Q}(0))/|\mathbf{P}'(0)||\mathbf{P}(0) - \mathbf{Q}(0)| = \cos \theta_0$$

where θ_0 is the angle between the initial direction of the bus and the direction of the line of motion. Similarly from (7)

$$\xi''(0) = \lambda^2 k_2 = (\alpha_1^2 + \alpha_2^2)/L^2 \text{ or } k_2 = 1.$$

Therefore,

(13) $$\xi(t) = \cosh \lambda t + k_3 \sinh \lambda t$$

where λ and k_3 are given by (11) and (12) respectively. Substituting this into (3) we derive the solution

(14) $$Q(t) = P(t) - \frac{P(0) - Q(0) + (P'(0)/\lambda)[\sinh \lambda t + k_3 \cosh \lambda t - k_3]}{\cosh \lambda t + k_3 \sinh \lambda t},$$

where λ and k_3 are given in (11), (12).

Circular motion. In the case of circular motion, we will take

(15) $$P(t) = (p_1(t), p_2(t)) = (\delta_1 + r \cos \omega t, \delta_2 + r \sin \omega t)$$

which are solutions of the differential equation $P''(t) + \omega^2 P(t) = d$ with $d = (\omega^2 \delta_1, \omega^2 \delta_2)$. Then (10) reduces to

(16) $$L^2 \xi'''(t) + (L^2 \omega^2 - r^2 \omega^2) \xi'(t) = 0.$$

There are three cases corresponding to the relative size of the length of the bus and the radius of the circle; $L > r, L = r, L < r$.

Case $L > r$. If $L > r$, then

(17) $$\xi(t) = k_1 + k_2 \cos \mu \omega t + k_3 \sin \mu \omega t$$

where

(18) $$\mu^2 = 1 - \left(\frac{r}{L}\right)^2.$$

From $\xi(0) = 1$, (5) and (7), we obtain linear equations for k_1, k_2, k_3 with solution

(19)
$$k_1 = 1 - r(\delta_1 - q_{1,0})/L^2 \mu^2,$$
$$k_2 = r(\delta_1 - q_{1,0})/L^2 \mu^2,$$
$$k_3 = r(\delta_2 - q_{2,0})/L^2 \mu.$$

In terms of the quantities (17), (18) and (19), the path $Q(t)$ is given by

(20)
$$
\begin{aligned}
Q(t) = P(t) - \frac{1}{\xi(t)} \Bigg\{ &P(0) - Q(0) + k_1(P(t) - P(0)) \\
&+ \frac{k_2}{2(1-\mu)} [P((1-\mu)t) - P(0)] \\
&+ \frac{k_2}{2(1+\mu)} [P((1+\mu)t) - P(0)] \\
&+ \frac{k_3}{2(1-\mu)\omega} [P'((1-\mu)t) - P'(0)]
\end{aligned}
$$

$$+ \frac{k_3}{2(1 + \mu)\omega} [\mathbf{P}'((1 + \mu)t) - \mathbf{P}'(0)] \bigg\}$$

where $\mathbf{P}'(\alpha t)$ is $d\mathbf{P}/dt$ evaluated at αt.

Case $L = r$. If $r = L$, then

(21) $$\xi(t) = k_1 + k_2 t + k_3 t^2$$

where

(22) $$k_1 = 1, \quad k_2 = L^{-2} r\omega(\delta_2 - q_{2,0}), \quad k_3 = -\tfrac{1}{2} L^{-2} r\omega^2(\delta_1 - q_{1,0}).$$

In this case the path of $\mathbf{Q}(t)$ has the description

(23)
$$\mathbf{Q}(t) = \mathbf{P}(t) - \frac{1}{\xi(t)} \bigg\{ \mathbf{P}(t) - \mathbf{Q}(0) + \frac{k_2}{\omega^2} (\mathbf{P}'(t) - \mathbf{P}'(0) - t\mathbf{P}''(t))$$
$$+ \frac{k_3}{\omega^2} (2t\mathbf{P}'(t) - t^2\mathbf{P}''(t) - 2(\mathbf{P}(t) - \mathbf{P}(0))) \bigg\}.$$

Observe that as $t \to \infty$, $\mathbf{Q}(t)$ approaches $\mathbf{P}(t) + 1/\omega^2 \, \mathbf{P}''(t) = (\delta_1, \delta_2)$.

Case $L < r$. In this case, $\xi(t)$ satisfies the differential equation $\xi'''(t) - \theta^2 \omega^2 \xi' = 0$ where

(24) $$\theta^2 = \left(\frac{r}{L}\right)^2 - 1.$$

Hence

(25) $$\xi(t) = k_1 + k_2 \cosh \theta\omega t + k_3 \sinh \theta\omega t,$$

where the constants are found to be

(26)
$$k_1 = 1 + r(\delta_1 - q_{1,0})/L^2\theta^2,$$
$$k_2 = -r(\delta_1 - q_{1,0})/L^2\theta^2,$$
$$k_3 = r(\delta_2 - q_{2,0})/L^2\theta.$$

Evaluating (3) in this case we find the equation of motion is

(27)
$$\mathbf{Q}(t) = \mathbf{P}(t) - \frac{1}{\xi(t)} \bigg\{ \mathbf{P}(0) - \mathbf{Q}(0) + k_1(\mathbf{P}(t) - \mathbf{P}(0))$$
$$+ \frac{k_2}{\omega^2(1 + \theta^2)} [-\cosh \theta\omega t \, \mathbf{P}''(t) + \mathbf{P}''(0) + \theta\omega \sinh \theta\omega t \, \mathbf{P}'(t)]$$
$$+ \frac{k_3}{\omega^2(1 + \theta^2)} [-\sinh \theta\omega t \, \mathbf{P}''(t) + \theta\omega [\cosh \theta\omega t \, \mathbf{P}'(t) - \mathbf{P}'(0)]] \bigg\}.$$

Note that as $t \to \infty$,

$$\mathbf{Q}(t) \to \mathbf{P}(t) - \frac{1}{\omega^2(1 + \theta^2)} (\cos \omega t - \theta \sin \omega t, \sin \omega t + \theta \cos \omega t),$$

or

$$\mathbf{Q}(t) \rightarrow \left(\delta_1 + \frac{r}{1 + \theta^2}\,(\theta \sin \omega t - \theta^2 \cos \omega t),\ \delta_2 + \frac{r}{1 + \theta^2}\,(-\theta^2 \sin \omega t - \theta \cos \omega t)\right),$$

which is a circle centered at (δ_1, δ_2) and radius $\sqrt{r^2 - L^2}$.

Other paths. Certain other motions satisfy simple second order linear equations, but the solutions for $\xi(t)$ and $\mathbf{Q}(t)$ will require numerical techniques. We list a few examples below.

Elliptic motion. If the point given by $\mathbf{P}$ traverses an ellipse, $\mathbf{P}$ may be parametrized by

$$\mathbf{P}(t) = (\xi_1 + \varepsilon_1 \cos \omega t,\ \delta_2 + \varepsilon_2 \sin \omega t)$$

which satisfies the same differential equation as circular motion. However in this case the equation for $\xi(t)$ is

$$L^2 \xi'''(t) + (L^2 - \varepsilon_1^2 \cos^2 \omega t - \varepsilon_2^2 \sin^2 \omega t)\omega^2 \xi'(t)$$
$$+\ [3\omega^3(\varepsilon_2^2 - \varepsilon_1^2)\cos \omega t \sin \omega t]\xi(t) = 0.$$

Parabolic motion. For parabolic motion we take $\mathbf{P}(t) = (\alpha_1 t^2 + \beta_1 t + \varepsilon_1,\ \beta_2 t + \varepsilon_2)$. Then $\mathbf{P}''(t) = \alpha,\ (\alpha_2 = 0)$, and the equation for $\xi(t)$ is

$$L^2 \xi'''(t) - [(2\alpha_1 t + \beta_1)^2 + \beta_2^2]\xi'(t) - 3\alpha_1(2\alpha_1 t + \beta_1)\xi(t) = 0.$$

Hyperbolic motion. Hyperbolic motion is given by

$$\mathbf{P}(t) = (\delta_1 + \varepsilon_1 \cosh \omega t,\ \delta_2 + \varepsilon_2 \sinh \omega t)$$

in which case $\mathbf{P}''(t) - \omega^2 \mathbf{P}(t) = -\mathbf{d} = (-\omega^2 \delta_1,\ -\omega^2 \delta_2)$. The equation for $\xi(t)$ in this case is

$$L^2 \xi'''(t) - [L^2 + \varepsilon_1^2 \sinh^2 \omega t + \varepsilon_2^2 \cosh^2 \omega t]\omega^2 \xi'(t)$$
$$-\ 3\omega^3(\varepsilon_1^2 + \varepsilon_2^2)[\cosh \omega t \sinh \omega t]\xi(t) = 0.$$

Applications. The above analysis can be applied to several specific common occurrences.

i) *Turning a corner.* The path of the rear wheels can be computed by having the point given by $\mathbf{P}$ move around a quarter circle followed by straight line motion. The path of the rear wheels is found by piecing together solutions from the two special cases above (see Figure 2b). Normally in making a right turn (left turn in Britain) $r < L$. Hence when the turn is initiated too close to the corner, a rear wheel will go over the curb (a common experience for bus riders).

ii) *Changing lanes.* Changing lanes or pulling out from a curb can be modelled by having the point given by $\mathbf{P}$ traverse an eighth of a circle followed by another eighth of a circle in the opposite direction so as to "straighten" out the front wheels (see Fig. 2a).

iii) *Traffic circles.* Traffic circles are quite common throughout Britain and in some places in North America (for example, Edmonton). The point given by $\mathbf{P}$ would traverse three circular arcs pieced together (see Fig. 2b) simulating entering, moving around, and exiting from the circle. In this case we would expect $r < L$ when entering and leaving the circle while $r > L$ in the remaining case. The figure shows that in a tight traffic circle, the wheels of a bus entering from the right lane, travelling around the center line of the circle and exiting into the right lane may go over the outside curb when entering and exiting the circle and over the inside curb as it traverses the circle. This explains why the driver will swing wide when entering and exiting the circle and will travel in the outside lane around the circle.

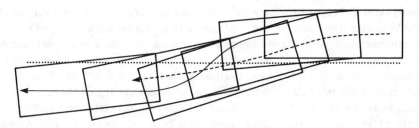

FIG 2a. *Changing lanes. Bus width is ⅚ of the lane width.* ___ *path of front axle,* ----- *path of rear axle.*

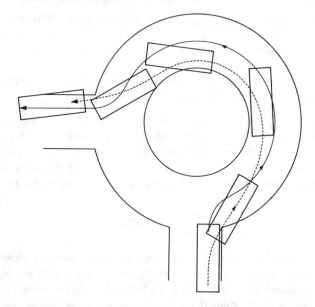

FIG 2b. *Bus entering a traffic circle from the right lane, traveling around the circle between lanes, and exiting into the right lane. Bus width is ⅚ of lane width. Inner circle radius = L.* ___ *path of front axle,* ----- *path of rear axle.*

Acknowledgment. The authors are indebted to Mr. Walter Aiello for providing computer drawn figures.

REFERENCES

[1] JOHN BAYLIS, *The mathematics of a driving hazard*, Math. Gaz., 57 (1973), pp. 23–26.
[2] EDWARD A. BENDER, *A driving hazard revisited*, SIAM Rev., 21 (1979), pp. 136–138.

ICE BREAKING WITH AN AIR CUSHION VEHICLE*

E. R. MULLER†

Abstract. The ice breaking mechanism under a slow moving air cushion vehicle is explored. The bending moments which cause the circumferential cracks in the ice are determined for the case where the pressure of the air cushion is sufficient to depress the water and force air below the ice field. References are given to the more difficult case where the pressure is insufficient to form an air cavity.

Introduction. Recent experiments[1] using self-powered air cushion vehicles have demonstrated that they can be used successfully to break up ice on rivers. This is particularly useful at ice break-up time, when the ice may pack in one section of the river hindering the flow of water and causing floods above the packed ice. Other experiments have used air cushion platforms tied to the front of ships. Such an experiment near Thunder Bay, Ontario, Canada, has demonstrated that an air cushion platform can be a relatively efficient means of ice breaking compared to conventional displacement vessels with their large power and crew requirements. The Alexander Henry, a coast guard vessel, has ice breaking capabilities of two knots in twelve inches of ice. When fitted with an air cushion platform across its bow the vessel was able to move through ice seventeen inches thick, at a speed of nine knots, through twenty-four inch ice at five knots, and thirty inch ice at two knots. When fitted with the air cushion platform the fuel consumption of the Alexander Henry was drastically reduced.

The question we wish to explore is how do the size and cushion pressure of the air cushion vehicle affect its ice breaking capacities.

Properties of the system. Consider an air cushion platform fitted to the front of a ship as shown in Fig. 1(a), and assume that the coupling between the ship and platform is such that the total weight of the platform is carried by the cushion pressure p (psi). At rest or at slow speeds the platform floats on the water and displaces its own weight of water. Measuring d, the mean depression of the water level below the air cushion, in feet we have

$$d = \frac{p \times 144}{62.3} = 2.3p$$

where the weight of one cubic foot of water is taken as 62.3 pounds. Now assume that the air cushion vehicle approaches an ice field, uniform in thickness, density ($0.9 \times$ density of water), and tensile strength. This ice field floats on the water. If h (in feet) is the thickness of the ice field, then $0.9\,h$ of this ice field lies below the water level. Therefore, if the mean depression of the water level below the air cushion exceeds $0.9\,h$, air will be forced below the ice field as shown in Fig. 1(a). If d does not exceed $0.9\,h$, then Fig. 2(a) is applicable.

To develop a model for the system we require to know how an ice sheet breaks. It is not difficult to demonstrate that when a weight is placed on a sheet of ice it will, if sufficiently heavy, cause radial cracks. These cracks emanate from the position of the weight outwards. If the weight is sufficient, circumferential cracks will then appear, and the ice sheet will fail. These observations are of primary importance in the development of any mathematical model since

1. they dictate the symmetry of the problem; that is, plane polar coordinates should be used, and

2. since ice breaking is of interest, that is failure of the ice sheet is being studied, the model can concentrate only on those conditions which cause the circumferential cracks. In other words, the model can assume that the load is sufficient to form radial

* Received by the editors December 5, 1977.

† Department of Mathematics, Brock University, St. Catharines, Ontario, Canada L2S 3A1. The author spent a year's leave with the Research and Development Centre, Transport, Canada, one of the Government agencies involved in this development.

[1] Recent reports have reached the media, for example, Toronto Globe and Mail, March 16, 1977, p. 7; The Financial Post, January 15, 1977, p. 10; Science Dimension, NRC, vol. 9, no. 6 (1977), p. 17.

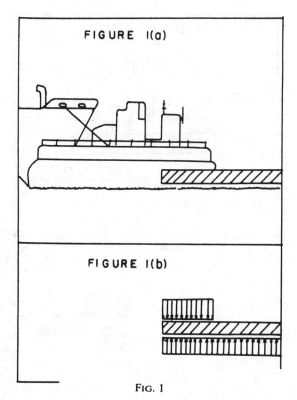

FIGURE I(a)

FIGURE I(b)

Fig. 1

cracks, for if these are not formed, the load will be insufficient to cause circumferential cracks and the ice sheet will not fail.

Basically, the difficult plate problem is reduced to one of determining bending moments in ice wedges. When these bending moments exceed those which can be borne by the ice, it breaks.

The model. Consider the case shown in Fig. 1(a) where the cushion pressure is sufficient to cause a mean water depression in excess of $0.9\,h$. The distribution of load is then as shown in Fig. 1(b), where underneath the air vehicle the total pressure on top of the ice wedge is equal to that below the ice wedge.

The maximum bending moment in the ice wedge is determined in a number of steps as follows:

Under a uniform load p per unit area, the total load up to a distance R in the wedge is given by

$$\int_0^R pb_0 r'\,dr' = \frac{pb_0 R^2}{2}$$

where b_0 is the width of wedge at unit distance.

The moment at R in the wedge is given by

$$\int_0^R pb_0 r'(R - r')\,dr' = \frac{pb_0 R^3}{6}.$$

In calculating moments it is possible to replace a distribution of load by a single point load acting through the center of gravity. Because of symmetry this point will lie on the bisector of the apex angle at a distance

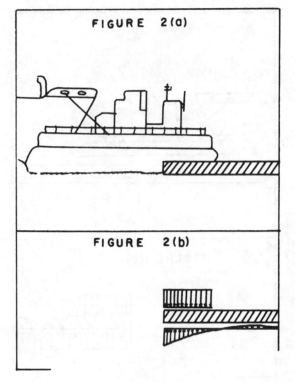

FIG. 2

$$\bar{r} = \int r'\,ds \Big/ \int ds = \frac{\int_{-b_0/2}^{b_0/2}\frac{2}{3}r\cos\alpha\cdot\frac{1}{2}r^2\,d\alpha}{\int_{-b_0/2}^{b_0/2}\frac{1}{2}r^2\,d\alpha} = \frac{2}{3}r\frac{\sin(b_0/2)}{b_0/2} = \frac{2}{3}\beta r$$

where for convenience we have defined

$$\beta = \frac{\sin(b_0/2)}{b_0/2}.$$

Consider a point $x > r$ (r size of cushion) and assume that the air cavity extends beyond x. The moment per unit width of an ice-wedge at the distance x under the distribution of load given in Fig. 1(b) is then, for $p > 0.4\,h$, given by

$$m = \frac{\frac{pr^2 b_0}{2}\left(x - \frac{2}{3}r\beta\right) - (p-0.4h)\frac{b_0 x^3}{6}}{b_0 x},$$

that is,

$$m = \frac{pr^2}{6}\left\{3 - 2\beta\frac{r}{x} - \left[1 - \frac{0.4h}{p}\right]\left[\frac{x}{r}\right]^2\right\}.$$

The maximum occurs when $dm/dx = 0$ or at

$$x = r\beta^{1/3}\left(1 - \frac{0.4h}{p}\right)^{-1/3}.$$

At this distance x, the maximum bending moment/unit width, is given by

$$m_{max} = \frac{pr^2}{2}\left\{1 - \beta^{2/3}\left(1 - \frac{0.4h}{p}\right)^{1/3}\right\}.$$

If this maximum exceeds the bending moment the ice can bear, the ice will fail.

A search of the literature to find the maximum bending moment ice can bear yields little result. Engineers usually measure the strength of a material by its tensile strength, namely, the maximum load a specimen will bear before it breaks. Engineers find the tensile strength more accessible to experimental measurement. The tensile strength (σ) for fresh water ice is 150 psi, and for sea water ice is 90 psi (see, for example, [4]). The relationship between bending moment and tensile strength is derived in most texts on strengths of materials, for example [3]. An explicit derivation for a wedge is given in [2a] and is

$$m = \frac{\sigma h^2}{6}.$$

Therefore, equating m_{max} to $\sigma h^2/6$ we find

(1)
$$\sigma = \frac{3pr^2}{h^2}\left\{1 - \beta^{2/3}\left(1 - \frac{0.4h}{p}\right)^{1/3}\right\}.$$

This equation relates the variables of interest:

r,   the radius of load or size of air cushion;
p,   the air pressure or cushion pressure;

and

h,   the thickness of the ice field.

Experiments have shown that the ice wedges tend to be narrow and therefore the factor

$$\beta = \frac{\sin(b_0/2)}{b_0/2}$$

is not expected to vary much from 1. In the illustrations which follow β_0 has been set at $\pi/6$.

Equation (1) will now be used to explore the following two questions:

(i) for a given cushion pressure, what is the minimum cushion size required to break a given thickness of ice, and

(ii) for an air cushion of a given size, what is the minimum cushion pressure required to break a given thickness of ice?

For the first case we write (1) as

$$r = h\sqrt{\frac{6}{3p}}[1 - \beta^{2/3}(1 - 0.4h)^{1/3}]^{-1/2}.$$

Figure 3 is a plot of $r(p, h)$ for fresh water ice. The corresponding radius of load required to break a given thickness of sea ice can easily be calculated from Fig. 3 by taking the corresponding value of r times 0.775.

Experience suggests that the model used here fails when p is close to $0.4h$, for no maximum in $r(p)$ is expected. In some sense this is no surprise, for when p is close to $0.4h$ two conditions are easily violated; firstly, any bending of the ice will cause it to touch the water, and secondly, it is unlikely that the air cavity will extend to large distances. It would therefore be reasonable to assume that the air cavity does not extend beyond some distance, for example $x = 3r$, three cushion lengths, as is indicated by a dashed line on Fig. 3.

For the second case we explore $p(r, h)$, and from (1) we obtain a cubic in p, namely

$$p^3(1-\beta^2)+\left(0.4h\beta^2-\frac{\sigma h^2}{r^2}\right)p^2+\frac{\sigma^2h^4}{3r^4}p-\left(\frac{\sigma h^2}{3r^2}\right)^3 = 0.$$

The most profitable way to solve for p is probably by some iterative procedure. Figure 4 shows a plot of p versus h for fixed values of r. We recall that the model is only valid for $p > 0.4h$. Therefore, even though the cubic has solutions for $p < 0.4h$, only those values of p which lie above the dashed line in Fig. 4 reflect the behavior of the system.

For the reader who is interested in following this problem further we note that

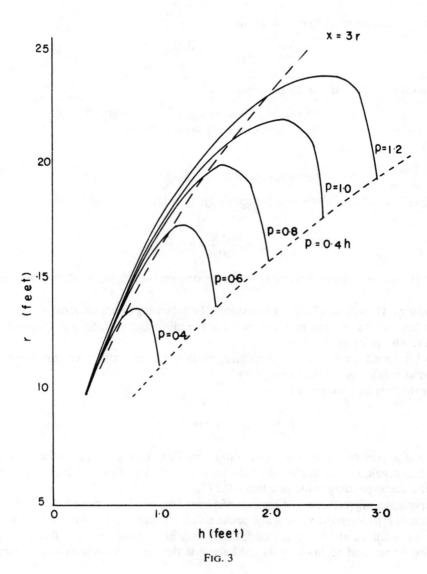

FIG. 3

existing air cushion vehicles tend to be large; 50 to 70 feet, with low cushion pressures from 0.2 to 0.4 psi. From the above model and solution we conclude that these vehicles break ice of any sizeable thickness without the formation of an air cavity. The distribution of load for such a case is shown in Fig. 2(b). Nevel [2], in a series of three reports, has studied the bending moments of an ice wedge on an elastic foundation, and more recently, Carter [1], has applied Nevel's work to uniformly distributed loads covering an area as large as air cushion vehicles. The development and solution of the fourth order partial differential equation for the model is beyond the scope of this classroom note.

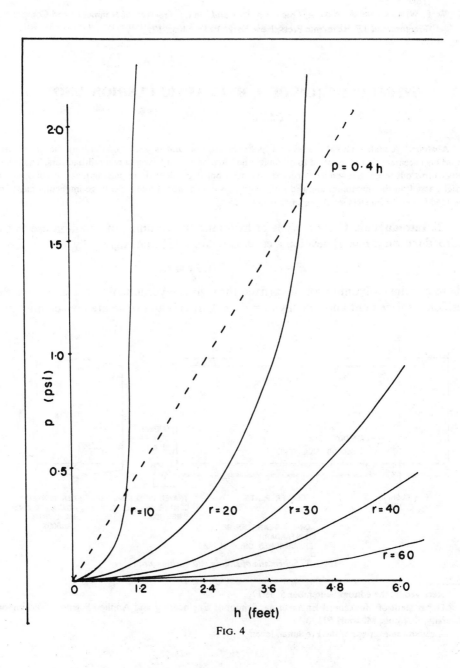

FIG. 4

REFERENCES

[1]   D. Carter, *Ice breaking mechanism under air cushion vehicles*, Draft report, Transportation Development Agency, Transport Canada, 1975.

[2a] D. E. Nevel, *The theory of a narrow infinite wedge on an elastic foundation*, Transactions, Engineering Inst. of Canada, 2 (1958).

[2b] ———, Tech. rep. 56, U.S. Army Snow, Ice and Permafrost Research Establishment, 1958.

[2c] ———, *The narrow free infinite wedge on an elastic foundation*, Res. rep. 79, U.S. Army Cold Regions Research and Engineering Laboratory, 1961.

[3a] S. Timoshenko, *Strength of Materials*, Van Nostrand, New York, 1955.

[3b] S. Timoshenko and S. Woinowsky-Krieger, *Theory of Plates and Shells*, McGraw-Hill, New York, 1959.

[4]   W. F. Weeks and A. Assur, *Fracture of lake and sea ice*, Fracture of Nonmetals and Composites, Fracture, vol 12, Academic Press, New York, 1972, Chap. 12.

OPTIMUM DESIGN OF A HYDRAULIC CUSHION UNIT*[1]

DAVID A. PETERS†

Abstract.[1] A mathematical model of a hydraulic cushion unit is used to determine the orifice area versus displacement that is required to minimize the force of impact between two railroad cars. The orifice design turns out to be independent of impact velocity but dependent on impacting mass. A suboptimum design, based on the maximum allowed rail weight, gives minimum force for a large impacting mass and bounded forces for impacts involving smaller masses.

1. Introduction. Certain types of hydraulic cusion units, of common use in the railroad industry, Fig. 1, develop a cushioning force, F, that is given by

$$(1) \qquad\qquad F = cv^2/a^2, \qquad 0 \leqq x \leqq x_m,$$

where c is the cushioning constant (fixed for a given hydraulic fluid), x and v are the cushion displacement and velocity ($v = dx/dt$), $a(x)$ is an orifice area which may vary

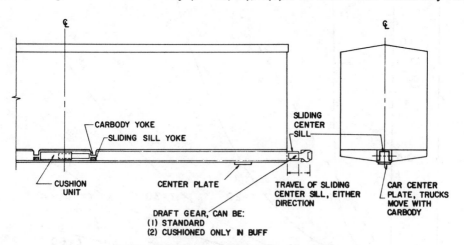

FIG. 1. *Schematic of sliding still cushion unit.*

---

* Received by the editors September 5, 1977.

† Department of Mechanical Engineering, School of Engineering and Applied Science, Washington University, St. Louis, Missouri 63130.

[1] A classroom exercise at the Freshman level.

with displacement, and x_m is the maximum allowable travel given the geometric constraints [1]. The optimum design of such a unit is obtained by choosing $a(x)$ in such a way that the maximum force is minimized for a given impact of the type shown in Fig. 2.

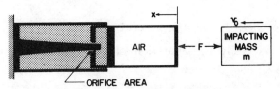

FIG. 2. *Schematic of cushion unit and orifice design.*

An impact is uniquely defined by two parameters: the impacting mass, m, and the initial velocity, v_0. Furthermore, the motion during impact is given by the work energy formula

(2)
$$\frac{1}{2}mv^2 = \frac{1}{2}mv_0^2 - \int_0^x F\,dx.$$

Thus, we wish to minimize max F for the system described by (1) and (2).

2. Optimum orifice area. The critical insight into the solution of this problem is found from (2) evaluated at the end of impact ($v = 0$, $x = x_f$) [2]:

(3)
$$\frac{1}{2}mv_0^2 = \int_0^{x_f} F\,dx.$$

Since the integral on the right hand side is fixed for a given impact (m, v_0), it is clear that the smallest max F is obtained when F maintains a constant value ($F = F_0$) over the maximum possible travel ($x_f = x_m$). It follows immediately from (3) that the desired force is

(4)
$$F_0 = \frac{mv_0^2}{2x_m}.$$

The next step in the solution is to determine the velocity behavior of the cushion unit under this constant force. From (2) and (4) we get

(5)
$$v^2 = v_0^2(1 - x/x_m).$$

Thus we see that the square of the velocity varies linearly under constant force. Now, if v^2 varies linearly, then (1) implies that a^2 must also vary linearly to provide a constant F or

(6)
$$a^2 = a_0^2(1 - x/x_m).$$

The only remaining unknown, a_0^2, can be obtained from (1) and (4), i.e.,

(7)
$$a_0^2 = \frac{2cx_m}{m}.$$

Thus, the optimum orifice area has been obtained; and, interestingly, it is not a function of impact speed.

3. Sub-optimal impacts. The results in (6) and (7) establish that, for a given impacting mass, a single orifice design provides optimum cushioning for all impact speeds. In practice, however, a cushion unit may encounter a variety of impacting masses. We shall now determine the resulting force when a mass different from the design mass impacts the cushion unit. For an impacting mass m', the differential of (2) gives

$$(8) \qquad\qquad m'v \, dv = -c(v^2/a^2) \, dx.$$

The solution follows easily by separation of variables and by use of the initial condition $v = v_0$ at $x = 0$:

$$(9) \qquad\qquad \frac{dv}{v} = -\frac{c}{m'a_0^2(1 - x/x_m)} \, dx,$$

$$(10) \qquad \ln\left|\frac{v}{v_0}\right| = \frac{cx_m}{m'a_0^2}\ln\left|1 - \frac{x}{x_m}\right| = \frac{1}{2}\frac{m}{m'}\ln\left|1 - \frac{x}{x_m}\right|,$$

$$(11) \qquad\qquad v^2 = v_0^2(1 - x/x_m)^{m/m'}.$$

The force can then be obtained from (1), (6), and (11):

$$F = c\frac{v_0^2}{a_0^2}(1 - x/x_m)^{m/m'-1}.$$

For $m' < m$, a relatively light impacting mass, the force decreases with x and reaches zero at maximum travel. For $m = m'$, the force remains constant for all x. For $m' > m$, a relatively heavy impacting mass, the force increases without bound as $x \to x_m$. Therefore, the cushion unit must be designed based on the heaviest expected impacting mass. This is quite feasible because of regulated limits on maximum rail weight.

4. Conclusion. A sub-optimal solution to the design problem is obtained by choosing a linear variation in a^2, which is the optimum solution for the greatest impacting mass. Smaller impacting masses will consequently give a sub-optimal but bounded cushioning force.

REFERENCES

[1] W. K. MacCurdy and R. M. Hermes, *The hydrocushion car*, Anthology of Rail Vehicle Dynamics, Vol. I., Freight Car Impact, ASME, New York, 1971, pp. 9–14.
[2] D. A. Peters, *The sliding still underframe in impact situations*, ASME paper No. 76-WA/RT-13, presented at the ASME Winter Annual Meeting, New York, Dec. 1976.

SLOWEST DESCENT TO THE MOON*

VITTORIO CANTONI† AND AMALIA ERCOLI FINZI‡

Abstract. The problem of a soft landing on the moon with minimum fuel consumption can be successfully dealt with by means of the Maximum Principle [3], [2]. The solution is also a minimum-time solution. However, the converse problem (descent in maximum time, with a given amount of fuel) is an

* Received by the editors February 16, 1979, and in final revised form February 28, 1980.
† Istituto Matematico dell'Università, via Saldini 50, Milano, Italy.
‡ Istituto di Ingegneria Aerospaziale del Politecnico, via Golgi 40, Milano, Italy.

extreme example of the possible ineffectiveness of the use of the Maximum Principle: indeed every solution turns out to satisfy Pontryagin's necessary conditions as a singular solution [1]. Actually every solution is optimal, for it can be shown that the descent time is independent of the control law, provided that the available fuel is entirely used.

1. Introduction. Suppose a spacecraft crew on a lunar mission, in order to carry out repairs or collect data, is confronted with the need of protracting for as long as possible the final stage of the descent to the lunar surface, while disposing of a maximum amount $\Delta m_{\max}$ of the fuel supply.

Assuming a vertical trajectory, and supposing that at time t_0 the height is h_0 and the velocity v_0, what is the control law which maximizes the time T required for a soft landing?

It must of course be assumed that $\Delta m_{\max}$ is greater than the minimum amount of fuel $\Delta m_{\min}$ required for a soft landing. Moreover the initial data h_0 and v_0 must be chosen, for a given maximum thrust of the engine, in such a way that the soft landing is possible. The appropriate conditions can be read from the exhaustive discussion of the moon-landing problem with minimum fuel consumption, based on Pontryagin's Maximum Principle, given in Fleming–Rishel [2, pp. 28–33, 35–37].

Leaving, at first, the performance index and part of the end conditions unspecified, we recall the mathematical formulation of the type of moon-landing optimization problem of interest to us, and state the Maximum Principle (§ 2).

Certain specific choices of the performance index and of the complete end conditions lead to problems with a unique solution actually determinable by means of the Maximum Principle: this is the case for the minimum fuel consumption problem referred to above, as well as for the minimum time problem discussed below by way of illustration (§ 3).

A completely different situation is exhibited by the attempt to apply the same technique to our slowest descent problem: any control law compatible with the end conditions turns out to be a candidate for optimality, so that Pontryagin's necessary conditions give no information whatever towards the solution of the control problem (§ 4).

By means of a first integral of the equations of motion, it can be shown that the descent time is a nondecreasing function of the amount of fuel consumed. Therefore, every control law compatible with the required end conditions on position and velocity is equivalent (and therefore optimal) for our problem, provided that the available fuel is entirely used (§ 5).

It is remarked that the two physically distinct performance indices (time and fuel consumption) are closely related in the maximization as well as in the minimization problem. In both cases they can be regarded as equivalent, in the sense that they lead to the same optimal solutions.

2. Soft landing optimization problems and Pontryagin's principle. Denote by $v(t)$ the velocity and by $m(t)$ the combined mass of the spacecraft and residual fuel at time t. At time $t + dt$ the velocity is $v(t) + dv$, the mass is $m(t) + dm$, while the mass of the ejected fuel increases by $-dm$ and has absolute velocity $v(t) + \sigma$ (σ = relative ejection velocity). Equating the infinitesimal change in momentum to the impulse $mg\,dt$ of the constant gravity force, one gets:

$$(m + dm)(v + dv) - dm(v + \sigma) - mv = mg\,dt.$$

Hence,

$$\frac{dv}{dt} = g + \frac{\sigma}{m}\frac{dm}{dt}.$$

Let the vertical axis x_1 be oriented downwards, with origin on the ground, and set:

$$x_1(t) = \text{position at time } t,$$

$$x_2(t) = v(t) = \frac{dx_1(t)}{dt},$$

$$x_3(t) = m(t).$$

The rate of ejection per unit time $-dm/dt$ can be controlled, and will be denoted by $u(t)$. Thus, with this notation, the system equations are:

$$\frac{dx_1}{dt} = x_2$$

(1)
$$\frac{dx_2}{dt} = g - \frac{\sigma}{x_3} u(t)$$

$$\frac{dx_3}{dt} = -u(t).$$

The maximum thrust of the braking engine is $\sigma\alpha$, so that $u(t)$ is a piecewise continuous function such that $0 \leq u(t) \leq \alpha$.

In each of the optimization problems examined here, the end conditions

(2)
$$x_1(t_0) = h_0, \qquad x_2(t_0) = v_0, \qquad x_3(t_0) = m_0, \qquad \text{(initial data),}$$
$$x_1(t_1) = 0, \qquad x_2(t_1) = 0, \qquad \text{(soft landing condition),}$$

are imposed, and the performance index to be minimized has the form

$$J = \int_{t_0}^{t_1} f[u(t)]\, dt.$$

When applied to such problems, Pontryagin's Maximum Principle states that with every optimal solution $u^*(t)$, $x^*(t)$, it is possible to associate a nonzero solution $\pi^*(t) = (\pi_0^*(t), \pi_1^*(t), \pi_2^*(t), \pi_3^*(t))$ of the adjoint system

$$\frac{d\pi_0}{dt} = 0,$$

$$\frac{d\pi_1}{dt} = 0,$$

(3)
$$\frac{d\pi_2}{dt} = -\pi_1,$$

$$\frac{d\pi_3}{dt} = -\frac{\sigma}{x_3^2}\pi_2 u,$$

such that $\pi_0 \leqq 0$, and the "Hamiltonian" function

(4)
$$H[\boldsymbol{\pi}, \mathbf{x}, u] = \pi_1 x_2 + \pi_2\left(g - \frac{\sigma}{x_3}u\right) - \pi_3 u + \pi_0 f(u)$$

satisfies the relations

(5)
$$H[\boldsymbol{\pi}^*(t), \mathbf{x}^*(t), u^*(t)] = 0$$

and

(6)
$$H[\boldsymbol{\pi}^*(t), \mathbf{x}^*(t), u^*(t)] \geqq H[\boldsymbol{\pi}^*(t), \mathbf{x}^*(t), u]$$

at any time.

Moreover, when the final mass $x_3(t_1)$ is not preassigned, the "transversality condition" $\pi_3(t_1) = 0$ must be satisfied.

From (3) one immediately deduces

(7)
$$\pi_0 = c_0, \qquad \pi_1 = c_1, \qquad \pi_2 = c_2 - c_1 t,$$

where c_0, c_1 and c_2 are constants.

3. The minimum time problem. Fleming and Rishel [2] apply the Maximum Principle to the minimum fuel consumption problem, corresponding to the performance index $J = \int_{t_0}^{t_1} u(t)\,dt$, i.e., to $f(u) = u$. Whenever the initial conditions and the amount of fuel available are compatible with a soft landing, the optimal control law consists of a final stage of braking with maximum thrust, generally preceded by an initial stage of free fall.

An analogous treatment can be applied to the minimum time problem, with performance index $J = \int_{t_0}^{t_1} dt$, $f = 1$, as we shall now show.

From the Hamiltonian (4), with $f = 1$, one sees that condition (6) can only be satisfied by control laws where $u = 0$ or $u = \alpha$, whenever the switching function

(8)
$$S = \pi_2\frac{\sigma}{x_3} + \pi_3$$

is respectively positive or negative. If S vanishes over some time interval, one is again led to the control value $u = 0$. This can be recognized from condition (5), which then yields $x_2 = -(\pi_2/\pi_1)g - (\pi_0/\pi_1)$, so that on account of (7) $dx_2/dt = g$, to be compared with the second equation of motion.

The final stage necessarily corresponds to braking $(u = \alpha)$, and since the transversality condition $\pi_3(t_1) = 0$ must be satisfied, there cannot be a time interval (τ_1, τ_2) with $u = 0$ preceded and followed by intervals with $u \neq 0$. In fact, in any time interval in which $u = 0$, x_3 is constant on account of $(1)_3$ and π_3 is constant on account of $(3)_4$, so that S is a linear function of t on account of (7). If the end-points τ_1 and τ_2 were switching points, the linear function S would have to vanish at τ_1 and τ_2, and therefore over the whole interval. But this would imply $\pi_2 = c_2 < 0$, as can be seen from condition (5) for time t_1.[1] Thus, for $\tau_1 \leqq t \leqq \tau_2$, from $S(t) = 0$ one would get $\pi_3(t) > 0$, while $(3)_4$

---

[1] In the Hamiltonian, and therefore in (5), π_0 must be strictly negative. The possibility $\pi_0 = 0$ must be excluded since it would imply $\pi_2 = 0$, as it can be seen from (5) written at time t_1, and $\pi_3 = 0$ on account of the assumption $S = 0$. Since we already know that $\pi_1 = 0$, the whole vector $\boldsymbol{\pi}$ would vanish, contrary to one of Pontryagin's conditions.

would imply that π_3 is always a nondecreasing function of t; but this is incompatible with the transversality condition $\pi_3(t_1) = 0$.

This reasoning means that whenever the smooth landing is possible, an optimal solution in minimum time can either correspond to a constant control law $u = \alpha$, or to a control law with a single switch from $u = 0$ to $u = \alpha$, exactly as in the minimum fuel consumption problem.

4. The maximum time problem and Pontryagin's Principle. Let us consider now the converse problem of the descent in maximum time.

At present we take it for granted that the entire amount of available fuel will have to be used in order to minimize the performance index, which is now $J = -\int_{t_0}^{t_1} dt$ ($f = -1$). Thus at time t_1 we have the additional condition

$$(9) \qquad\qquad x_3(t_1) = m_0 - \Delta m_{\max}.$$

In the next section the correctness of this obvious assumption will be formally justified.

The Hamiltonian now becomes

$$(10) \qquad\qquad H = \pi_1 x_2 + \pi_2\left(g - \frac{\sigma}{x_3} u\right) - \pi_3 u - \pi_0,$$

and differs only by a sign from the Hamiltonian of the minimum time problem.

The essential technical difference between the present discussion and the preceding one lies in the end condition (9), which now allows for singular solutions, i.e., for solutions satisfying Pontryagin's necessary conditions and such that the switching function is identically zero over some time interval [1, Chap. 6].

Indeed, if the coefficient of u in H is equated to zero, one gets

$$(11) \qquad\qquad \pi_2\frac{\sigma}{x_3} + \pi_3 = 0,$$

i.e., recalling (7),

$$(c_2 - c_1 t)\frac{\sigma}{x_3} + \pi_3 = 0,$$

so that

$$\frac{d\pi_3}{dt} = \frac{c_1\sigma}{x_3} + (c_2 - c_1 t)\frac{\sigma}{x_3^2}\frac{dx_3}{dt} = \frac{c_1\sigma}{x_3} - (c_2 - c_1 t)\frac{\sigma u}{x_3^2},$$

where $(1)_3$ has been taken into account. Replacing in the last equation $d\pi_3/dt$ with its expression $(3)_4$, one gets, after simplification, $(c_1\sigma/x_3) = 0$, so that $c_1 = 0$. Equation (5) thus becomes

$$(12) \qquad\qquad c_2 g - \pi_0 = 0.$$

The possibility $\pi_0 = 0$ must be excluded. In fact, it would imply $c_2 = 0$ and, from $(3)_4$, $d\pi_3/dt = 0$, i.e., $\pi_3 = c_3$. But on one hand the constant c_3 cannot vanish (for this would imply $\pi = 0$); on the other hand, since (5) now yields $H = uc_3 = 0$, the nonvanishing of c_3 would imply $u = 0$ at any time, which is incompatible with the soft landing condition.

Now if $\pi_0 \neq 0$ it can be assumed (by suitable normalization) that $\pi_0 = -1$, and (12) yields $c_2 = -1/g$, i.e., $\pi_2 = -1/g$. Inserting this value in (11), we obtain $\pi_3 = \sigma/gx_3$, which is immediately seen to be compatible with $(3)_4$. We can therefore conclude that, whatever the actual form of the control law $u(t)$ (and of the related solutions of (1)), the adjoint vector with components

$$\pi_0 = -1, \qquad \pi_1 = 0, \qquad \pi_2 = -\frac{1}{g}, \qquad \pi_3 = \frac{\sigma}{gx_3}$$

satisfies all the requirements of Pontryagin's Principle and provides a singular solution of the problem.

This is an extreme example where no conclusion at all can be drawn from the application of the Maximum Principle, since for every control $u(t)$ the necessary conditions of optimality can be satisfied.

5. Solution of the slowest descent problem. Eliminating u from $(1)_2$ and $(1)_3$ we obtain the relation

$$\frac{dx_2}{dt} = g + \frac{\sigma}{x_3}\frac{dx_3}{dt},$$

and hence the integral

(13) $$x_2 - gt - \sigma \log x_3 = \text{const.}$$

The constant value of this function is immediately determined from the initial data.

Solving (13) with respect to t and setting $T = t_1 - t_0$ and $\Delta m = x_3(t_0) - x_3(t_1)$, one gets

$$T = \frac{1}{g}\left(\sigma \log \frac{m_0}{m_0 - \Delta m} - v_0\right).$$

The duration T of the descent is therefore an increasing function of the fuel consumed Δm, independent of $u(t)$ (which can be freely chosen from among the control laws ensuring the smooth landing conditions).

Thus in order to maximize T, the maximum amount $\Delta m_{\max}$ of fuel available must be used. Moreover, the descent laws with end conditions (2) and (9), which all satisfy Pontryagin's necessary conditions (as shown in § 4), are all optimal, since T only depends on the total amount of fuel used, and not on the distribution of the ejection in time interval $(t_0, t_0 + T)$.

We conclude by noting that the integral (13) connects the performance indices $\int_{t_0}^{t_1} u \, dt = \Delta m$ and $\int_{t_0}^{t_1} dt = T$ by a relation which explains the identity of the respective optimal solutions, exhibited in § 3.

REFERENCES

[1] M. ATHANS AND P. L. FALB, *Optimal Control*, McGraw-Hill, New York, 1966.
[2] W. H. FLEMING AND R. W. RISHEL, *Deterministic and Stochastic Optimal Control*, Springer-Verlag, Heidelberg, 1975.
[3] L. S. PONTRYAGIN, V. G. BOLTYANSKII, R. V. GAMKRELIDZE AND E. F. MISCENKO, *The Mathematical Theory of Optimal Processes*, John Wiley & Sons, New York, 1962.

ON EXTREME LENGTH FLIGHT PATHS*

M. S. KLAMKIN†

In this note we extend the following problem proposed by M. F. Gardner[1]:

"A swimmer can swim with speed v in still water. He is required to swim for a given length of time T in a stream whose speed is $w < v$. If he is also required to start and finish at the same point, what is the longest path (total arc length) that he can complete? Assume the path is continuous with piecewise continuous first derivatives."

It is physically intuitive that the longest and shortest paths must be perpendicular and parallel, respectively, to the velocity of the stream. We generalize and prove these results by considering an aeroplane flying with a speed v with respect to ground in a bounded irrotational wind field given by $\mathbf{W} = \nabla \varphi(x, y, z)$ (and $w = |\mathbf{W}|$).

If $\alpha(t), \beta(t), \gamma(t)$ denote the direction angles of the direction of the heading of the aeroplane at time t, then

$$\dot{x} = v \cos \alpha + \varphi_x \qquad \left(\dot{x} = \frac{dx}{dt}, \quad \varphi_x = \frac{\partial \varphi}{\partial x} \right),$$

$$\dot{y} = v \cos \beta + \varphi_y,$$

$$\dot{z} = v \cos \gamma + \varphi_z.$$

The length L of a closed path flown in time T is then given by[1]

$$L = \int_0^T \{\dot{x}^2 + \dot{y}^2 + \dot{z}^2\}^{1/2} \, dt = \int_0^T \{v^2 + w^2 + 2v \sum \varphi_x \cos \alpha\}^{1/2} \, dt$$

$$= \int_0^T \{v^2 - w^2 + 2 \sum \varphi_x (v \cos \alpha + \varphi_x)\}^{1/2} \, dt$$

$$= \int_0^T \{v^2 - w^2 + 2\dot{\varphi}\}^{1/2} \, dt.$$

Applying the Schwarz–Buniakowski inequality, we have

(1) $$L^2 \leq \int_0^T \{v^2 - w^2 + 2\dot{\varphi}\} \, dt \cdot \int_0^T dt \leq T^2 \{v^2 - w^2_{\min}\}$$

(note that $\int_0^T \dot{\varphi} \, dt = 0$ because the path is closed). The upper bound $T\{v^2 - w^2_{\min}\}^{1/2}$ is the best possible, and there exist paths whose length is arbitrarily close to this bound. These can be achieved by flying in an arbitrary small closed path around a point where w is least and in a plane perpendicular to $\mathbf{W}$ at this minimum point. For the special case when $\mathbf{W}$ is constant, the latter can be achieved by flying in arbitrary closed paths (satisfying the time constraint) in a plane perpendicular to $\mathbf{W}$. (To have equality in (1), $\dot{\varphi}$ or equivalently $\sum \varphi_x \cos \alpha$ must be constant.)

We now consider the closed path of minimum length in a given time T. Since

---

* Received by the editors June 1, 1975.

† Department of Mathematics, University of Alberta, Edmonton, Alberta, Canada T6G 2G1.

[1] Here and elsewhere, $\sum$ denotes a cyclic sum, e.g.,

$$\sum \varphi_x \cos \alpha = \varphi_x \cos \alpha + \varphi_x \cos \beta + \varphi_y \cos \gamma.$$

$$v^2 + w^2 + 2v \sum \varphi_x \cos \alpha \geqq (v + \sum \varphi_x \cos \alpha)^2$$

(note that $\max_{\alpha,\beta,\gamma} |\sum \varphi_x \cos \alpha| = w$),

$$L \geqq \int_0^T \{v + \sum \varphi_x \cos \alpha\} \, dt = vT + \frac{1}{v} \int_0^T \{\sum (\varphi_x \dot{x} - \varphi_x^2)\} \, dt$$

or

$$L \geqq vT - \frac{1}{v} \int_0^T w^2 \, dt \geqq (v^2 - w_{\max}^2) T / v$$

(note that $\sum \varphi_x \dot{x} = \dot{\varphi}$). The latter lower bound is also the best possible, and there exist paths whose length is arbitrarily close to this bound. These are achieved by flying back and forth along an arbitrary small segment containing a point where w is a maximum and whose direction is parallel to the wind velocity at this maximum point. For the special case when $\mathbf{W}$ is constant, the latter bound can be achieved by flying back and forth along an arbitrary segment (satisfying the time constraint) in a direction parallel to the wind velocity.

A related result is that if the aeroplane flies any closed path in a nonconstant irrotational wind field, the time of flight is greater than the time of flight over the same path without the wind field [2].

For the two-dimensional case in which $\mathbf{W}$ is constant, both extreme length paths are up and back segments. It is easy to show that the length of any up and back segment of the same time duration T is a monotonic function of the angle the segment makes with $\mathbf{W}$. (See Fig. 1.)

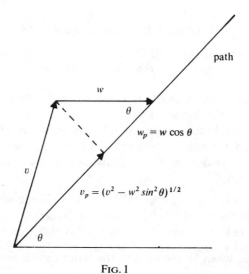

FIG. 1

The component of v along the path is $(v^2 - w^2 \sin^2 \theta)^{1/2}$. If T_1 denotes the time for the downwind flight, then

$$\{(v^2 - w^2 \sin^2 \theta)^{1/2} + w \cos \theta\} T_1 = \{(v^2 - w^2 \sin^2 \theta)^{1/2} - w \cos \theta\}(T - T_1).$$

Whence

$$T_1 = \{(v^2 - w^2 \sin^2 \theta)^{1/2} - w \cos \theta\} T / 2(v^2 - w^2 \sin^2 \theta)^{1/2}$$

and

$$L(\theta) = \frac{T(v^2 - w^2)}{(v^2 - w^2 \sin^2 \theta)^{1/2}}.$$

A related and more difficult problem for the two-dimensional case with constant $\mathbf{W}$ is Chaplygin's problem [3, pp. 206–208]. Here we want to determine the closed path one should fly in a given time such that the enclosed area is a maximum. Using the calculus of variations, it has been shown that the path is an ellipse whose major axis is perpendicular to $\mathbf{W}$ and whose eccentricity is w/v. This result contains the isoperimetric theorem for circles (just set $\mathbf{W} = 0$). For other wind field problems, see [4].

REFERENCES

[1] *Problem* 926, Math. Mag., 48 (1975), p. 51.
[2] *Problem* 61-4, *Flight in an irrotational wind field*, this Review, 4 (1962), pp. 155–156.
[3] N. I. AKHIEZER, *The Calculus of Variations*, Blasidell, New York, 1962.
[4] M. S. KLAMKIN AND D. J. NEWMAN, *Flying in a wind field I, II*, Amer. Math. Monthly, 76 (1969), pp. 16–23, pp. 1013–1019.

POSTSCRIPT

A simpler complete proof for Chaplygin's problem, using Wulff's construction for the equilibrium shape of crystals, appears in SIAM J. Math. Anal., 8 (1977), pp. 288–289.

Flight in an Irrotational Wind Field

Problem 61-4, by M. S. KLAMKIN (University of Alberta)
AND D. J. NEWMAN (Temple University).

If an aircraft travels at a constant air speed, and traverses a closed curve in a horizontal plane (with respect to the ground), the time taken is always less when there is no wind, then when there is any constant wind. Show that this result is also valid for any irrotational wind field and any closed curve (the constant wind case is due to T. H. Matthews, Amer. Math. Monthly, Dec. 1945, Problem 4132).

Solution by G. W. Veltkamp (Technical University, Eindhoven, Netherlands).

Let the speed of the aircraft (with respect to the air) be 1 and let at any point of the (oriented) path the tangential and normal components of the velocity of the wind be u and v, respectively. Then the velocity of the aircraft along its path is easily seen to be

(1) $w = u + \sqrt{1 - v^2}.$

Of course the desired flight can be performed only if

(2) $w > 0$

everywhere along the path. We have to prove that for any closed curve C along which (2) is satisfied

(3) $\int_C \frac{ds}{w(s)} > \int_C ds,$

unless $u = v = 0$ everywhere on C. It follows from (1) by some algebra that

$$\frac{1}{w} = \frac{1}{2}\left(w + \frac{1}{w}\right) - u + \frac{u^2 + v^2}{2w}.$$

Hence (since (2) is assumed to hold) we have everywhere on C

(4)
$$\frac{1}{w} \geq 1 - u$$

with equality only if $u = v = 0$. Since the irrotationality of the wind field implies that $\int_C u\,ds = 0$, the assertion (3) directly follows from (4).

Solution by the proposers.

If we let

$\mathbf{W}$ = wind velocity,

$\mathbf{V}$ = actual plane velocity (which is tangential to the path of flight), then $|\mathbf{V} - \mathbf{W}|$ is the constant air speed of the airplane (without wind) and will be taken as unity for convenience.

We now have to show that

(1)
$$\oint \frac{ds}{|\mathbf{V}|} \geq \oint \frac{ds}{1}.$$

By the Schwarz inequality,

(2)
$$\oint |\mathbf{V}|\,ds \cdot \oint \frac{ds}{|\mathbf{V}|} \geq \left\{\oint ds\right\}^2.$$

Since

$$\oint |\mathbf{V}|\,ds = \oint \mathbf{V}\cdot d\mathbf{R} = \oint (\mathbf{V} - \mathbf{W})\cdot d\mathbf{R} + \oint \mathbf{W}\cdot d\mathbf{R},$$

and

$$\oint \mathbf{W}\cdot d\mathbf{R} = 0 \qquad\qquad (\mathbf{W} \text{ is irrotational}),$$

(3)
$$\oint |\mathbf{V}|\,ds \leq \oint |\mathbf{V} - \mathbf{W}|\,|dR| = \oint ds.$$

(1) now follows from (2) and (3).

Flight in an Irrotational Wind Field II

Problem 82-15, by M. S. KLAMKIN (University of Alberta).

It is a known result (see *Problem 61-4, SIAM Rev.*, 4 (1962), p. 155) that if an aircraft traverses a closed curve at a constant air speed with respect to the wind, the time taken is always less when there is no wind than when there is any bounded irrotational wind field.

(i) Show more generally that if the wind field is $k\mathbf{W}$ ($\mathbf{W}$ bounded and irrotational and k is a constant), then the time of traverse is a monotonic increasing function of k ($k \geq 0$).

(ii) Let the aircraft be subject to the bounded irrotational wind field $\mathbf{W}_i$, $i = 1, 2$, and let T_i denote the time of flight over the same closed path. If $|\mathbf{W}_1| \leq |\mathbf{W}_2|$ at every point of the traverse, does it follow that $T_1 \leq T_2$?

Solution by the proposer.

(i) Let the arc length s denote the position of the plane on its path and let $w(s)$, $\theta(s)$ denote, respectively, the speed and the direction of the wind with respect to the tangent line to the path at position s. It is assumed that the wind field is continuous and that $1 > kw$ where the plane's speed is taken as 1. By resolving $k\mathbf{W}$ into components along and normal to the tangent line of the plane's path, the aircraft's ground speed is

$$\sqrt{1 - k^2 w^2 \sin^2 \theta} + kw \cos \theta,$$

and then the time of flight is given by

$$T(k) = \oint \frac{ds}{\sqrt{1 - k^2 w^2 \sin^2 \theta} + kw \cos \theta}.$$

From Problem 61-4, it is known that $T(k) \geq T(0)$ with equality iff $k\mathbf{W} = 0$. We now show that $T(k)$ is a strictly convex function of k which implies the desired result. Differentiating $T(k)$, we get

$$\frac{dT}{dk} = -\oint \left\{ w \cos \theta - \frac{kw^2 \sin^2 \theta}{\sqrt{1 - k^2 w^2 \sin^2 \theta}} \right\} \left\{ \sqrt{1 - k^2 w^2 \sin^2 \theta} + kw \cos \theta \right\}^{-2} ds.$$

Then $T'(0) = -\oint w \cos \theta \, ds = 0$ since $\mathbf{W}$ is irrotational. On differentiating again, $T''(k) > 0$ since the integrand consists of positive terms. Thus $T(k)$ is strictly convex (for $\mathbf{W} \neq 0$).

(ii) The answer here is negative. Just consider two constant wind fields, both having the same wind speeds. Since the times of the traverses will in general be different, we cannot have both $T_1 \leq T_2$ and $T_2 \leq T_1$.

A problem related to part (i) is that the aircraft flies the same closed path twice with the second time around in the reverse direction. All the other conditions of the problem are the same as before except that the wind field need not be irrotational. Then the total time of flight is an increasing function of k ($k\mathbf{W} \neq 0$). In this case if the aircraft only flew one loop, the time of flight could be less then the time of flight without wind (just consider a whirlwind). Here the total time of flight is

$$T(k) = \oint \frac{ds}{\sqrt{1 - k^2 w^2 \sin^2 \theta} + kw \cos \theta} + \oint \frac{ds}{\sqrt{1 - k^2 w^2 \sin^2 \theta} - kw \cos \theta}.$$

By the A.M.–G.M. inequality, the sum of the integrands is $\geq 2(1 - k^2 w^2)^{-1/2} \geq 2$ which shows that $T(k) \geq T(0)$ with equality iff $k\mathbf{W} = 0$. Then as before $T''(k) > 0$.

Escape Velocity with Drag

Problem 64-3, by D. J. NEWMAN (Temple University).

If we assume that the frictional force (or drag) retarding a missile is proportional to the density of the air, $\rho(x)$, at altitude x (above the earth) and to the square of the velocity, then the differential equation of the motion of the missile can be written in the form

$$\ddot{x} + \rho(x)\dot{x}^2 + (x + 1)^{-2} = 0; \qquad x(0) = 0, \qquad \dot{x}(0) = V_0$$

(after a proper normalization of the constants).

  a) Show that escape is not always possible (e.g., if $\rho(x) \geq (2x + 2)^{-1}$).
  b) Find the necessary and sufficient condition on $\rho(x)$ in order to allow escape.
  c) Give an explicit formula for the escape velocity when it exists.

  Solution by J. ERNEST WILKINS, JR., (General Dynamics/General Atomic Div.)

  Let v be the velocity $\dot{x}$, and let E be the kinetic energy $v^2/2$. Then

$$\frac{dE}{dx} = \frac{dv}{dt} = -2\rho(x)E - (x + 1)^{-2},$$

so that

(1) $$E = \frac{1}{\phi(x)} \left\{ V_0^2 - \int_0^x \frac{\phi(\lambda)\, d\lambda}{(\lambda + 1)^2} \right\},$$

where

$$\phi(x) = 2 \exp \int_0^x 2\rho(y)\, dy.$$

Suppose that

$$I \equiv \int_0^\infty \frac{\phi(\lambda)\, d\lambda}{(\lambda + 1)^2} = \infty.$$

Then for any positive finite V_0, there exists a unique positive finite value ξ such that

$$V_0^2 = \int_0^\xi \frac{\phi(\lambda)\, d\lambda}{(\lambda + 1)^2}.$$

The kinetic energy E, and hence the velocity v, vanishes when $x = \xi$. Since the acceleration $\ddot{x}$ is always negative (the air density $\rho(x)$ is assumed to be nonnegative) it follows that ξ is the maximum value attained by x and that the missile falls back to earth after reaching the height ξ.

  If, on the other hand, the integral I is finite, and the initial velocity V_0 is not less than $I^{1/2}$, then it follows from (1) that $E > 0$, and hence that $v > 0$, throughout the missile trajectory. It then follows that $x \to \infty$ as $t \to \infty$ and the missile will escape the earth.

  We conclude that (a) escape will not occur if $\rho(x) \geq (2x + 2)^{-1}$, for in that case $I = \infty$, (b) the finiteness of the integral I is a necessary and sufficient condition for escape to occur, and (c) the minimum initial velocity V_0 which the missile must possess in order to escape is $I^{1/2}$.

  Editorial Note: The term $P(x)\dot{x}^2$ in the differential equation should have been $k\rho(x)\dot{x}|\dot{x}|$ since the drag force changes sign with $\dot{x}$ and also one cannot normalize all the proportionality constants away.

  Morduchow notes that the above analysis has been based on a strictly "vertical" (radial) motion of the missile and raises the question of finding the escape condition when the initial velocity V_0 is not in a radial direction. To answer that question one would have to consider the following system of equations:

$$\ddot{r} - r\dot{\theta}^2 = -gr^{-2}k^2 - k\rho(r)r\dot{\theta}\sqrt{\dot{r}^2 + r^2\dot{\theta}^2},$$

$$r\ddot{\theta} + 2\dot{r}\dot{\theta} = -k\rho(r)\dot{r}\sqrt{\dot{r}^2 + r^2\dot{\theta}^2}.$$

In connection with this, there are some interesting remarks with regard to the general effect of a resisting medium on the motion of a comet (e.g., Encke's Comet) in *A Treatise on Dynamics of a Particle*, by E. J. Routh, Dover, New York, 1960, pp. 246–249.

Also, it would be of interest to determine the escape condition in the original problem if the drag force is replaced by $\rho(x)\dot{x}^n$, $n > 0$. For the special case of $\rho(x) = k(x + 1)^{-2}$, we have escape if and only if

$$\int_0^\infty \frac{v\,dv}{kv^n + 1} > 1.$$

This holds for all $n \leqq 2$. For $n > 2$, the condition reduces to

$$\pi > nk^{2/n} \sin \frac{2\pi}{n}.$$

A Minimum Time Path

Problem 59-7 (Corrected), by D. J. NEWMAN (Temple University).

Determine the minimal time path of a jet plane from take-off to a given point in space. We assume the highly idealized situation in which the total energy of the jet (kinetic + potential + fuel) is constant and, what is reasonable, that the jet burns fuel at its maximum rate which is also constant. We also assume that $v = y = x = t = 0$ at take-off which gives

$$v^2 + 2gy = at$$

as the energy equation.

Solution by the proposer.

The problem here is to minimize $\int dt$ subject to the constraints

$$\dot{x}^2 + \dot{y}^2 + 2gy = at,$$

$$\int \dot{x}\,dt = X, \qquad \int \dot{y}\,dt = Y.$$

Defining $\tan \gamma = \dot{y}/\dot{x} = \dfrac{dy}{dx}$, the problem becomes: minimize $\int dt$ subject to the constraints

(1) $$\dot{y}^2/\sin^2\gamma + 2gy = at,$$

$$\int \dot{y} \cot \gamma \, dt = X, \qquad \int \dot{y}\,dt = Y,$$

(note: by using γ instead of x, we obtain a simplification).

The usual Lagrange multiplier technique yields the variational problem:

Minimize $\int \{\lambda(t)(\dot{y}^2/\sin^2\gamma + 2gy - at) - C\dot{y} \cot \gamma\} \, dt.$

The resulting Euler equations are

(2) $$\frac{d}{dt}(2\lambda\dot{y}/\sin^2\gamma - C \cot \gamma) = 2gy,$$

(3) $$2\lambda\dot{y} = C \tan \gamma,$$

(note: (3) has already been integrated and simplified). Substituting (3) into (2) yields

(4) $$2\lambda g = C \dot{\gamma} \sec^2\gamma$$

and then (3) becomes

(5) $$\dot{y} = gt' \sin \gamma \cos \gamma, \qquad \left(t' = \frac{dt}{d\gamma}\right).$$

From (1) and (5), we obtain

(6) $$g^2 t'^2 \cos^2\gamma + 2gy = at,$$

which on differentiating gives

(7) $$2g^2 t' t'' \cos^2\gamma - g^2 t'^2 \sin 2\gamma + 2g\dot{y}t' = at'.$$

Replacing $\dot{y}$ from (5) into (7) leads to

$$2g^2 t'' = a \sec^2\gamma,$$

which on integration is

(8) $$2g^2 t' = a(\tan \gamma - \tan \gamma_0).$$

Since $\dfrac{dy}{d\gamma} = y' = gt'^2 \sin \alpha \cos \alpha, \; x' = y' \cot \gamma$, we obtain y' and x' parametrically;

(9) $$4g^3 y' = a^2 \tan \gamma \, (\sin^2\gamma + k \sin 2\alpha + k^2 \cos^2\gamma),$$

(10) $$4g^3 x' = a^2 \, (\sin^2\gamma + k \sin 2\alpha + k^2 \cos^2\gamma),$$

where $k = -\tan \gamma_0$. Noting that initially $x = y = t = 0, \gamma = -\tan^{-1} k$, we can integrate (8), (9), and (10) to obtain

(16) $$\frac{2g^2 t}{a} = \ln (\sec \gamma)/(\sec \gamma_0) - \gamma \tan \gamma_0 + \gamma_0 \tan \gamma_0 ,$$

(17) $$\frac{16g^3 x}{a^2} = 2(1 + k^2)(\gamma - \gamma_0) + (k^2 - 1)(\sin 2\gamma - \sin 2\gamma_0)$$
$$- 2k(\cos 2\gamma - \cos 2\gamma_0),$$

(18) $$\frac{16g^3 y}{a^2} = 4 \ln \frac{\sec \gamma}{\sec \gamma_0} + 4k(\gamma - \gamma_0) - (k^2 - 1)(\cos 2\gamma - \cos 2\gamma_0)$$
$$- 2k(\sin 2\gamma - \sin 2\gamma_0).$$

If γ (final) and γ_0 are chosen properly, we can insure that $x = X$ and $y = Y$ (finally). Then from (16), (17), (18) we have the parametric equations for the optimal path plus the minimum time of flight.

Editorial note: For more realistic and more complicated minimum time-to-climb problems, it is very improbable that the variational equations can be explicity integrated (as in the above equations). In general, it would be more practical to use the techniques of dynamic programming. In particular, see R. E. Bellman and S. E. Dreyfus, "Applied Dynamic Programming", Princeton University Press, New Jersey, 1962, pp. 209–219.

HANGING ROPE OF MINIMUM ELONGATION*

GHASI R. VERMA† AND JOSEPH B. KELLER‡

Abstract. It is shown how to taper a heavy rope, hanging vertically, to minimize the elongation due to its own weight plus a load at its lower end. Hooke's law is used to determine the elongation, and the calculus of variations is used to find that taper which minimizes it.

Let a heavy rope (or beam or string or chain) hang vertically from a fixed support and carry a load of weight W at its lower end. It is stretched elastically by the load and by its own weight. To minimize its total elongation, we can taper the rope so that its upper part, which carries the greatest load, is thickest while its lower part which carries the least load, is thinnest. Its unstretched length L, its volume V, its mass density ρ, and its elastic properties are assumed to be given. Thus we can vary only its cross-sectional area distribution $A(x)$, subject to the volume condition

$$(1) \qquad \int_0^L A(x)dx = V.$$

Here x denotes distance from the upper end in the unstretched state. We seek that area distribution $A(x)$ satisfying (1) which minimizes the total elongation.

To solve this problem we first determine the downward displacement $y(x)$ of the point at position x in the unstretched state. We assume that the strain $dy(x)/dx$ is small, and then Hooke's law of elasticity applies. It states that the strain at each point is proportional to the stress there. The stress at x is just the total downward force at x divided by the cross-sectional area at x. Thus Hooke's law yields

$$(2) \qquad \frac{dy}{dx} = \left[W + \rho g \int_0^L A(x')dx' \right] \bigg/ EA(x).$$

In (2) g is the acceleration of gravity and the proportionality factor E, called Young's modulus, is a characteristic of the rope material. Since the top of the rope is fixed, the displacement there is zero, so $y(0) = 0$. The solution of (2) satisfying this condition is

$$(3) \qquad y(x) = \int_0^x \left[W + \rho g \int_{x'}^L A(x'')dx'' \right] \bigg/ EA(x')dx'.$$

Now we must find $A(x)$ satisfying (1) to minimize $y(L)$.

To take account of (1), we shall minimize $y(L) - \lambda[\int_0^L A(x)dx - V]$, where λ is a Lagrange multiplier. Since the integrand in (3) contains an integral of A, it is convenient

---

*Received by the editors July 1, 1983, and in revised form November 16, 1983.

†Department of Mathematics, University of Rhode Island, Kingston, Rhode Island 02881.

‡Departments of Mathematics and Mechanical Engineering, Stanford University, Stanford, California 94305.

to let that integral be the unknown function. Thus we introduce $B(x)$ defined by

(4)
$$B(x) = \frac{W}{\rho g} + \int_x^L A(x')dx'.$$

Then the quantity to be minimized becomes

(5)
$$y(L) - \lambda \left[\int_0^L A\, dx - V \right] = \int_0^L \left\{ \frac{\rho g}{EA} \left[\frac{W}{\rho g} + \int_x^L A(x')\, dx' \right] - \lambda A \right\} dx + \lambda V$$
$$= \int_0^L \left\{ \frac{-\rho g B}{EB'} + \lambda B' \right\} dx + \lambda V.$$

In order that the last integral in (5) be stationary, with respect to variations in B, B must satisfy the Euler equation of the calculus of variations, which is

(6)
$$\left(\frac{d}{dx} \frac{\partial}{\partial B'} - \frac{\partial}{\partial B} \right) \left(\frac{-\rho g B}{EB'} + \lambda B' \right) = 0.$$

Upon carrying out the differentiations in (6) and simplifying, we obtain

(7)
$$BB'' = (B')^2.$$

By writing (7) as $B''/B' = B'/B$ and integrating, we get $B' = - KB$ where K is a constant. The solution of this equation, which is also the general solution of (6), is

(8)
$$B(x) = B(L)e^{K(L-x)}.$$

By setting $x = L$ in (8) and in (4), and equating the results, we get $B(L) = W/\rho g$. Then (4) shows that $A(x) = -B'(x)$, so (8) yields

(9)
$$A(x) = \frac{KW}{\rho g} e^{K(L-x)}.$$

To determine K we substitute (9) into (1) and find that

(10)
$$K = L^{-1} \log \left(1 + \frac{\rho g V}{W} \right).$$

Then (9) becomes

(11)
$$A(x) = \frac{W}{\rho g L} \log \left(1 + \frac{\rho g V}{W} \right) \cdot \left(1 + \frac{\rho g V}{W} \right)^{1-(x/L)}.$$

By our construction, this is the unique function $A(x)$ which satisfies (1) and makes $y(L)$ stationary. Thus if $y(L)$ has a minimum, this function yields it.

By using (11) in (3) we find

(12)
$$y(L) = \frac{\rho g L^2}{E \log (1 + \rho g V/W)}.$$

For a uniform rope $A = VL^{-1}$ and the displacement is $y_u(L)$ given by

(13)
$$y_u(L) = \frac{\rho g L^2}{E} \left(\frac{1}{2} + \frac{W}{\rho g V} \right).$$

Thus the ratio of the extension of the optimum rope to that of the uniform rope is

(14)
$$\frac{y(L)}{y_u(L)} = \frac{1}{(1/2 + W/\rho gV) \log (1 + \rho gV/W)}.$$

This ratio decreases monotonically as the ratio of the weight of the rope to the load increases. It is unity at $\rho gV/W = 0$ and falls to zero at $\rho gV/W = \infty$.

Substitution of (4) into (2) gives dy/dx in terms of B. Then we use (8) for B to obtain

(15)
$$\frac{dy(x)}{dx} = \frac{-\rho gB}{EB'} = \frac{\rho g}{EK}.$$

This shows that the strain is constant for the optimal rope, and therefore so is the stress.

Paul Bunyan's Washline

Problem 78-17, by J. S. LEW (IBM T. J. Watson Research Center).

It is well known that if a uniform thin flexible cord is suspended freely from its endpoints in a uniform gravitational field, then the shape of the cord will be an arc of a catenary. Determine the shape of the cord if we use a very long one which requires the replacement of the uniform gravitational field approximation by the inverse square field.

Solution by the proposer.
We may determine a unique plane through the earth's center via the two fixed supports, and may suppose the required curve in this plane by symmetry arguments. Introduce polar coordinates (r, θ) in this plane, and locate the origin at the earth's center. Let s denote the cord arclength, l the constant linear density, $U(r)$ the gravitational potential, P and Q the fixed supports. Then $\int_P^Q ds$ is the same for any admissible curve, and $\int_P^Q lU(r) ds$ is a minimum for the equilibrium curve. However if the arclength is constant while the potential energy is either minimal or maximal, then clearly $\delta \int_P^Q L \, d\theta = 0$, where

(1)
$$L(r, r', \theta) = [lU(r) - F](r^2 + r'^2)^{1/2}, \qquad r' = \frac{dr}{d\theta}.$$

Here F is an unknown constant having the dimensions of a force, a Lagrange multiplier representing the tension in the cord. We might now derive the Euler–Lagrange equation for this problem,

(2)
$$(d/d\theta)(\partial L/\partial r') = \partial L/\partial r,$$

but we observe its translation-invariance in the variable θ [1, p. 262], and we obtain a first integral with no intermediate steps:

(3)
$$r'(\partial L/\partial r') - L = E = \text{constant}.$$

Here E has the dimensions of energy, whence E/F has the dimensions of length. Direct substitution into (3) gives a first-order equation for the unknown curve:

(4)
$$E(r^2 + r'^2)^{1/2} = r^2[F - lU(r)].$$

Moreover we expect a "turning point" on this curve where we achieve minimal or maximal distance from the origin. Thus we let (r_0, θ_0) be the coordinates of this point, and we note that $dr/d\theta = 0$ at (r_0, θ_0). Evaluating (4) at this point yields a relation among these unknown constants:

(5) $$lU(r_0) = F - (E/r_0).$$

If we introduce the dimensionless quantities

(6) $$\rho = r/r_0, \qquad \gamma = 1 - (Fr_0/E),$$

then we obtain the differential equation

(7) $$\rho^{-2}(\rho^2 + \rho'^2)^{1/2} = 1 - \gamma + \gamma U(r_0\rho)/U(r_0).$$

Recalling the standard formula $U(r) = -k/r$ for a suitable constant k, we can rewrite (7) as

(8) $$(\rho^2 + \rho'^2)^{1/2} = (1 - \gamma)\rho^2 + \gamma\rho.$$

Setting $u = 1/\rho$, we can simplify (8) to

(9) $$(u^2 + u'^2)^{1/2} = 1 - \gamma + \gamma u.$$

We square equation (9) and differentiate the result, cancel the factor $2u'$ and obtain

(10) $$u'' + (1 - \gamma^2)u = \gamma(1 - \gamma).$$

Since the solution curve in the original variables has a turning point at (r_0, θ_0), the corresponding function in these new variables satisfies the conditions

(11) $$u = 1, \qquad u' = 0, \quad \text{at } \theta = \theta_0.$$

To find the unknown curve we need only solve (10)–(11); to recover the original variables we need only use (6). The form of the solution depends on the value of γ. The results are:

(12) $\gamma = 1$: $\rho \equiv 1$,

(13) $\gamma = -1$: $\rho = [1 - (\theta - \theta_0)^2]^{-1}$,

(14) $|\gamma| < 1$: $\rho = (1 + \gamma)[\gamma + \cos((1 - \gamma^2)^{1/2}(\theta - \theta_0))]^{-1}$,

(15) $|\gamma| > 1$: $\rho = (1 + \gamma)[\gamma + \cosh((\gamma^2 - 1)^{1/2}(\theta - \theta_0))]^{-1}$.

Thus the solution is a circular arc for $\gamma = 1$, a straight line for $\gamma = 0$, and an elementary function for all values of γ.

Moreover the standard expression for curvature in polar coordinates is

(16) $$\kappa(\theta) = (\rho^2 + \rho'^2)^{-3/2}[\rho^2 + 2\rho'^2 - \rho\rho''];$$

whence the value of this curvature at the turning point is

(17) $$\kappa(\theta_0) = 1 - \rho''(\theta_0) = 1 + u''(\theta_0) = \gamma.$$

If γ is positive then the solution curve is concave at θ_0 and the potential energy is maximal for this curve. If γ is negative then the solution curve is convex at θ_0 and the potential is minimal for this curve. Thus the desired solutions have negative γ.

REFERENCE

[1] R. COURANT AND D. HILBERT, *Methods of Mathematical Physics I*, Interscience, New York, 1953.

PLATE FAILURE UNDER PRESSURE*

JOSEPH B. KELLER†

Let us consider a flat plate occupying a plane domain P and rigidly supported along its boundary ∂P. Let a uniform pressure p act on one of its faces. We suppose that the plate fails or yields or breaks along a curve if the shear force per unit length along that curve exceeds a certain constant σ. The shear force is just the force normal to the plate exerted by one part of the plate upon an adjacent part across their common boundary. What is the largest value of p which can be applied before the plate fails? This question was considered and solved by Gilbert Strang of MIT [1], [2]. His method, which differs from that given here, is described at the end of this note.

To answer this question, we consider the balance of forces on any subdomain D of P. The total force normal to the plate exerted on D by the pressure p is pA, where A is the area of D. If the plate has not failed, the total opposing force exerted on D by the surrounding portion of the plate is at most σL, where L is the length of the boundary of D. Thus if failure has not occurred, we must have

$$(1) \qquad pA \leq \sigma L.$$

From this equation, we obtain

$$(2) \qquad p \leq \sigma L/A.$$

Failure will occur when this inequality is violated for some subdomain $D \subseteq P$. The largest pressure at which failure will not occur is p^*, defined by

$$(3) \qquad p^* = \sigma \min_{D \subseteq P} L/A.$$

Thus to find p^* we must determine the minimum value of L/A among all subdomains $D \subseteq P$. This is analogous to the isoperimetric problem of maximizing A for fixed L, or equivalently of minimizing L/A for fixed L. It has been solved for certain general domains by Steiner [3], and for more specific polygonal domains by DeMar [4] and Lin [5]. We shall call the subdomain D^* which yields the minimum the weakest subdomain of P. See Fig. 1.

If the boundary ∂D lies in the interior of P, it follows from the isoperimetric analogy that D must be a circle of some radius r. Then $L = 2\pi r$, $A = \pi r^2$, and $L/A = 2/r$, which is least for the largest possible value of r. However, a still smaller value of L/A may be obtained if ∂D is not entirely in the interior of P, but consists of arcs of the boundary ∂P and arcs in the interior of P. By geometrical arguments from the calculus of

* Received by the editors May 15, 1979, and in revised form October 4, 1979.

† Departments of Mathematics and Mechanical Engineering, Stanford University, Stanford, California 94305. This work was supported by the National Science Foundation, by the Office of Naval Research, by the Air Force Office of Scientific Research and by the Army Research Office.

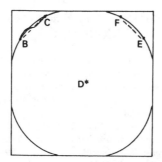

FIG. 1. *A square domain P with its weakest subdomain D* bounded by four circular arcs and four straight line segments. The heavy arc BC has the same length as the circular arc BC.*

variations, we shall show below that the interior arcs must be arcs of circles with equal radii which are tangent to ∂P at their endpoints. Therefore to find the weakest subdomain bounded by n arcs of ∂P and n interior arcs, it is necessary to determine the $2n$ endpoints. Thus for each n, the problem of finding p^* becomes one of ordinary calculus, but not necessarily easy, and we shall call its solution p_n^*. Then the value of n which minimizes p_n^* must be found, and finally $p^* = \min_n p_n^*$.

As an example, let us suppose that P is a square. Then the minmizing subdomain D^* will have the same symmetry as the square. Therefore its boundary will consist of 4 identical line segments on the 4 sides of ∂P, joined by 4 identical circular arcs. See Fig. 1.

To prove that each arc of D^* in the interior must be a circle, we consider any sufficiently short segment of it such as BC in Fig. 1. If that segment is not circular, we can increase A and keep L fixed by replacing the segment by a circular arc with the same length and the same endpoints. To prove that any two interior arcs of D^* must have equal radii, we interchange a short segment of one with a short segment of the other having the same chord length, such as BC and EF in Fig. 1. This does not change L or A. Since the resulting interior arcs must be circles, as we have shown above, the two arcs must have equal radii. To prove that each interior arc of D^* must be tangent to the boundary, we suppose that some arc is not tangent, but intersects the boundary at an angle different from π. Then we cut off from D^* a little "triangle" near the intersection point with a line segment of length ε. This will reduce L by a positive amount $O(\varepsilon)$ of order ε, and reduce A by an amount of order ε^2. Thus L/A is changed to $[L - O(\varepsilon)]/[A - O(\varepsilon^2)] = L/A - O(\varepsilon) + O(\varepsilon^2)$, which is less than L/A for ε sufficiently small. Therefore the interior arcs must be tangent to the boundary.

Professor Strang formulated the problem of static deformation on the assumption that the plate was made of an elastic-plastic material. This formulation involved a partial differential equation for the displacement of the plate. Then by finding the dual problem, he was led to study the constrained isoperimetric problem (3).

REFERENCES

[1] GILBERT STRANG, *A family of model problems in plasticity*, Computing Methods in Applied Sciences and Engineering, Springer-Verlag Lecture Notes 704, R. Glowinski and J. L. Lions, eds., Springer-Verlag, New York, 1979, 292–308.
[2] ——, *A minimax problem in plasticity theory*, Functional Analysis Methods in Numerical Analysis, Springer-Verlag Lecture Notes, 701, M. Z. Nashed, ed., Springer-Verlag, New York, 1979.
[3] J. STEINER, *Sur le maximum et le minimum des figures dans le plan, sur le sphere, et dans l'esplace en general, I and II*, J. Reine Angew. Math. (Crelle), 24 (1842), pp. 93–152, 189–250.
[4] R. F. DEMAR, *A simple approach to isoperimetric problems in the plane*, Math. Mag., 48 (1975), pp. 1–12.
[5] T. P. LIN, *Maximum area under constraint*, Ibid., 50 (1977), pp. 32–34.

Supplementary References
Mechanics

[1]   R. N. ARNOLD AND L. MAUNDER, *Gyrodynamics and its Engineering Applications*, Academic, London, 1961.

[2]   K. L. ARORA AND N. X. VINH, *"Maximum range of ballistic missiles,"* SIAM Rev. (1965) pp. 544–550.

[3]   C. D. BAKER AND J. J. HART, *"Maximum range of a projectile in a vacuum,"* Amer. J. Phys. (1955) pp. 253–255.

[4]   R. M. L. BAKER, JR. AND M. W. MAKEMSON, *An Introduction to Astrodynamics*, Academic, N.Y., 1967.

[5]   S. BANACH, *Mechanics*, Hafner, N.Y., 1951.

[6]   R. R. BATE, D. D. MUELLER AND J. E. WHITE, *Fundamentals of Astrodynamics*, Dover, N.Y., 1971.

[7]   D. C. BENSON, *"An elementary solution of the brachistochrone problem,"* AMM (1969) pp. 890–894.

[8]   A. BERNHART, *"Curves of general pursuit,"* Scripta Math. (1959) pp. 189–206.

[9]   ———, *"Curves of pursuit,"* Scripta Math. (1954) pp. 125–141.

[10] ———, *"Curves of pursuit,"* Scripta Math. (1957) pp. 49–65.

[11] ———, *"Polygons of pursuit,"* Scripta Math. (1959) pp. 23–50.

[12] L. BLITZER AND A. D. WHEELON, *"Maximum range of a projectile in vacuum on a spherical earth,"* Amer. J. Phys. (1957) pp. 21–24.

[13] R. L. BORRELLI, C. S. COLEMAN AND D. D. HOBSON, *"Poe's pendulum,"* MM (1985) pp. 78–83.

[14] B. BRADEN, *"Design of an oscillating sprinkler,"* MM (1985) 29–38.

[15] F. BRAUER, *"The nonlinear simple pendulum,"* AMM (1972) pp. 348–354.

[16] M. N. BREARLY, *"Motorcycle long jump,"* Math. Gaz. (1981) pp. 167–171.

[17] W. BURGER, *"The yo-yo: A toy flywheel,"* Amer. Sci. (1984) pp. 137–142.

[18] D. N. BURGHES, *"Optimum staging of multistage rockets,"* Int. J. Math. Educ. Sci. Tech. (1974) pp. 3–10.

[19] D. N. BURGHES AND A. M. DOWNS, *Modern Introduction to Classical Mechanics and Control*, Horwood, Chichester, 1975.

[20] W. E. BYERLY, *An Introduction to the Use of Generalized Coordinates in Mechanics and Physics*, Dover, N.Y., 1944.

[21] A. C. CLARKE, *Interplanetary Flight, An Introduction to Astronautics*, Harper, N.Y.

[22] J. M. A. DANBY, *Fundamentals of Celestial Mechanics*, Macmillan, N.Y., 1962.

[23] D. E. DAYKIN, *"The bicycle problem,"* MM (1972) p. 1 (also see (1973) pp. 161–162).

[24] R. W. FLYNN, *"Spacecraft navigation and relativity,"* Amer. J. Phys. (1985) pp. 113–119.

[25] G. GENTA, *Kinetic Energy Storage. Theory and Practice of Advanced Flywheel Systems*, Butterworths, Boston, 1985.

[26] H. GOLDSTEIN, *Classical Mechanics*, Addison-Wesley, Reading, 1953.

[27] A. GREY, *A Treatise on Gyrostatics and Rotational Motion*, Dover, N.Y., 1959.

[28] B. HALPERN, *"The robot and the rabbit—a pursuit problem,"* AMM (1969) pp. 140–144.

[29] V. G. HART, *"The law of the Greek catapult,"* BIMA (1982) pp. 58–63.

[30] M. S. KLAMKIN, *"Dynamics: Putting the shot, throwing the javelin,"* UMAP J. (1985) pp. 3–18.

[31] ———, *"Moving axes and the principle of the complementary function,"* SIAM Review (1974) pp. 295–302.

[32] ———, *"On a chainomatic analytical balance,"* AMM (1955) pp. 117–118.

[33] ———, *"On some problems in gravitational attraction,"* MM (1968) pp. 130–132.

[34] M. S. KLAMKIN AND D. J. NEWMAN, *"Cylic pursuit or the three bugs problem,"* AMM (1971) pp. 631–638.

[35] ———, *"Flying in a wind field I, II,"* AMM (1969) pp. 16–22, 1013–1018.

[36] ———, *"On some inverse problems in potential theory,"* Quart. Appl. Math. (1968) pp. 277–280.

[37] ———, *"On some inverse problems in dynamics,"* Quart. Appl. Math. (1968) pp. 281–283.

[38] J. M. J. KOOY AND J. W. H. UYENBOGAART, *Ballistics of the Future*, Stam, Haarlem.

[39] C. LANCZOS, *The Variational Principles of Mechanics*, University of Toronto Press, Toronto, 1949.

[40] D. F. LAWDEN, *"Minimal rocket trajectories,"* Amer. Rocket Soc. (1953) pp. 360–367.

[41] ———, *Optimal Trajectories for Space Navigation*, Butterworths, London, 1963.

[42] ———, *"Orbital transfer via tangential ellipses,"* J. Brit. Interplanetary Soc. (1952) pp. 278–289.

[43] J. E. LITTLEWOOD, *"Adventures in ballistics,"* BIMA (1974) pp. 323–328.

[44] S. L. LONEY, *Dynamics of a Particle and of Rigid Bodies*, Cambridge University Press, Cambridge, 1939.

[45] W. F. OSGOOD, *Mechanics*, Dover, N.Y., 1965.

[46] W. M. PICKERING AND D. M. BURLEY, *"The oscillation of a simple castor,"* BIMA (1977) pp. 47–50.

[47] H. PRESTON-THOMAS, *"Interorbital transport techniques,"* J. Brit. Interplanetary Soc. (1952) pp. 173–193.

[48] D. G. MEDLEY, *An Introduction to Mechanics and Modelling*, Heinemann, London, 1982.

[49] A. MIELE, *Flight Mechanics I, Theory of Flight Paths*, Addison-Wesley, Reading, 1962.

[50] N. MINORSKY, *Introduction to Nonlinear Mechanics*, Edwards, Ann Arbor, 1946.

[51] F. R. MOULTON, *An Introduction to Celestial Mechanics*, Macmillan, N.Y., 1962.

[52] ———, *Methods in Exterior Ballistics*, Dover, N.Y., 1962.

[53] A. S. RAMSEY, *Dynamics I, II*, Cambridge University Press, Cambridge, 1946.

[54] R. M. ROSENBERG, *"On Newton's law of gravitation,"* Amer. J. Phys. (1972) pp. 975–978.

[55] L. I. SEDOV, *Similarity and Dimensional Methods in Mechanics*, Academic, N.Y., 1959.

[56] M. J. SEWELL, *"Mechanical demonstration of buckling and branching,"* BIMA (1983) pp. 61–66.

[57] D. B. SHAFFER, *"Maximum range of a projectile in a vacuum,"* Amer. J. Phys. (1956) pp. 585–586.

[58] S. K. STEIN, *"Kepler's second law and the speed of a planet,"* AMM (1967) pp. 1246–1248.

[59] J. J. STOKER, *Nonlinear Vibrations*, Interscience, New York, 1950.

[60] J. L. SYNGE AND B. A. GRIFFITH, *Principles of Mechanics*, McGraw-Hill, N.Y., 1959.

[61] J. L. SYNGE, *"Problems in mechanics,"* AMM (1948) pp. 22–24.

[62] H. S. TSIEN, *"Take-off from satellite orbit,"* Amer. Rocket Soc. (1953) pp. 233–236.

[63] W. G. UNRUH, *Instability in automobile braking,"* Amer. J. Phys. (1984) pp. 903–908.

[64] P. VAN DE KAMP, *Elements of Astromechanics*, Freeman, San Francisco, 1964.

[65] J. WALKER, *"The amateur scientist: In which simple questions show whether a knot will hold or slip,"* Sci. Amer. (1983) pp. 120–128.

[66] L. B. WILLIAM, *"Fly around a circle in the wind,"* AMM (1971) pp. 1122–1125.

[67] E. T. WHITTAKER, *A Treatise on the Analytical Dynamics of Particles and Rigid Bodies*, Dover, N.Y., 1944.

[68] C. WRATTEN, *"Solution of a conjecture concerning air resistance,"* MM (1984) pp. 225–228.

[69] W. WRIGLEY, W. M. HOLLISTER AND W. G. DENHARD, *Gyroscopic Theory, Design and Instrumentation*, M.I.T. Press, Cambridge, 1969.

[70] J. ZEITLIN, *"Rope strength under dynamic loads: The mountain climber's surprise,"* MM (1978) pp. 109–111.

2. Heat Transfer and Diffusion

THE TEMPERATURE DISTRIBUTION WITHIN A HEMISPHERE EXPOSED TO A HOT GAS STREAM*

L. M. CHIAPPETTA† AND D. R. SOBEL†

Abstract. The temperature distribution in the spherical tip of a gas sampling probe is obtained. The boundary condition representing the flux of heat from a hot gas stream to the tip is of the mixed Dirichlet–Neumann type. This solution is expressed as a series of Legendre polynomials.

Introduction. The purpose of this Classroom Note is to present the solution to the steady-state heat conduction problem representing a hemisphere whose flat surface is maintained at a constant temperature while the curved surface is exposed to a gaseous stream maintained at a second constant temperature. This solution was obtained during a thermal integrity study of the tip region of a combustion-gas sampling probe.

Problem description. A schematic diagram of a combustion-gas sampling probe is shown in Fig. 1. A small sample of the gas to be analyzed is admitted through the center of the probe tip. The curved surface of the tip is exposed to the hot combustion gas while the flat, downstream surface is maintained at essentially constant temperature by water impingement cooling. It is known that the heat flux on the interior wall of the hemispherical tip, i.e., the surface exposed to the gas sample, is negligible compared with that on the curved upstream and flat downstream surfaces. Therefore, the simplified physical model shown in Fig. 2 was used in the development of an approximate solution for the temperature distribution in the probe tip region.

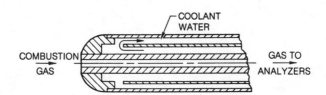

FIG. 1. *Combustion-gas sampling probe.*

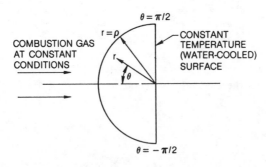

FIG. 2. *Physical model of tip region.*

---
*Received by the editors November 15, 1983, and in revised form January 15, 1984.
†United Technologies Research Center, East Hartford, Connecticut 06108.

The physical model shown in Fig. 2 can be represented mathematically by the heat conduction equation in spherical coordinates with axial symmetry, i.e.,

(1) $\dfrac{\partial}{\partial r}\left(r^2 \dfrac{\partial T}{\partial r}\right) + \dfrac{1}{\sin\theta}\dfrac{\partial}{\partial\theta}\left(\sin\theta\cdot\dfrac{\partial T}{\partial\theta}\right) = 0,\qquad 0\le r\le\rho,\quad -\dfrac{\pi}{2}\le\theta\le\dfrac{\pi}{2},$

and the associated boundary conditions, viz.,

(2) $T\left(r,\dfrac{\pi}{2}\right) = T\left(r,-\dfrac{\pi}{2}\right) = T_c,\qquad 0\le r\le\rho,$

where T_c is the coolant temperature,

(3) $T(r,\theta)$ finite   for $0\le r\le\rho,\quad -\dfrac{\pi}{2}\le\theta\le\dfrac{\pi}{2},$

and

(4) $k\dfrac{\partial T}{\partial r}(\rho,\theta) = h[T_G - T(\rho,\theta)],\qquad -\dfrac{\pi}{2}\le\theta\le\dfrac{\pi}{2},$

where the heat transfer coefficient, h, the thermal conductivity k and the gas temperature T_G are known. These equations, after the substitutions $z = \cos\theta$ and $u(r,z) = T(r,z) - T_c$, become

(1') $\dfrac{\partial}{\partial r}\left[r^2\dfrac{\partial u}{\partial r}\right] + \dfrac{\partial}{\partial z}\left[(1-z^2)\dfrac{\partial u}{\partial z}\right] = 0,\qquad 0\le r\le\rho,\quad 0\le z\le 1,$

(2') $u(r,0) = 0,\qquad 0\le r\le\rho,$

(3') $u(r,z)$ finite   for $0\le r\le\rho,\quad 0\le z\le 1,$

(4') $k\dfrac{\partial u}{\partial r}(\rho,z) = h[T_G - T_c - u(\rho,z)],\qquad 0\le z\le 1.$

Application of the method of separation of variables yields

(5) $u(r,z) = \displaystyle\sum_{m=0}^{\infty} C_m r^{2m+1} P_{2m+1}(z),$

or

(6) $T(r,z) = T_c + \displaystyle\sum_{m=0}^{\infty} C_m r^{2m+1} P_{2m+1}(z),$

where $P_n(Z)$ represents the Legendre polynomial of order n. Equation (6) represents a solution of (1), (2) and (3). The constants C_m can be determined by satisfying the boundary condition, (4), as follows:
     Substitution of (6) into (4) yields

(7) $\displaystyle\sum_{m=0}^{\infty} C_m\{k(2m+1)\rho^{2m} + h\rho^{2m+1}\} P_{2m+1}(z) = h(T_G - T_c),\qquad 0 < z \le 1.$

Letting

$$(8) \qquad B_m = C_m \left\{ \frac{(2m+1)k}{\rho h} + 1 \right\} \rho^{2m+1},$$

we have

$$(9) \qquad \sum_{m=0}^{\infty} B_m P_{2m+1}(z) = T_G - T_c.$$

Since the Legendre polynomials form a complete set of orthogonal functions,

$$(10) \qquad B_m = \frac{\int_0^1 (T_G - T_c) P_{2m+1}(z) \, dz}{\int_0^1 P_{2m+1}^2(z) \, dz}.$$

It follows that

$$(11) \qquad B_m = \frac{(-1)^m (2m)! (T_G - T_c)(4m+3)}{2^{2m+1}(m+1)(m!)^2},$$

and

$$(12) \qquad C_m = \frac{B_m}{\{(2m+1)k/\rho h + 1\}\rho^{2m+1}},$$

which, along with (6), represents a solution of (1) through (4), thereby permitting the temperature distribution within the hemispherical probe tip to be calculated.

POST-FURNACE DRAWDOWN OF GLASS FIBER*

J. A. LEWIS†

Abstract. Fiber for optical waveguide is commonly made by drawing from a softened glass billet in a furnace. Here we calculate the small secondary drawdown which occurs after the hot fiber leaves the furnace. We model the fiber as a viscous, heat-transferring fluid under uniaxial tension. Use of the rapid variation of viscosity with temperature, typical of glasses, allows a simplification giving the maximum allowable fiber exit temperature explicitly in a wide variety of cases.

1. The model. Glass fiber for optical waveguide is commonly made by feeding a carefully fabricated glass billet (the so-called "preform") into a tube furnace where it is softened and drawn into fiber, as shown schematically in Fig. 1. As in most glass forming processes, the rapid change of viscosity with temperature prevents any appreciable drawdown after the fiber leaves the furnace. Nevertheless, for fiber guide even this small secondary drawdown is of interest.

We estimate this drawdown by modeling the fiber as a viscous, incompressible fluid with a free surface. (See [1] for a general discussion of glass viscosity and [2] for the fused silica used for fiber guides.) For a fiber of slowly varying radius $a(z)$ the fiber velocity $v(z)$, axial stress $t_{zz}(z)$, and fiber temperature $T(z)$ satisfy the equations

$$(1) \qquad Q = \pi a^2 v = \text{const.},$$

---

* Received by the editors November 1, 1977, and in final revised form April 21, 1978.

† Bell Laboratories, Murray Hill, New Jersey 07974.

(2) $$F = \pi a^2 t_{zz} = \text{const.},$$

(3) $$\pi a^2 v c \, dT/dz = -2\pi a H,$$

where Q is the volume flux, F the draw force, constant when fluid inertia and surface tension are neglected, c the heat capacity per unit volume, H the surface cooling rate, and axial heat conduction is neglected. (See § 3.)

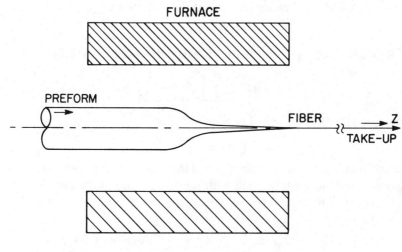

FIG. 1 *Fiber drawing.*

Now the axial stress t_{zz} and radial stress t_{rr}, which must vanish in a fiber with its surface free and its radius slowly varying, are given in general by the relations

$$t_{zz} = -p + 2\mu \, \partial v/\partial z, \qquad t_{rr} = -p + 2\mu \, \partial u/\partial r,$$

for pressure p, viscosity $\mu = \mu(T)$, and radial velocity u. With $v = v(z)$ the continuity equation

$$\partial u/\partial r + u/r + \partial v/\partial z = 0$$

implies that

$$u = -(r/2)(dv/dz),$$

so that

$$t_{rr} = -p - \mu \, dv/dz.$$

With $t_{rr} \equiv 0$, $p = -\mu \, dv/dz$, and

(4) $$t_{zz} = 3\mu \, dv/dz.$$

(3μ is the viscous analogue of the Young's modulus $3G$ for an incompressible elastic medium with shear modulus G.) Elimination of v and t_{zz} then yields the pair of equations for the free surface $r = a(z)$ and fiber temperature $T(z)$

(5) $$da/dz = -Fa/(6Q\mu),$$

(6) $$dT/dz = -2\pi Ha/(cQ),$$

strongly nonlinear because μ is a rapidly varying function of T, typically of the form

(7) $$\mu = \mu_0 \exp [\beta_0(T_0/T - 1)],$$

with $\beta_0 \gg 1$. On the other hand, they are autonomous, H typically having the form

(8) $$H = H_r(T) + (a/a_0)^m H_c(T),$$

giving the combined effect of radiation and convection. Thus, we finally obtain

(9) $$da/dT = cF/(12\pi\mu(T)H(T, a)),$$

giving $a = a(T)$, with the initial condition at the furnace exit

$$a(T_0) = a_0.$$

2. Radiative cooling. For radiative cooling $H = H_r(T)$, (9) gives

$$a = a_0 - \frac{cF}{12\pi} \int_T^{T_0} \frac{dT}{\mu(T)H_r(T)}.$$

Furthermore, with $H_r \sim T^4$, the change of variable

$$\theta = \beta_0(T_0/T - 1)$$

reduces the integral to elementary form. However, we shall not make use of this special result. Instead we note that, with $\mu \sim \exp(\theta)$, the integrand is exponentially small for large θ. Only for $\theta \ll \beta_0$ is there an appreciable contribution to the integral. In this range, however,

$$H_r = H_0[1 + O(1/\beta_0)], \qquad dT = -(T_0 \, d\theta/\beta_0)[1 + O(1/\beta_0)],$$

so that, for $\beta_0 \gg 1$, to a first approximation,

(10) $$a = a_0 - \Delta a(1 - e^{-\theta}),$$

where

(11) $$\Delta a = cFT_0/(12\pi\mu_0 H_0\beta_0).$$

This is the desired result, giving the radius change during post-furnace cooling. It is obviously valid for *any* slowly varying function $H_r(T)$. It is a simple exercise to show that it is also valid for the cooling rate given by (8), provided that both $H_r(T)$ and $H_c(T)$ are slowly varying and $\Delta a \ll a_0$.

3. An example. Figure 2 shows the relative radius change $\Delta = \Delta a/a_0$ as a function of fiber exit temperature T_0 for a silica fiber, with $a_0 = 50 \ \mu m$, $F = 10^4$ dyne, $c = 0.6 \ \text{cal/cm}^3\text{-}^\circ C$,

$$\mu_0 = (3.4 \times 10^5 \ \text{Poise}) \exp [30.56(2300/T_0 - 1)],$$

$$\beta_0 = 30.56(2300/T_0),$$

(see [2]), and radiative cooling, with

$$H = (1.37 \times 10^{-13})T_0^4 \ \text{cal/cm}^2\text{-sec.}$$

Drawdown increases rapidly with increasing exit temperature T_0. For $T_0 > 2250^\circ K$, the exit viscosity μ_0 is too small to support the given draw force F and no steady state exists ($\Delta > 1$).

Direct experimental confirmation of this result is lacking, because of the difficulty of measuring T_0 and a_0 during the drawing process. However, numerical solution of

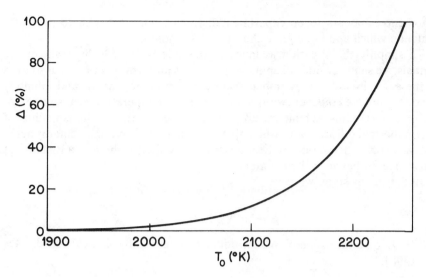

FIG. 2. *Secondary drawdown as a function of fiber exit temperatures.*

(1) to (4) in the furnace, using crudely measured values of furnace heating rate—H_f in place of H, gives a fiber profile $a = a(z)$ which agrees well with a measured profile "frozen" by severing the fiber at the takeup end and rapidly withdrawing it from the furnace. It also gives a value of fiber exit temperature T_0 consistent with the observed small secondary drawdown.

Finally, to estimate the effect of axial conduction, we differentiate (6) and use (5) to obtain $(\pi a^2 K d^2 T / dz^2)/(cQ\, dT/dz) = -(k/v)(F/(6\mu Q) + 8\pi a H/(cQT))$. For thermal diffusivity $k = K/c \sim 10^{-2}$ cal/sec-cm-°C and $v \sim 10^2$ cm/sec, $k/v \sim 10^{-4}$ cm, while both the drawing length $6\mu Q/F$ and the cooling length $8\pi a H/(cQT)$ are a few centimeters. Thus, except in a tiny region near $z = 0$, where a conduction boundary layer must be inserted to make dT/dz continuous, axial conduction is negligible.

REFERENCES

[1] G. W. MOREY, *The Properties of Glass*, Reinhold, New York, 2nd ed., 1954.
[2] J. F. BACON, A. A. HASAPIS AND J. W. WHOLLEY, JR., *Viscosity and density of molten silica and high silica content glasses*, Tech. Rep. RAD-TR-9(7)-59-35, AVCO Corp., Wilmington, MA.

A BUNDLING PROBLEM*

M. L. GLASSER† AND S. G. DAVISON‡

It is a primeval observation of most warm blooded creatures that by huddling together they can keep each other warm. However, most of the work on this

* Received by the editors July 1, 1975, and in revised form April 17, 1977.
† Department of Applied Mathematics, Clarkson University, Potsdam, N.Y. 13676.
‡ Department of Applied Mathematics, University of Waterloo, Waterloo, Ontario, Canada N2L 3G1.

phenomenon appears to be experimental. In this note, we wish to present a calculation which applies, e.g., to the case of armadillos.

To simplify the situation as much as possible without losing sight of the essentials, we shall consider sleeping armadillos as uniform spherical heat sources and the earth in which they imbed themselves as homogeneous and infinitely extended. Thus we consider two spherical constant temperature heat sources, of the same (unit) radius, at temperature $T = T_0$ above that of an infinite medium of uniform thermal diffusivity, in which they are embedded. We shall determine, in the steady state, the amount of heat given off by either of them as a function of the distance $2a$ between their centers.

In the steady state, we have

$$\nabla^2 T(\mathbf{r}) = 0 \tag{1}$$

with $T = T_0$ on either sphere. We introduce a system of bispherical coordinates [1, p. 1298].

$$-\infty < \eta < \infty, \quad 0 \le \theta \le \pi, \quad 0 \le \phi \le 2\pi$$

with scale parameter $c = (a^2 - 1)^{1/2}$. Then the spheres are the coordinate surfaces $\eta = \pm\eta_1$, with $\cosh \eta_1 = a$. The general solution of Laplace's equation, which is even in η and independent of ϕ, is

$$T(\eta, \theta) = (\cosh \eta - \cos \theta)^{1/2} \sum_{n=0}^{\infty} A_n \cosh (n + \tfrac{1}{2})\eta P_n(\cos \theta). \tag{2}$$

By noting the generating formula for the Legendre polynomials in the form

$$(\cosh \eta - \cos \theta)^{-1/2} = \sqrt{2} \sum_{n=0}^{\infty} e^{-(n+1/2)\eta} P_n(\cos \theta), \tag{3}$$

we see that the temperature distribution is given by

$$T(\eta, \theta) = 2^{1/2} T_0 (\cosh \eta - \cos \theta)^{1/2} \sum_{n=0}^{\infty} e^{-(n+1/2)\eta_1} \frac{\cosh (n + 1/2)\eta}{\cosh (n + 1/2)\eta_1} P_n(\cos \theta). \tag{4}$$

We require the heat flux from one of the spheres, given by

$$Q = \kappa \iint_{\text{sphere}} dS \, \hat{n} \cdot \nabla T$$

$$= (a^2 - 1)\kappa \int_0^{2\pi} d\phi \int_0^{\pi} d\theta \frac{\sin \theta}{(\cosh \eta_1 - \cos \theta)^2} \hat{n} \cdot \nabla T|_{\eta=\eta_1}, \tag{5}$$

where κ is a constant whose value depends on the set of units used and $\hat{n}$ is the outward normal to the sphere. In the present case, $\hat{n} = \hat{\imath}_\eta$ so

$$\hat{n} \cdot \nabla T\Big|_{\eta=\eta_1} = \frac{(\cosh \eta_1 - \cos \theta)}{(a^2 - 1)^{1/2}} \frac{\partial T}{\partial \eta}\Big|_{\eta=\eta_1}$$

(6)
$$= \frac{2^{-1/2} T_0}{(a^2-1)^{1/2}} \sinh \eta_1 \left(\cosh \eta_1 \right.$$

$$\left. - \cos \theta \right)^{1/2} \sum_{n=0}^{\infty} e^{-(n+1/2)\eta_1} P_n(\cos \theta)$$

$$+ \frac{2^{1/2} T_0}{(a^2-1)^{1/2}} (\cosh \eta_1 - \cos \theta)^{3/2}$$

$$\cdot \sum_{n=0}^{\infty} (n+1/2) e^{-(n+1/2)\eta_1} \tanh (n+1/2)\eta_1 P_n(\cos \theta).$$

The series in (6) are easily evaluated by using (3) and the expansion

(7)
$$\tanh x = 1 + 2 \sum_{k=1}^{\infty} (-1)^k e^{-2kx}.$$

The angular integrations required are elementary and, after some simplification, we find

(8)
$$Q(a) = 4\pi\kappa T_0 \sum_{k=1}^{\infty} \frac{(-1)^{k+1} \sinh \eta_1}{\sinh k\eta_1}.$$

The quantity η_1 can be eliminated by means of the relation $\cosh \eta_1 = a$, so alternatively, we have

(9)
$$Q(a) = 4\pi\kappa T_0 \sum_{k=0}^{\infty} (-1)^k U_k^{-1}(a)$$

where $U_k^{-1}(a)$ is the reciprocal of the Chebyshev polynomial of the second kind [2, Chap. 22]. The first term of (9) is precisely the result for an isolated sphere

(10)
$$Q_0 = 4\pi\kappa T_0 = Q(\infty).$$

The remainder of the series, which is negative, represents the effect of the second sphere and shows that its presence, no matter how remote, reduces the heat output of the first sphere. Physical intuition suggests that $Q(a)$ should increase steadily from the value $Q(1) = 4\pi\kappa T_0 \ln 2$ to the value Q_0 as the distance between the spheres increases. This is supported by a numerical calculation which shows

TABLE 1

a	$Q(a)/Q_0$	a	$Q(a)/Q_0$
1.00	0.693147	2.0	0.802656
1.01	0.694849	2.5	0.834407
1.02	0.695882	3.0	0.857699
1.03	0.697725		
1.04	0.699207	4.0	0.990113
1.05	0.700629	5.0	0.909184
1.1	0.707721	10.0	0.952381
1.2	0.721281	20.0	0.975610
1.3	0.734090	30.0	0.983611
1.4	0.745882	40.0	0.987656
1.5	0.757311	50.0	0.990100

that the series in (8) is rapidly convergent even for values as small as $a = 1.01$. It was with some surprise that we were unable to find a mathematical proof of the monotonic increase of $Q(a)$ for $a > 1$ which is an interesting open problem [3]. The numerical calculation (see Table 1) supports the empirical result that the greatest warming effect occurs for tangency. Since $Q(1)/Q(\infty) = \ln 2$, the heat output of each sphere is reduced by nearly a third. The extension of this calculation to three (or more) spheres should be valuable in the study of Armadillo colonies.

REFERENCES

[1] P. H. MORSE AND H. FESHBACH, *Methods of Theoretical Physics*, McGraw-Hill, New York,. 1953.
[2] M. ABRAMOWITZ AND I. STEGUN, *Handbook of Mathematical Functions*, (N.B.S. AMS #55), American Mathematical Society, Providence, RI, 1964.
[3] M. L. GLASSER, *Problem 77-5*, this Review, 19 (1977), p. 148.

Steady-State Plasma Arc

Problem 60-6, by JERRY YOS (AVCO Research and Advanced Development Division).

In studying the positive column of an electric arc, a model is considered in which the arc strikes between two plane electrodes in an infinite channel. The sides of the channel are held at a fixed temperature $T = 0$ and are perfect electrical insulators. The electrodes are held at fixed potentials and are perfect thermal insulators (Fig. 1). The steady-state distributions of temperature and electrical potential in the arc are then determined by the equations for the con-

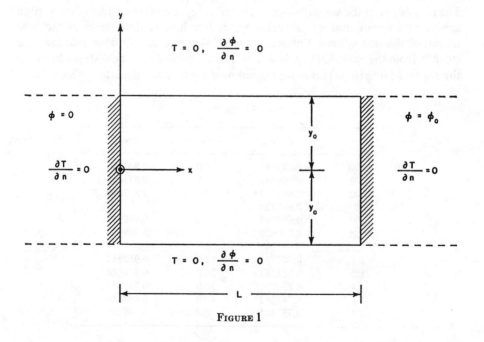

FIGURE 1

servation of current and for the energy balance between the electrical heating of the gas and the cooling due to thermal conduction to the walls, i.e.

(1) $$\nabla \cdot (\sigma \nabla \phi) = 0,$$

(2) $$\sigma (\nabla \phi)^2 = -\nabla \cdot (k \nabla T).$$

Here,
$$\sigma = \sigma(T), \qquad k = k(T)$$

are the electrical and thermal conductivities of the gas, respectively. The boundary conditions at the electrodes are

(3)
$$\phi = 0, \qquad \frac{\partial T}{\partial n} = 0, \quad \text{for} \quad x = 0,$$

$$\phi = \phi_0, \qquad \frac{\partial T}{\partial n} = 0, \quad \text{for} \quad x = L.$$

The boundary conditions at the walls of the channel are

(4) $$T = 0, \qquad \frac{\partial \phi}{\partial n} = 0, \quad \text{for} \quad y = \pm Y_0.$$

One solution of this problem can be readily found in the form

$$\phi = \frac{\phi_0 x}{L}, \qquad T = T(y),$$

where $T(y)$ is determined implicitly from

(5)
$$\frac{\phi_0 y}{L} = \int_T^{T_m} \frac{k \, dT}{\sqrt{2 \int_T^{T_m} k\sigma \, dT}}$$

and where the maximum temperature T_m is given by

(6)
$$\frac{\phi_0 Y_0}{L} = \int_0^{T_m} \frac{k \, dT}{\sqrt{2 \int_T^{T_m} k\sigma \, dT}}.$$

Is this solution unique?

A Free Boundary Problem

Problem 60-7, by CHRIS SHERMAN (AVCO Research and Advanced Development Division).

In analyzing the phenomena occurring in the column of an electric arc operating with forced convection, a set of coupled partial differential equations involving the dependent variables $\mathbf{J}$, $\mathbf{E}$ (electric current density and field strength), $\mathbf{U}$, p (velocity and pressure of the gas stream), and T (gas temperature) arises (see Problem 60-6). This set may be reduced to a single equation by the following drastic simplifying assumptions: $\mathbf{U}$ is taken to be a constant directed along the x-axis of a rectangular coordinate system, $\mathbf{E}$ a constant directed along the z-axis, and T is assumed to be a function of x and y only. $\mathbf{J}$ is related to $\mathbf{E}$ by $\mathbf{J} = \sigma(T)\mathbf{E}$ where $\sigma(T)$, the electrical conductivity, is a known function of temperature. The equation which results is the conservation of energy equation

$$(1) \qquad \frac{\partial^2 T}{\partial x^2} + \frac{\partial^2 T}{\partial y^2} - 2a\frac{\partial T}{\partial x} + b^2\sigma(T) = 0,$$

where a and b are constants. If

$$\sigma(T) = cT \qquad\qquad (T \geq T_0) \quad \text{in region 1,}$$

$$\sigma(T) = 0 \qquad\qquad (T \leq T_0) \quad \text{in region 2,}$$

where c and T_0 are known constants, a solution of equation (1) which includes the determination of the shape of the boundary separating regions 1 and 2 is sought. The boundary conditions to be satisfied are

    1. the separating boundary B passes through the point $(x_0, 0)$,
    2. T and $\partial T/\partial n$ are continuous across B,
    3. T bounded,
    4. $\lim_{\sqrt{x^2+y^2}\to\infty} T = 0$.

The following questions are of particular interest:

    (a) Does a stable solution exist?
    (b) If it does, is the solution unique?
    (c) Is the separating boundary B closed?

A similar alternative but simpler problem in which

$$\sigma(T) = c, \qquad\qquad (T \geq T_0) \quad \text{in region 1,}$$

$$\sigma(T) = 0, \qquad\qquad (T \leq T_0) \quad \text{in region 2,}$$

is also of interest.

Ohmic Heating

Problem 69-11, by J. A. LEWIS (Bell Telephone Laboratories).

Consider an isotropic, homogeneous, conducting body, ohmically heated by the passage of direct current between perfectly conducting electrodes on its surface, the rest of the surface being electrically and thermally insulated. Show that the maximum temperature in the body depends only on the potential difference between the electrodes, the electrode temperature, and the electrical and thermal conductivities, in general functions of the temperature, and is independent of the size and shape of the body and of the electrode configuration.

    Solution by the proposer.

    The potential V and temperature T in a body with electrical and thermal conductivities $\sigma(T)$, $k(T)$ satisfy the equations

$$(1) \qquad \nabla\cdot(\sigma\nabla V) = 0, \qquad \nabla\cdot(k\nabla T) + \sigma|\nabla V|^2 = 0$$

and the boundary conditions on the electrodes

$$V = V_0 \ \text{ on } S_0, \qquad V = V_1 \ \text{ on } S_1, \qquad T = T_0 \ \text{ on } S_0 + S_1,$$

while on the insulated surface,

$$\partial V / \partial n = \partial T / \partial n = 0.$$

One may verify by direct substitution that the pair of functions V, $T(V)$ satisfy the equations and boundary conditions provided that $T(V)$ is given by

(2) $$\int_{T_0}^{T(V)} \frac{k(T)}{\sigma(T)} dT = \frac{1}{2}(V - V_0)(V_1 - V).$$

The maximum temperature, obtained by differentiation with respect to V, is given by

$$\int_{T_0}^{T_{max}} \frac{k(T)}{\sigma(T)} dT = \frac{1}{8}(V_1 - V_0)^2,$$

which is the desired result.

Editorial note. The result is also valid if the boundary condition $T = T_0$ on $S_0 + S_1$ is extended to

$$T = T_0 \quad \text{on } S_0 \quad \text{and} \quad T = T_1 \quad \text{on } S_1.$$

Assuming that $V_1 > V_0$, $T_1 > T_0$, we replace (2) by

(2)′ $$\int_{T_0}^{T(V)} \frac{k(T)}{\sigma(T)} dT = \frac{1}{2}(V - V_0)(V_1 - V) + \frac{V - V_0}{V_1 - V_0} \int_{T_0}^{T_1} \frac{k(T)}{\sigma(T)} dT.$$

It still follows easily that the pair of functions V, $T(V)$ satisfy (1) and the boundary conditions.

The maximum temperature, obtained by differentiation of the right-hand side of (2)′ (or by completing the square), occurs when $V = a + (V_0 + V_1)/2$ and is now given by

$$\int_{T_0}^{T_{max}} \frac{k(T)}{\sigma(T)} dT = \frac{1}{8}(V_1 - V_0 + 2a)^2,$$

where

$$a = \frac{1}{V_1 - V_0} \int_{T_0}^{T_1} \frac{k(T)}{\sigma(T)} dT,$$

again giving the desired result.

Since $V_0 \leqq V \leqq V_1$, the previous equation for T_{max} is only valid if $2a < V_1 - V_0$. If $2a \geqq V_1 - V_0$, then $T_{max} = T_1$ and occurs on the boundary S_1. A similar argument applies for the case $T_1 < T_0$. [M.S.K.]

A Heat Transfer Problem

Problem 70-23, by J. ERNEST WILKINS, JR. (Howard University).

Solve the partial differential equation

$$k\nabla^2 T \equiv k\left\{ \frac{1}{r} \frac{\partial}{\partial r}\left(r \frac{\partial T}{\partial r}\right) + \frac{1}{r^2} \frac{\partial^2 T}{\partial \theta^2} \right\} = -q, \qquad 0 < r < a, \quad 0 \leqq \theta \leqq 2\pi,$$

subject to the boundary condition

$$-k\frac{\partial T}{\partial r} = h(\theta)T, \qquad r = a, \quad 0 \le \theta \le 2\pi,$$

which describes the temperature distribution T in an infinitely long, uniformly heated rod of radius a and thermal conductivity k, when the heat transfer co-efficient h varies around its mean value h_0 over the circumference in such a manner that

$$h(\theta) = h_0(1 + \varepsilon \cos n\theta)$$

for some nonzero integer n and some number ε such that $0 < \varepsilon < 1$.

Solution by G. W. VELTKAMP (Technological University, Eindhoven, Netherlands).

Since $T = -qr^2/(4k)$ is a solution of the inhomogeneous differential equation, we assume that the full solution can be represented as

$$(1) \qquad T = \frac{qa^2}{4k}\left[1 - \frac{r^2}{a^2} + \sum_{j=0}^{\infty} {}' c_j\left(\frac{r}{a}\right)^{jn} \cos jn\theta \right]$$

(where the $'$ indicates that the term with $j = 0$ should be taken half). This "ansatz" is motivated by separation of variables together with the regularity condition at $r = 0$ and the observation that the mode-coupling mechanism constituted by the boundary condition will not excitate other modes than those present in (1).

We write the boundary condition as

$$(2) \qquad -\frac{a}{n}\frac{\partial T}{\partial r} = (v + \lambda \cos n\theta)T \quad \text{at } r = a,$$

where

$$v = h_0 a/(nk), \qquad \lambda = \varepsilon v.$$

Substitution of (1) into (2) gives

$$\frac{2}{n} - \sum_{j=1}^{\infty} jc_j \cos jn\theta = (v + \lambda \cos n\theta) \sum_{j=0}^{\infty} {}' c_j \cos jn\theta$$

$$= v \sum_{j=0}^{\infty} {}' c_j \cos jn\theta + \tfrac{1}{2}\lambda\left[c_1 + \sum_{j=1}^{\infty} (c_{j+1} + c_{j-1}) \cos jn\theta \right].$$

Equating coefficients of $\cos jn\theta$ gives

$$(3) \qquad vc_0 + \lambda c_1 = 4/n,$$

$$(4) \qquad (v + j)c_j + \tfrac{1}{2}\lambda(c_{j+1} + c_{j-1}) = 0.$$

Comparing the difference equation (4) with the recurrence relation for the Bessel functions, and observing that convergence of (1) at $r = a$ implies boundedness of the c_j, we find

$$c_j = A(-1)^j J_{j+v}(\lambda).$$

Substitution into (3) gives

$$\frac{4}{n} = A(\nu J_\nu(\lambda) - \lambda J_{\nu+1}(\lambda)) = A\lambda J_\nu'(\lambda).$$

It is well known that $J_\nu'(z) \neq 0$ for $0 < |z| < \nu$. Hence the condition $0 < \varepsilon < 1$ ensures that $J_\nu'(\lambda) = J_\nu'(\varepsilon\nu)$ is different from zero. Therefore the solution is

$$T = \frac{qa^2}{4k}\left[1 - \frac{r^2}{a^2} + \frac{4}{n}\sum_{j=0}^{\infty}{}' (-1)^j \frac{J_{j+\nu}(\varepsilon\nu)}{\varepsilon\nu J_\nu'(\varepsilon\nu)}\left(\frac{r}{a}\right)^{jn} \cos jn\theta \right],$$

where $\nu = h_0 a/(nk)$.

A Steady-State Temperature

Problem 62-1, by ALAN L. TRITTER (Data Processing Inc.)
AND A. I. MLAVSKY (Tyco, Inc.).

Consider the steady-state temperature $(T(r, z))$ distribution boundary-value problem for an infinite solid bounded by two parallel planes:

(1) $$\frac{\partial^2 T}{\partial r^2} + \frac{1}{r}\frac{\partial T}{\partial r} + \frac{\partial^2 T}{\partial z^2} = 0, \qquad 0 < z < H, \qquad r \geq 0,$$

$$\left\{ -k\frac{\partial T}{\partial z} = \begin{matrix} Q, r < R \\ 0, r > R \end{matrix} \right\}_{z=0},$$

$$\{ T = 0 \}_{z=H},$$

$$|T| < M \text{ (boundedness condition)},$$

(all the parameters involved are constants). Determine the temperature at the point $r = z = 0$.

The solutions by E. DEUTSCH (Institute of Mathematics, Bucharest, Rumania), THOMAS ROGGE (Iowa State University), J. ERNEST WILKINS JR. (General Dynamics Corporation) and M. S. KLAMKIN (University of Buffalo) were essentially the same and are given by the following:

Letting

$$\phi(\lambda, z) = \int_0^\infty r J_0(\lambda r) T(r, z) \, dr,$$

it follows by integration by parts that the Hankel transform of Eq. (1) is

$$\{ D^2 - \lambda^2 \}\phi = 0,$$

subject to the boundary conditions

$$k\frac{\partial\phi}{\partial z}\bigg]_{z=0} = \int_0^R Q r J_0(\lambda r) \, dr = \frac{QR J_1(\lambda R)}{\lambda},$$

$$\{ \phi = 0 \}_{z=H}.$$

Consequently,

$$\phi(\lambda, z) = \frac{QR}{k}\frac{J_1(\lambda R)}{\lambda^2}\frac{\sinh \lambda(H - z)}{\cosh H}.$$

---

* H. T. Davis, *Introduction to Nonlinear Differential and Integral Equations*, U. S. Atomic Energy Commission, 1960, pp. 405–407.

Inverting the latter transform:

$$(2) \qquad T(r, z) = \frac{QR}{k} \int_0^\infty \frac{\sinh \lambda(H - z)}{\lambda \cosh \lambda H} J_0(\lambda r) J_1(\lambda R) \, d\lambda.$$

On letting $H \to \infty$, we obtain

$$\lim_{H \to \infty} T(r, z) = \frac{QR}{k} \int_0^\infty e^{-\lambda z} J_0(\lambda r) J_1(\lambda R) \frac{d\lambda}{\lambda}$$

which corresponds to a result given in Carslaw and Jaeger, Conduction of Heat in Solids, Oxford University Press, London, 1959, p. 215.

In particular, the temperature at $r = 0$, $z = 0$, is given by

$$(3) \qquad T(0, 0) = \frac{QR}{k} \int_0^\infty \lambda^{-1} J_1(\lambda R) \tanh \lambda H \, d\lambda.$$

The series expansion

$$(4) \qquad T(0, 0) = \frac{QR}{k} \left\{ 1 - \frac{R}{H} \sum_{m=1}^\infty \frac{(-1)^{m+1}}{m + \sqrt{m^2 + R^2/4H^2}} \right\}$$

is obtained by expanding $\tanh \lambda H$ into the exponential series

$$\tanh \lambda H = 1 - 2 \sum_{m=1}^\infty (-1)^{m+1} e^{-2m\lambda H}$$

and employing the integral

$$\int_0^\infty \lambda^{-1} e^{-a\lambda} J_1(\lambda R) \, d\lambda = (\sqrt{a^2 + R^2} - a)/R$$

(Watson, Theory of Bessel Functions, Cambridge University Press, London, 1952, p. 386).

Deutsch also obtains the alternate series expansion

$$(5) \qquad T(0, 0) = \frac{QH}{k} \left\{ 1 - \frac{4R}{\pi H} \sum_{n=1,3,5,\cdots} \frac{1}{n} K_1 \left(\frac{\pi n R}{2H} \right) \right\}.$$

This latter expansion is suitable for numerical calculations for large values of the parameter R/H. It shows that $\lim_{H \to 0} T(0, 0) = 0$ which is what one would expect physically. It also implies that

$$\sum_{m=1}^\infty \frac{(-1)^{m+1}}{m + \sqrt{m^2 + \lambda^2}} = \frac{1}{2\lambda} - \frac{1}{4\lambda^2} \quad \text{as} \quad \lambda \to \infty.$$

For small values of R/H, Wilkins expanded

$$\sum_{m=1}^\infty \frac{(-1)^{m+1}}{m + \sqrt{m^2 + \lambda^2}} = \sum_{m=1}^\infty \frac{(-1)^{m+1}}{\lambda^2} \left\{ \sqrt{m^2 + \lambda^2} - m \right\}$$

into powers of λ using the binomial theorem. The coefficients depend on the Zeta function $\zeta(2m + 1)$. In particular,

$$\lim_{H \to \infty} T(0, 0) = \frac{QR}{k}.$$

By a superposition integral, we can also find the temperature distribution from (2) for the more general flux condition

$$k \frac{\partial T}{\partial z}\bigg]_{z=0} = Q(r), \qquad r \geqq 0.$$

Amos also gives the following series expansion for $T(r, z)$:

$$T(r, z) = \begin{cases} \dfrac{Q}{k}(H - z) - \dfrac{4QR}{\pi k} \sum_{n=1}^{\infty} \dfrac{(-1)^{n+1}}{2n - 1} I_0(\lambda_n r) K_1(\lambda_n R) \sin \lambda_n(H - z), \\ \hspace{8cm} r < R, \\ \dfrac{4QR}{\pi k} \sum_{n=1}^{\infty} \dfrac{(-1)^{n+1}}{2n - 1} K_0(\lambda_n r) I_1(\lambda_n R) \sin \lambda_n(H - z), \hspace{0.5cm} r \geqq R, \end{cases}$$

where $\lambda_n = (2n - 1)\pi/2H$. This reduces to (5) for $r = z = 0$.

For extensions of this problem to the unsteady-state in finite or infinite cylinders see *Unsteady Heat Transfer into a Cylinder Subject to a Space- and Time-Varying Surface Flux,* by M. S. Klamkin, Tr-2-58-5, AVCO Research and Advanced Development Division, May, 1958.

Steady-State Diffusion-Convection

Problem 60-1, by G. H. F. GARDNER (Gulf Research & Development Company).

When a homogeneous fluid flows through a porous material, such as sandstone or packings of small particles, its molecules are scattered by the combined action of molecular diffusion and convective mixing. Thus a sphere of tagged fluid particles expands as it moves and its deformation may be resolved into the longitudinal and transverse component. The transverse mixing, which is many times less than the longitudinal mixing, has been investigated at turbulent rates of flow but has received little attention when the flow rate is low as for subterranean fluid movement. The following experiment was set up to investigate transverse mixing at low flow rates.

A rectangular porous block with impermeable sides was mounted with one side horizontal. An impermeable horizontal barrier AB divided the block into equal parts for about one-third of its length. One fluid was pumped at a constant rate into the block above AB and another was pumped at an equal rate below AB. After passing B the fluids mingled and a steady-state distribution was attained.

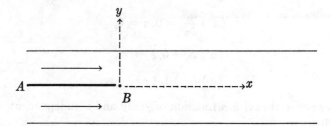

The fluids have approximately equal densities and the heavier was flowed under AB so that the equilibrium would be stable.

The steady-state distribution of the fluids is assumed to be given by

(1) $$\frac{\partial S}{\partial x} = \alpha \frac{\partial^2 S}{\partial x^2} + \beta \frac{\partial^2 S}{\partial y^2},$$

where $S(x, y)$ denotes the fractional amount of the lower fluid present at the point (x, y). The equation was obtained under the following assumptions (for an analogous problem, see H. Bateman, *Partial Differential Equations*, p. 343):

(1) Streamline motion and molecular diffusion cause the dispersion.
(2) The coefficient α reflects the dispersion of flow caused by velocity variations along each streamline and molecular diffusion in the direction of flow.
(3) The coefficient β reflects the dispersion perpendicular to the direction of flow caused by diffusion between steamlines.

The boundary conditions to be satisfied are

(a) $\frac{\partial S}{\partial y} = 0$ on the impermeable boundaries,

(b) $S \rightarrow 0$ for $y > 0$, $x \rightarrow -\infty$, and

(c) $S \rightarrow 1$ for $y < 0$, $x \rightarrow -\infty$.

Solve equation (1) for an infinite medium. Here boundary condition (a) is replaced by

(a') $\frac{\partial S}{\partial y} = 0$ for $y = 0$, $x < 0$.

Solution by the proposer.

The use of parabolic coordinates (ξ, η) instead of rectangular coordinates (x, y) is more appropriate because the semi-infinite boundary $y = 0$, $x < 0$ is simply given by $\eta = 0$. Thus, writing

(2) $$\frac{x}{\alpha} = \frac{1}{2}(\eta^2 - \xi^2), \frac{y}{\sqrt{\alpha\beta}} = \xi\eta,$$

the differential equation is transformed to

(3) $$\frac{\partial^2 S}{\partial \xi^2} + \frac{\partial^2 S}{\partial \eta^2} + \xi \frac{\partial S}{\partial \xi} - \eta \frac{\partial S}{\partial \eta} = 0,$$

and the boundary conditions become

(a") $\frac{\partial S}{\partial \eta} = 0, \eta = 0,$

(b') $S \rightarrow 0, \xi \rightarrow \infty,$

(c') $S \rightarrow 1, \xi \rightarrow -\infty.$

If we now assume that S is a function of ξ only and is independent of η, equation (3) reduces to

(4) $$\frac{d^2 S}{d\xi^2} + \xi \frac{dS}{d\xi} = 0.$$

The general solution of equation (4) may be written

(5)
$$S = A \int_0^{\xi/\sqrt{2}} e^{-u^2}\, du + B.$$

Boundary condition (a″) is obviously satisfied. Conditions (b′) and (c′) can be satisfied by choosing the constants A and B appropriately. Hence the required solution may be written,

(6)
$$S = \tfrac{1}{2}[1 - erf\ \xi/\sqrt{2}].$$

The curves of constant concentration are given by constant values of ξ and therefore are parabolas confocal with the end of the barrier.

One simple result is perhaps noteworthy. The vertical concentration gradient at points on the x-axis is given by

(7)
$$\left(\frac{\partial S}{\partial y}\right)_{y=0} = \frac{1}{\sqrt{4\pi\beta x}}\ .$$

It is independent of α and hence gives a convenient way of measuring β experimentally.

Generalization of the solution for an infinite medium, given by equation (6), to certain bounded regions is easily accomplished by use of the method of images.

Resistance of a Cut-Out Right Circular Cylinder

Problem 62-4, by ALAN L. TRITTER (Data Processing, Inc.).

Determine the resistance of the cut-out right circular cylinder (Fig. 1) between the two perfectly conducting cylindrical electrodes E_1 and E_2. All electrical properties are assumed to be constant and the two small circles of radius r are orthogonal to the large circle of radius R.

Solution by D. E. AMOS (Sandia Corporation).

The potential problem for the interior region bounded by the three circles can be solved by conformal mapping. We assign the potentials $V = 0$ and $V = V_0$ to the arcs of E_1 and E_2, respectively, which border this region. Since it is assumed that the remainder of the conductor is insulated on its surface, $\partial V/\partial n = 0$ on the arcs of the center circle between E_1 and E_2. The transformation is constructed by noticing that the points labeled $b + R \equiv ar$ and $b - R \equiv r^2/ar$ (Fig. 2) are images of one another in circles E_1 and E_2, where

$$a = R/r + \sqrt{(R/r)^2 + 1}$$

and

$$b = \sqrt{R^2 + r^2}.$$

We, therefore, form the transformation by adding logarithmic potentials of opposite signs, and normalize so that the image of E_1 is on the imaginary axis. Thus, we verify that

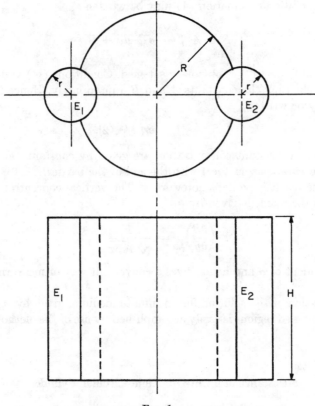

FIG. 1

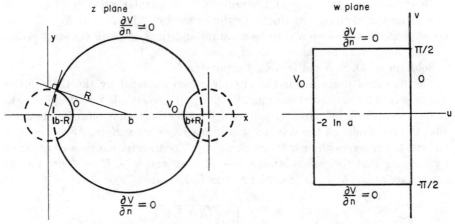

FIG. 2

$$w = \ln \frac{z - ar}{a(z - r^2/ar)}$$

$$= \ln \frac{1}{a} \left| \frac{z - ar}{z - r^2/ar} \right| + i \tan^{-1} \frac{2Ry}{(x - b)^2 + y^2 - R^2}$$

is the required transformation which takes the two-dimensional problem in the z plane into a one-dimensional problem in the w plane.

Image of E_1, $x^2 + y^2 = r^2$:

$$\frac{1}{a}\left|\frac{z - ar}{z - r^2/ar}\right| = \frac{1}{a}\sqrt{\frac{r^2 + R^2 - 2x(b + R) + b^2 + 2Rb}{r^2 + R^2 - 2x(b - R) + b^2 - 2Rb}}$$

$$= \frac{1}{a}\sqrt{\frac{b + R}{b - R}}$$

$$= 1.$$

Therefore, the image of E_1 is on the imaginary axis, $u = 0$.

Image of E_2, $(x - 2b)^2 + y^2 = r^2$:

$$\frac{1}{a}\left|\frac{z - ar}{z - r^2/ar}\right| = \frac{1}{a}\sqrt{\frac{2x(b - R) + r^2 - 4b^2 + (R + b)^2}{2x(b + r) + r^2 - 4b^2 + (b - R)^2}}$$

$$= \frac{1}{a}\sqrt{\frac{2x(b-R) + 2Rb - 2b^2}{2x(b+R) - 2Rb - 2b^2}}$$

$$= \frac{1}{a}\sqrt{\frac{b - R}{b + R}}$$

$$= \frac{1}{a^2}.$$

Therefore, the image of E_2 is on the line $u = \ln 1/a^2$.

Image of E_3, $(x - b)^2 + y^2 = R^2$:

$$v = \tan^{-1}\frac{2Ry}{(x - b)^2 + y^2 - R^2}.$$

For $y > 0$, $v \to -\pi/2$ as a point interior to E_3 approaches the boundary of E_3.

For $y < 0$, $v \to +\pi/2$ as a point interior to E_3 approaches the boundary of E_3.

Therefore, the image of the upper arc is on the line $v = -\pi/2$ and the image of the lower arc is on the line $v = +\pi/2$.

The solution of the problem in the w plane is

$$V = \frac{V_0}{\ln 1/a^2}u$$

and the total current in a cylinder of conductivity k and height H is

$$I_0 = \frac{-k(\pi H)\, V_0}{\ln 1/a^2}.$$

Finally, the resistance between E_1 and E_2 becomes

$$\frac{V_0}{I_0} = \frac{2\ln a}{\pi k H} = \frac{2}{\pi k H}\ln\left[\frac{R}{r} + \sqrt{\left(\frac{R}{r}\right)^2 + 1}\right].$$

THE MONITORING OF AIRBORNE DUST AND GRIT*

A. W. BUSH†, M. CROSS‡ AND R. D. GIBSON§

Abstract. The collection efficiency of a dust gauge is analysed using the inviscid theory of hydrodynamics. The classical methods of conformal mapping and the Milne-Thomson circle theorem

* Received by the editors November 25, 1975.
† Department of Mathematics and Statistics, Teesside Polytechnic, Cleveland, England.
‡ British Steel Corporation Research Laboratory, Middlesbrough, Cleveland, England.
§ Department of Mathematics and Statistics, Teesside Polytechnic, Cleveland, England.

are used to obtain the airflow around the dust gauge. Then, using the Stokes drag equations, the dust particle paths are found. The variation of the collection efficiency of the gauge with particle size is obtained and used to determine the true dust particle size distribution from an experimental distribution.

For a number of years, dust gauges have been used to monitor the deposition of dust from industrial chimneys (Lucas and Moore [2]; [1]). An obvious feature of dust pollution which causes complaint is the fouling of objects at ground level so that surfaces which would otherwise be clean and bright are masked by a film of dust.

As a dusty airstream passes around an object, the bulk of the dust, especially if it is fine, is carried around the object with the airstream, but some of the dust reaches the object by inertia. The dust from a modern plant has typically a free-falling speed of $0.3 \text{ m} \cdot \text{s}^{-1}$ and is carried by the wind at a speed which is on the average $5 \text{ m} \cdot \text{s}^{-1}$. Its path is therefore very nearly horizontal. For this reason, the collecting orifice of certain gauges (see, for example, Lucas and Moore [2] and Fig. 1) is a vertical slit in a vertical cylinder. Such a gauge is not 100 percent effective in collecting dust which is moving in its direction, and its collection efficiency will be greater for coarser particles and less for finer particles, the finer particles being swept around the gauge without leaving the airstream. Further-more, the collection efficiency of the gauge is dependent on the wind direction. It will be maximum when the wind is blowing straight into the opening and will fall to zero as the wind direction swings through a right angle in either direction. Here we examine a simple theoretical model of the flow past the gauge and examine the variation of collection efficiency with particle and wind characteristics.

The airflow past the gauge is modeled by considering the horizontal irrota-tional motion of an incompressible fluid past an infinite vertical cylinder of radius a with a vertical slit subtending an angle $2\pi - 4\alpha$ at the center. Boundary-layer effects and those due to the ends of the gauge are ignored. In the z-plane, where $z = x + iy$, the gauge has cross section, C say, which consists of the circular arc $z = a e^{i\theta}$ $(-2\alpha < \theta < +2\alpha)$. Now the conformal relation

$$(1) \qquad \frac{z - a e^{2i\alpha}}{z - a e^{-2i\alpha}} = \left(\frac{Z - ia\, e^{i\alpha}}{Z + ia\, e^{-i\alpha}}\right)^2$$

transforms the domain outside the circle $|Z| = a$ in the Z-plane into the domain outside C in the z-plane (Milne-Thomson [3, p. 207]). Simplifying (1), we obtain

$$(2) \qquad Z = \frac{(z - a) \pm \sqrt{(z - a)^2 + 4az \sin^2 \alpha}}{2 \sin \alpha},$$

where the positive sign in (2) applies if both $|z| > a$ and $x > -\cos \alpha$; otherwise the negative sign applies. Also, at infinity, z tends to $Z \sin \alpha$. In the Z-plane $(Z = X + iY)$, the complex potential, w, for the flow of a uniform stream with speed $V \sin \alpha$ inclined at an angle β to the negative X-axis past the cylinder $|Z| = a$ is

$$(3) \qquad w = -V \sin \alpha (Z e^{-i\beta} + (a^2/Z) e^{i\beta})$$

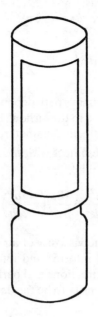

FIG. 1. *The dust gauge and collection jar*

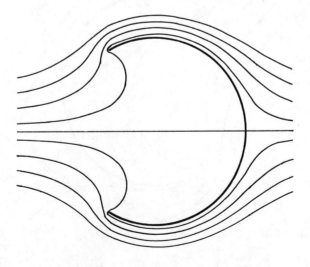

FIG. 2. *The streamlines for $\alpha = \pi/3$ and $\beta = 0$*

(Milne-Thomson [3, (3), p. 158]). The stream function $\psi = \text{Im}\,(w)$ can be found from (3) and streamlines in the cases $\alpha = \pi/3$, $\beta = 0$ and $\pi/4$ are plotted in Figs. 2 and 3, respectively. Furthermore if $\mathbf{v} = (v_x, v_y)$ is the fluid velocity, then

$$(4) \qquad v_x - iv_y = -\frac{dw}{dz} = V \sin \alpha \left(e^{-i\beta} - \frac{a^2}{Z^2} e^{i\beta} \right)\left(\frac{Z + a \sin \alpha}{a - z + 2Z \sin \alpha} \right).$$

The trajectory equation for a particle's path depends only on inertia forces and the drag due to the movement relative to the airflow. For the size range of particles of interest here (i.e., those whose effective diameter, d, is less than 100 microns) the Reynolds number is small. Thus we assume that the drag on the particle is the Stokes drag and the trajectory equation is

$$(5) \qquad \frac{d^2\mathbf{r}}{dt^2} = \frac{18\mu}{\rho\,d^2}\left(\mathbf{v} - \frac{d\mathbf{r}}{dt} \right)$$

where $\mathbf{r} = (x, y)$, t is time, μ is the viscosity of air and ρ is the particle density.

Substituting for $\mathbf{v}$ from (4) into (5) and integrating numerically gives the particle paths for various wind directions and particle diameters (see Figs. 4 and 5). We have taken the particle density to be the density of water; clearly from (5),

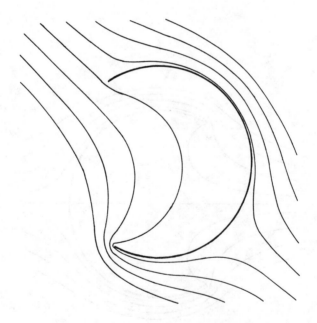

FIG. 3. The streamlines for $\alpha = \pi/3$ and $\beta = \pi/4$

varying ρ is equivalent to varying d^2. Figure 4 shows the outermost paths a particle may take and still be captured, and we let γ be the perpendicular distance between these paths. Clearly γ provides a measure of how efficiently the particles are captured. If the motion is rectilinear (i.e., uninfluenced by the fluid streamlines) then $\gamma = 2a \sin 2\alpha \cos \beta$. We define the collection efficiency, δ, by

$$\delta = \frac{\gamma}{2a \sin 2\alpha},$$

and Fig. 6 is a plot of δ as a function of β and d. As $d \to 0$, the particles stay with the airstream and $\delta \to 0$. As d increases, the particles are less influenced by the airstream, and $\delta \to \cos \beta$ as $d \to \infty$.

The variation of collection efficiency with particle size can be used to determine the true distribution of particles from an experimental distribution of particles. In Fig. 7, we plot an assumed experimental distribution of particles and the corresponding true distributions for wind speeds of $10 \text{ m} \cdot \text{s}^{-1}$ and $100 \text{ m} \cdot \text{s}^{-1}$ in the case $\beta = 0$. These results show how important the collection efficiency factor is, especially for small particles, and low wind speeds.

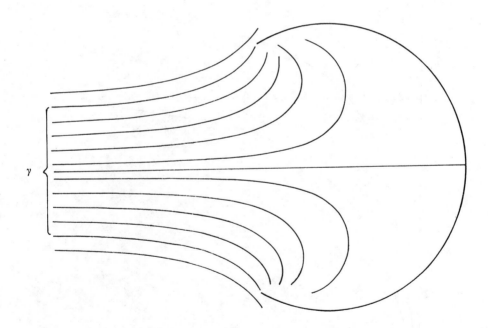

FIG. 4. *The particle paths in the case $a = 0.1$ m, $d = 40$ microns and airspeed $= 10 \text{ m} \cdot \text{s}^{-1}$ ($\beta = 0$)*

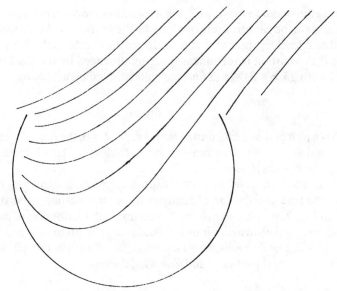

FIG. 5. *The particle paths in the case a = 0.1 m, d = 57 microns and airspeed = 10 m · s⁻¹ (β = π/4)*

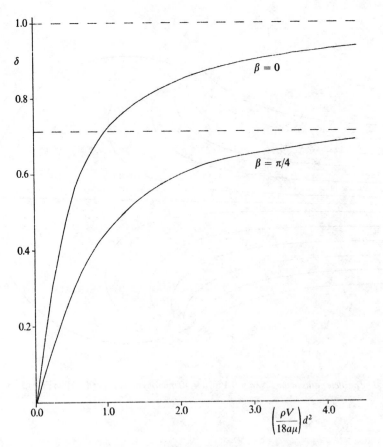

FIG. 6. *The variation of the collection efficiency, δ, with particle size, d, for β = 0 and π/4*

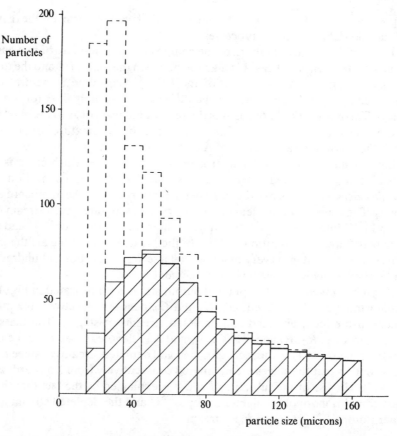

FIG. 7. *An assumed experimental distribution of particles* ▨ *and the corresponding true distributions for wind speeds of* $10 \text{ m} \cdot \text{s}^{-1}$ *(– – – –) and* $100 \text{ m} \cdot \text{s}^{-1}$ *(———)*

REFERENCES

[1] *Directional Dust Gauges*, Methods for the Measurement of Air Pollution, B.S. 1747: Part 5, British Standards Institution, 1972.
[2] D. M. LUCAS AND D. J. MOORE, *The measurement in the field of pollution by dust*, Internat. J. Air Water Pollution, 8 (1964), pp. 441–453.
[3] L. M. MILNE-THOMSON, *Theoretical Hydrodynamics*, Macmillan, London, 1968.

DESIGN OF AN ASYMMETRIC IDEAL CASCADE FOR ISOTOPE ENRICHMENT*

G. GELDENHUYS†

Abstract. Linear algebra can be used in the design of asymmetric cascades for isotope enrichment.

1. Introduction. Natural uranium contains approximately 0.71% of the light ^{235}U isotope. The rest is mainly the heavier ^{238}U isotope with traces of other isotopes such as ^{234}U. The ^{235}U isotope is used in nuclear reactors. Enrichment is required to increase

* Received by the editors March 17, 1978.
† Department of Applied Mathematics, University of Stellenbosch, Stellenbosch, South Africa.

the presence of ^{235}U in uranium, which is used in the form of the gas uranium hexafluoride (UF$_6$) in certain processes.

It is useful to have definitions of the composition of a mixture of two gases, A and B. The *mass fraction* X_A of gas A is the ratio between the mass of A and the total mass of the mixture. The mass fraction X_l of the light ^{235}U isotope in natural uranium is approximately 0.0071. This fraction should be approximately 0.03 for use in nuclear reactors. The *mass ratio* R_A of gas A is the ratio of the mass of A to the mass of B in the mixture, and $R_A = X_A/(1-X_A)$. The composition of two mixtures of A and B is the same if they have the same values of X_A (or R_A).

Large-scale uranium enrichment is done in a *cascade* which consists of a large number of interconnected stages. Each stage is a parallel connection of similar *separating elements*, the size of the stage being proportional to the number of elements it contains. The function of an element or a stage is to separate a feed stream containing ^{235}U and ^{238}U into an enriched stream with a higher mass ratio for ^{235}U and a depleted stream with a lower mass ratio for ^{235}U. In the present state of the art the enrichment achieved in a single stage is very small, so that a series connection of hundreds of stages must be used to achieve meaningful enrichment.

A typical stage i with its input and output streams is illustrated in Fig. 1, in which the following symbols are used. G_0^i, G_l^i and G_h^i denote respectively the mass flow of uranium in the feed, enriched and depleted streams of stage i. The mass flows are measured in kg/s. R_0^i, R_l^i and $R_h^i(X_0^i, X_l^i$ and $X_h^i)$ denote respectively the mass ratios (mass fractions) of ^{235}U in the feed, enriched and depleted streams of stage i. We follow the convention that the stages are numbered consecutively from 1 onwards so that $i > j$ if $X_l^i > X_l^j$. The use of the subscripts l and h is explained by the fact that the enriched stream contains more of the lighter isotope ^{235}U and the depleted stream more of the heavier isotope ^{238}U than the feed stream.

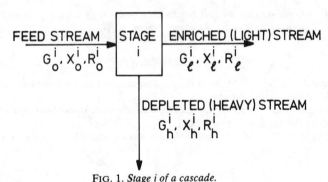

FIG. 1. *Stage i of a cascade.*

2. **Separation factors.** The *separation factor* between two streams containing ^{235}U and ^{238}U is defined as the ratio between the mass ratios of the desired isotope ^{235}U in the two streams. The separation factor between the enriched stream and the feed stream of stage i is $\alpha_{l0} = R_l^i/R_0^i$, between the depleted and the feed streams is $\alpha_{h0} = R_h^i/R_0^i$, and between the enriched and depleted streams is $\alpha_{lh} = R_l^i/R_h^i = \alpha_{l0}/\alpha_{h0}$. We regard these three separation factors as constants for a particular cascade. The value of α_{l0} is only slightly larger than 1 and the value of α_{h0} only slightly less than 1, so that

$$(1) \qquad\qquad\qquad \alpha_{h0} < 1 < \alpha_{l0} < \alpha_{lh}.$$

3. **Conservation equations.** The mass of uranium (mass of ^{235}U) entering stage i

must equal the total mass of uranium (total mass of ^{235}U) leaving stage i. The corresponding equations are

(2a) $$G_0^i = G_l^i + G_h^i,$$

(2b) $$X_0^i G_0^i = X_l^i G_l^i + X_h^i G_h^i.$$

The *mass flow ratio* μ_i of stage i is defined as the ratio between the mass flows of uranium in the enriched and depleted streams, i.e.,

(3) $$\mu_i = G_l^i / G_h^i.$$

If we use the relations $G_l^i = \mu_i G_h^i$ and $G_0^i = G_l^i + G_h^i = (1 + \mu_i)G_h^i$ in (2b) we can derive the equation

$$\mu_i = \frac{X_0^i - X_h^i}{X_l^i - X_0^i}.$$

The definitions of the separation factors α_{l0} and α_{lh} and the mass ratio R can be used to express X_0^i and X_h^i in terms of X_l^i, as follows:

$$X_0^i = \frac{X_l^i}{X_l^i + \alpha_{l0}(1 - X_l^i)}, \qquad X_h^i = \frac{X_l^i}{X_l^i + \alpha_{lh}(1 - X_l^i)}.$$

The last three equations can be combined to yield the equation

(4) $$\mu_i = \frac{\alpha_{lh} - \alpha_{l0}}{(\alpha_{l0} - 1)[(\alpha_{lh} - 1)(1 - X_l^i) + 1]}.$$

In a cascade X_l^i increases with increasing stage number i. This, taken together with (1) and (4), implies that μ_i increases with increasing i.

4. The interconnection of stages in a cascade. The systematic interconnection of the stages of a cascade can be done as follows. The enriched stream from stage r is taken forwards to stage $r + i$ for further enrichment. As r increases, the depleted stream from stage r contains more of the desired isotope. Rather than letting the depleted stream

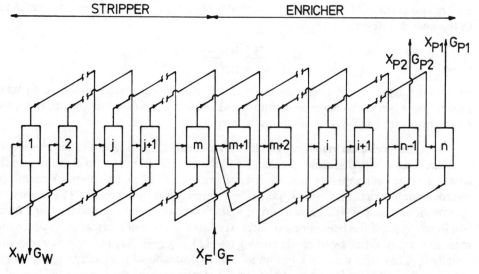

FIG. 2. *Arrangement for a* (2, 1) *cascade.*

from stage r go to waste, it is taken backwards to stage $r - j$ for further enrichment. The numbers i and j are constants for a given cascade, and the cascade is called an (i, j) cascade. We shall consider a $(2, 1)$ cascade, as shown in Fig. 2.

The positions in the cascade where the most highly enriched streams are removed are called the *product ends* of the cascade. The position where the worst depleted stream is removed is called the *waste end*. Somewhere between these extreme points is the *feed point*, where natural uranium is fed into the cascade. The portion of the cascade from the feed point to the product ends is called the *enricher*. The portion to the left of the feed point is called the *stripper*. The purpose of the enricher is to make product of the desired grade and that of the stripper is to reduce the amount of feed required to make a given amount of product.

We shall consider the *static behavior* of a cascade, in which the isotopic compositions of the various streams do not change in time. The *design* of such a cascade involves, amongst other things, the determination of the total number n of stages in the cascade, the number m of stages in the stripper, and the mass ratio of ^{235}U as well as the mass flow in every stream of the cascade. The known quantities are the separation factor α_{10}, the mass ratio R_F of ^{235}U and the mass flow G_F in the feed stream at the feed point, the mass ratio R_W of ^{235}U at the waste end, and, for a $(2, 1)$ cascade, the desired mass ratios R_{P1} and R_{P2} of ^{235}U at the two product ends. Note that, for example, $R_W = R_h^1$, $R_l^{n-1} = R_{P2}$ and $R_l^n = R_{P1}$.

A very important design consideration is that enrichment work that has already been done should not be destroyed. This implies that the isotopic composition of streams that are mixed should be the same. If this is true for the entire cascade, it is called an *ideal cascade*. For stage $i + 1$ of an ideal $(2, 1)$ cascade the requirement is that

$$(5) \qquad\qquad R_h^{i+2} = R_l^{i-1}, \qquad i = 2, 3, \cdots, n - 2.$$

If $\beta = \alpha_{10}$, it can be shown that (5) leads to the equations $\alpha_{lh} = \beta^{3/2}$,

$$(6) \qquad\qquad R_l^j = \beta^{1/2j+1} R_W, \qquad j = 1, 2, \cdots, n,$$

and $R_{P2} = \beta^{-1/2} R_{P1}$. Since R_W and β are known, it follows from (6) and (5) that the isotopic composition of all the streams in the cascade can be determined. Since R_{P1} and R_F are also known, (6) can be used to compute the total number of stages in the cascade, to the nearest integer,

$$n = \frac{2 \log (R_{P1}/R_W)}{\log \beta} - 2.$$

The total number m of stages in the stripper can be calculated in the same way. Furthermore, since the isotopic compositions of all the streams are known, the mass flow ratio μ_i for each stage i can be calculated.

5. A mathematical model of mass flows. An ideal $(1, 1)$ cascade is called *symmetric* because, if α_{10} is only slightly larger than 1 and the mass fraction of ^{235}U in the streams is small, a stage will separate a feed stream into enriched and depleted streams of approximately equal mass flows, i.e., $\mu_i \approx 1$ throughout the cascade. The theory of symmetric cascades is well known and its mass flows can be determined in closed form [4]. Recently there has been renewed interest in ideal *asymmetric* cascades, in which the mass flow ratios differ significantly from 1 (see [1], [3], [6]). Asymmetric cascades are usually (i, j) cascades with $i \neq j$. For an ideal $(2, 1)$ cascade with α_{10} only slightly larger than 1 and small mass fractions of ^{235}U in the streams, $\mu_i \approx \frac{1}{2}$ throughout the cascade.

We now show that the calculation of the mass flows in a $(2, 1)$ cascade can be reduced to the solution of a system of linear equations. Similar results apply to other asymmetric (i, j) cascades. We concentrate on the calculation of the various G_h^i. Once these are known, the known values of the μ_i can be used to calculate the G_l^i from (3) and the G_0^i from (2a).

From the conservation of the mass flow of uranium in stages 1, 2, and 3 we have

(7) $$G_l^1 + G_h^1 = G_h^2,$$

(8) $$G_l^2 + G_h^2 = G_h^3,$$

(9) $$G_l^n + G_h^n = G_h^{n-2}.$$

For stage $i(3 \leq i \leq n-1)$ we have

(10) $$G_l^i + G_h^i = G_l^{i-2} + G_h^{i+1} - \delta_{m+1,i}G_F,$$

where $\delta_{i,j} = 1$ if $i = j$ and $\delta_{i,j} = 0$ if $i \neq j$. Note that $G_l^i = G_{P2}$ if $i = n-1$.

So far we have not used the conservation equations for the ^{235}U isotope. As an alternative we follow a suggestion in [6] and use equation (3) to transform equations (7) to (10) to the following n equations for the determination of the n mass flows $G_h^1, G_h^2, \cdots, G_h^n$.

$$(1 + \mu_1)G_h^1 - G_h^2 = 0,$$

$$(1 + \mu_2)G_h^2 - G_h^3 = 0,$$

$$-\mu_{i-2}G_h^{i-2} + (1 + \mu_i)G_h^i - G_h^{i+1} = \delta_{m+1,i}G_F, \qquad 3 \leq i \leq n-1,$$

$$-\mu_{n-2}G_h^{n-2} + (1 + \mu_n)G_h^n = 0.$$

These equations can be written in the matrix form $AG_h = c$, where G_h is a column vector with the required mass flows as components, where the column vector c contains only one nonzero component, and where the matrix A is shown in Fig. 3.

Since the mass flow ratio μ_i is positive and increases with increasing i, the matrix A is strictly diagonally dominant[1] and therefore nonsingular [5, p. 23], so that there exists a unique solution for the mass flows in the streams.

6. The numerical calculation of mass flows. Because of the large number of stages in a practical cascade and the very real danger of error propagation, a numerical method for the solution of the final system of equations in the previous section should be chosen with care. Of course, the special banded structure of the matrix A can be used to advantage. If we use the techniques in [2, p. 56] we can show that Gauss–Jordan elimination of the system of equations is equivalent to a decomposition of the matrix A into matrix factors L and U, that the decomposition can be achieved by a simple and stable recursion, and that forward elimination and back substitution can then be done economically and in a stable manner, even for large n.

The numerical analysis indicated above and the actual design of a large asymmetric cascade for isotope enrichment should be valuable exercises for students who are

---

[1] An $n \times n$ matrix $A = [a_{ij}]$ is strictly diagonally dominant if

$$|a_{ii}| > \sum_{\substack{j=1 \\ j \neq i}}^{n} |a_{ij}| \qquad for\ i = 1, 2, \cdots, n.$$

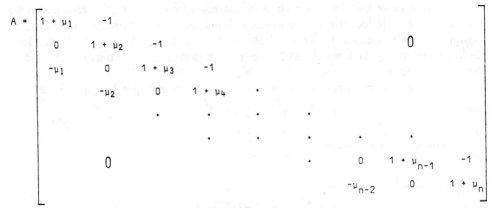

$$
A = \begin{bmatrix}
1 + \mu_1 & -1 & & & & & & \\
0 & 1 + \mu_2 & -1 & & & & 0 & \\
-\mu_1 & 0 & 1 + \mu_3 & -1 & & & & \\
& -\mu_2 & 0 & 1 + \mu_4 & \cdot & & & \\
& & \cdot & \cdot & \cdot & & & \\
& & & \cdot & \cdot & \cdot & & \\
& 0 & & & \cdot & 0 & 1 + \mu_{n-1} & -1 \\
& & & & & -\mu_{n-2} & 0 & 1 + \mu_n
\end{bmatrix}
$$

FIG. 3. *The matrix A.*

interested in the application of linear algebra. The study of asymmetric (i, j) cascades also provides insight into the reasons for the entirely new arrangement described in [1] for cascades with small mass flow ratios.

REFERENCES

[1] P. C. HAARHOFF, *Die helikontegniek vir isotoopverryking,* Tydskrif vir Natuurwetenskappe, 16 (1976), pp. 68–126. (Translation: *The helikon technique for isotope enrichment,* Uranium Enrichment Corporation of South Africa, Pretoria, Report VAL 1, 1976.)

[2] E. ISAACSON AND H. B. KELLER, *Analysis of Numerical Methods,* John Wiley, New York, 1966.

[3] D. R. OLANDER, *Two-up, one-down ideal cascades for isotope separation,* Nuclear Tech., 62 (1976), pp. 108–112.

[4] H. R. C. PRATT, *Countercurrent Separation Processes,* Elsevier, Amsterdam, 1967.

[5] R. S. VARGA, *Matrix Iterative Analysis,* Prentice-Hall, Englewood Cliffs, NJ, 1962.

[6] D. WOLF, J. L. BOROWITZ, A. GABOR AND Y. SHRAGE, *A general method for the calculation of an ideal cascade with asymmetric separation units,* Industrial and Engineering Chemistry, Fundamentals, 15 (1976), pp. 15–19.

Supplementary References
Heat Transfer and Diffusion

[1] R. ARIS, *The Mathematical Theory of Diffusion and Reaction in Permeable Catalysts,* Clarendon Press, Oxford, 1977.

[2] T. R. AUTON AND J. H. PICKLES, *"Deflagration in heavy flammable vapours,"* BIMA (1980) pp. 126–133.

[3] M. AVRAMI AND V. PASCHKIS, *"Application of an electric model to the study of two-dimensional heat flow,"* Trans. Amer. Inst. Chem. Eng. (1942) pp. 631–652.

[4] R. M. BARRER, *Diffusion in and through Solids,* Cambridge University Press, Cambridge, 1941.

[5] A. J. BROWN AND S. M. MARCO, *Introduction to Heat Transfer,* McGraw-Hill, N.Y., 1942.

[6] J. D. BUCKMASTER, *Theory of Laminar Flames,* Cambridge University Press, Cambridge, 1982.

[7] J. D. BUCKMASTER AND G. D. LUDFORD, *Lectures on Mathematical Combustion,* SIAM, 1983.

[8] H. S. CARSLAW AND J. C. JAEGER, *Conduction of Heat in Solids,* Clarendon Press, Oxford, 1959.

[9] A. J. CHAPMAN, *Heat Transfer,* Macmillan, N.Y., 1974.

[10] J. CRANK, *"Diffusion mathematics in medicine and biology,"* BIMA (1976) pp. 106–112.

[11] ———, *The Mathematics of Diffusion,* Clarendon Press, Oxford, 1956.

[12] J. B. DIAZ, *"Mathematical prolegomena to every theory of homogeneous heat engines,"* SIAM. Rev. (1978) pp. 265–277.

[13] R. J. DUFFIN, *"Optimum heat transfer and network programming,"* J. Math. Mech. (1968) pp. 759–768.

[14] R. J. DUFFIN AND D. K. McLAIN, *"Optimum shape of a cooling fin on a convex cylinder,"* J. Math. Mech. (1968) pp. 769–784.

[15] G. M. DUSINBERRE, *Numerical Analysis of Heat Flow,* McGraw-Hill, N.Y., 1949.

[16] E. ECKERT, *Introduction to the Transfer of Heat and Mass*, McGraw-Hill, N.Y., 1950.

[17] W. E. FITZGIBBON AND H. F. WALKER, eds., *Nonlinear Diffusion*, Pitman, London, 1977.

[18] C. M. FOWLER, *"Analysis of numerical solutions of transient heat-flow problems,"* Quart. Appl. Math. (1945) pp. 361–376.

[19] F. W. GEHRING, *"The boundary behavior and uniqueness of solutions of the heat equation,"* Trans. AMS (1960) pp. 337–364.

[20] L. GREEN, JR., *"Gas cooling of a porous heat source,"* Trans. ASME (1952) pp. 173–178.

[21] J. P. HOLMAN, *Heat Transfer*, McGraw-Hill, N.Y., 1976.

[22] L. R. INGERSOLL, O. J. ZOBEL AND A. C. INGERSOLL, *Heat Conduction with Engineering and Geological Applications*, McGraw-Hill, N.Y., 1948.

[23] M. JACOB, *Heat Transfer I, III*, Wiley, N.Y., 1959, 1962.

[24] W. JOST, *Diffusion in Solids, Liquids and Gases*, Academic Press, N.Y., 1952.

[25] C. F. KAYAN, *"An electrical geometric analogue for complex heat flow,"* Trans. ASME (1945) pp. 713–716.

[26] D. Q. KERN AND A. D. KRAUS, *Extended Surface Heat Transfer*, McGraw-Hill, N.Y., 1972.

[27] M. S. KLAMKIN, *"Asymptotic heat conduction in arbitrary bodies,"* Philips Res. Reports (1975) pp. 31–39.

[28] ———, *"On cooking a roast,"* SIAM Rev. (1961) pp. 167–169.

[29] D. I. LAWSON AND J. H. McGUIRE, *"The solution of transient heat-flow problems by analogous electrical networks,"* Proc. Inst. Mech. Eng. (1953) pp. 275–287.

[30] R. W. LEWIS, K. MORGAN AND B. A. SCHREIFLER, *Numerical Methods in Heat Transfer*, Wiley, N.Y., 1983.

[31] E. A. MARSHALL AND D. N. WINDLE, *"A moving boundary diffusion model in drug therapeutics,"* Math. Modelling (1982) pp. 341–369.

[32] M. H. MARTIN, *"A mathematical study of the transport of auxin,"* Math. Modelling (1980) pp. 141–166.

[33] W. H. McADAMS, *Heat Transmission*, McGraw-Hill, N.Y., 1954.

[34] J. R. OCKENDON AND W. R. HODGKINS, eds., *Moving Boundary Problems in Heat Flow and Diffusion*, Oxford University Press, London, 1975.

[35] K. H. OKES, *"Modelling the heating of a baby's milk bottle,"* Int. J. Math. Educ. Sci. Tech (1979) pp. 125–136.

[36] D. B. OULTON, D. J. WOLLKIND AND R. N. MAURER, *"A stability analysis of a prototype moving boundary problem in heat flow and diffusion,"* AMM (1979) pp. 175–186.

[37] D. T. PIELE, M. W. FIREBAUGH AND R. MANULK, *"Applications of conformal mapping to potential theory through computer graphics,"* AMM (1977) pp. 677–692.

[38] D. ROSENTHAL, *"The theory of moving sources of heat and its application to metal treatments,"* Trans. ASME (1946) pp. 849–866.

[39] P. J. SCHNEIDER, *Conduction Heat Transfer*, Addison-Wesley, Reading, 1955.

[40] F. H. SCOFIELD, *"The heat loss from a cylinder embedded in an insulating wall,"* Philo. Mag. (1931) pp. 329–349.

[41] A. SHITZER AND R. C. EBERHART, eds., *Heat Transfer in Medicine and Biology. Analysis and Applications I*, Plenum, N.Y., 1985.

[42] A. D. SNIDER, *"Mathematical techniques in extended surface analysis,"* Math. Modelling (1982) pp. 191–206.

[43] J. L. SYNGE, *"Thermostatic control,"* AMM (1933) pp. 202–215.

[44] D. V. WIDDER, *The Heat Equation*, Academic Press, N.Y., 1975.

[45] ———, *"Some analogies from classical analysis in the theory of heat conduction,"* Arch. Rational Mech. Anal. (1966) pp. 108–119.

[46] G. B. WILKES, *Heat Insulation*, Wiley, N.Y., 1950.

3. Traffic Flow

Malevolent Traffic Lights

Problem 82-16, by J. C. Lagarias (Bell Laboratories, Murray Hill, NJ).

Can the red-green pattern of traffic lights separate two cars, originally bumper-to-bumper, by an arbitrary distance? We suppose that:

(1) Two cars travel up a semi-infinite street with traffic lights set at one block intervals. Car 1 starts up the street at time 0 and car 2 at time $t_0 > 0$.

(2) Both cars travel at a constant speed 1 when in motion. Cars halt instantly at any intersection with a red light, and accelerate instantly to full speed when the light turns green. If the car has entered an intersection as the light turns red, it does not stop.

(3) Each light cycles periodically, alternately red and green with red time μ_j, green time λ_j and initial phase θ_j (i.e., phase at time 0) at intersection j.

Can one define triplets $(\lambda_j, \mu_j, \theta_j)$ ($j = 1, 2, \cdots$) so that $\{(\lambda_j, \mu_j): j = 1, 2, \cdots\}$ is a finite set and so that car 1 gets arbitrarily far ahead of car 2?

Solution by O. P. Lossers (Eindhoven University of Technology, Eindhoven, The Netherlands).

The answer is yes as we shall now prove. We consider two types of traffic lights (I) $\lambda = \mu = a/2$, (II) $\lambda = \mu = b/2$, where a and b are chosen in such a way that a/b is irrational. The phases of the lights are chosen in such a way that the first car never has to stop. This is easily accomplished. However, we can manipulate more with the phases. Suppose that the second car arrives at a light of type I with a delay of $ra + \theta$ seconds ($r \in N$, $0 < \theta < a$). The phase is chosen in such a way that the light changes $\frac{1}{2}\theta$ seconds after car 1 passes. So the delay of car 2 increases to $(r + \frac{1}{2})a + \frac{1}{2}\theta$. If we had only used lights of type I the delay would monotonically increase to $(r + 1)a$. A similar statement is true for lights of type II. At each intersection we still have the choice of the type of light. We choose this type in such a way that the increase is maximal. Since a/b is irrational, no multiple of a is a multiple of b and the time delay will tend to ∞.

Editorial note. The proposer shows that if all the cycle times are commensurable then any two cars remain within a bounded distance of each other, no matter how the lights are specified. [C.C.R.]

AVERAGE DISTANCES IN l_p DISKS*

C. K. WONG AND KAI-CHING CHU†

Abstract. In traffic flow studies as well as computer mass storage problems, two quantities of interest are the average distance between any two points in a certain region and that between any point and a fixed point of the region. In this note the regions are assumed to be l_p disks with uniform distribution of locations. For $p = 1, 2$, we have the street traffic flow problem. For $p = \infty$, we have the computer mass storage problem.

* Received by the editors February 2, 1976.
† IBM Thomas J. Watson Research Center, Yorktown Heights, New York 10598.

1. The model. In [1], [2], [4] and the references therein, a common model for studying urban street traffic is represented by a circular city uniformly inhabited with distance measured either by the rectilinear metric or the Euclidean metric, i.e., from point (u, v) to point (x, y), the distance can be calculated as $|x - u| + |y - v|$ or $((x - u)^2 + (y - v)^2)^{1/2}$. The quantities of interest are the average distance between any two points and that between any point and a fixed location in the city.

Recently, in a study of computer mass storage [3], [5], the above quantities came up naturally but in a different setting. The storage system can be represented by the lattice points (x, y) in a plane where x, y are integers and $-N \leq x, y \leq N$. N is usually very large. At each lattice point, a file of records is stored. These files are accessed by an electromechanical fetching mechanism which has equal and constant moving speeds along the x-axis and the y-axis. Suppose the fetching mechanism moves from point (u, v) to point (x, y); then the path it takes consists of two segments: a straight line at 45° to the x-axis followed by either a horizontal or a vertical line segment. (See Fig. 1.) The time it takes to complete the journey is therefore proportional to max $(|x - u|, |y - v|)$. To see this, assume without loss of generality that $u = v = 0$, $y > x > 0$. Then the path consists of two line segments: one at 45° and of length $\sqrt{2}x$; the other being vertical and of length $y - x$. If the speed in the horizontal and vertical directions is S, then the diagonal movement has speed $\sqrt{2}S$. The total time is therefore

$$(\sqrt{2}x)/(\sqrt{2}S) + (y - x)/S = y/S.$$

Depending on the actual system installation, two kinds of operations to access a sequence of files are available: (i) after accessing a file, the fetching mechanism moves directly to the location of the next file; (ii) after accessing a file, the fetching mechanism always returns to the center of the storage system, i.e., point $(0, 0)$. The choice of the specific access operation depends on the expected time between the arrivals of consecutive access requests as well as other system requirements.

If we assume every file is equally likely to be accessed, the quantity of interest in case (i) is the average distance between any two lattice points in the $(2N + 1) \times$

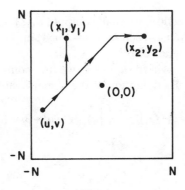

FIG. 1

$(2N+1)$ square, since it is a measure of the average time spent between two consecutive accesses. For case (ii), the quantity of interest is the average distance from any lattice point to the center.

Interestingly, the traffic flow problem [1], [2], [4] and the above-mentioned computer mass storage problem can all be put in a unified mathematical framework as will be seen in the next section.

2. The l_p metric. Let $\zeta = (u, v)$ and $\eta = (x, y)$ be any two points in a plane Q. For $p = 1, 2, \cdots$, define

$$(1) \qquad d_p(\zeta, \eta) = (|x - u|^p + |y - v|^p)^{1/p},$$

and for $p = \infty$, define

$$(2) \qquad d_\infty(\zeta, \eta) = \max(|x - u|, |y - v|).$$

As is well-known, d_p and d_∞ are indeed metrics and can be defined on the n-dimensional generalization R^n of $Q = R^2$.

Let D_p be the unit l_p disk, $1 \leq p \leq \infty$, i.e.

$$(3) \qquad D_p = \{\eta \mid \eta \in Q \text{ and } d_p(0, \eta) \leq 1\},$$

where 0 is the point $(0, 0)$ in Q. Clearly, D_1 is a diamond with edge $\sqrt{2}$, D_2 is a circular disk with radius 1, and D_∞ is a square with edge 2.

Define $\bar{d}_p(\zeta)$ as the average distance between any point in D_p and point ζ, and $\bar{d}_p$ as the average distance between any two points in D_p. Clearly, assuming uniform distribution of points,

$$(4) \qquad \bar{d}_p(\zeta) = A(D_p)^{-1} \int_{\eta \in D_p} d_p(\zeta, \eta) \, d\eta,$$

$$(5) \qquad \bar{d}_p = A(D_p)^{-2} \iint_{\zeta, \eta \in D_p} d_p(\zeta, \eta) \, d\eta \, d\zeta,$$

where $A(D_p)$ is the Euclidean area of D_p.

Cases $p = 1, 2$ correspond to traffic flow problems in diamond-shaped cities and circular cities. Case $p = \infty$ corresponds to the mass storage problem, with the square D_∞ serving as a continuous approximation to the discrete set of points $(x/N, y/N)$ with x and y integers, $-N \leq x, y \leq N$, and integer N very large.

3. Results. In this section, we shall compute $\bar{d}_1(\zeta), \bar{d}_\infty(\zeta), \bar{d}_1, \bar{d}_\infty$, and $\bar{d}_p(0)$, for all $1 \leq p \leq \infty$. $\bar{d}_2$ is given as $128/(45\pi)$ $(= .9054)$ in [1], [2], and $\bar{d}_2(\zeta)$ is given in [4] as certain elliptic integrals of ζ.

To compute $\bar{d}_1(u, v)$, where $(u, v) = \zeta$, we first compute $\bar{d}_1(u, v)$ for $(u, v) \in \Delta_1 = \{(u, v) \mid 0 \leq u, v \leq 1\}$. By definition of the l_1 metric and (4),

$$(6) \qquad \bar{d}_1(u, v) = A(D_1)^{-1} \iint_{(x, y) \in D_1} (|x - u| + |y - v|) \, dx \, dy.$$

As $A(D_1) = 2$, we have

$$A(D_1)^{-1} \iint\limits_{(x,y)\in D_1} |x-u|\, dx\, dy$$

(7)
$$= \frac{1}{2}\int_{x=-1}^{0}\int_{y=-1-x}^{x+1}(u-x)\,dy\,dx + \frac{1}{2}\int_{x=0}^{u}\int_{y=x-1}^{1-x}(u-x)\,dy\,dx$$

$$+ \frac{1}{2}\int_{x=u}^{1}\int_{y=x-1}^{1-x}(x-u)\,dy\,dx$$

$$= \tfrac{1}{3}+u^2-\tfrac{1}{3}u^3.$$

Similarly,

(8)
$$A(D_1)^{-1} \iint\limits_{(x,y)\in D_1} |y-v|\, dx\, dy = \tfrac{1}{3}+v^2-\tfrac{1}{3}v^3.$$

Substituting (7) and (8) into (6), we have for any $(u,v)\in D_1$,

(9)
$$\bar{d}_1(u,v) = \tfrac{2}{3}+u^2+v^2-\tfrac{1}{3}(|u|^3+|v|^3).$$

The minimum is $\bar{d}_1(0,0)=\tfrac{2}{3}$, the maximum is $\bar{d}_1(\pm 1,0)=\bar{d}_1(0,\pm 1)=\tfrac{4}{3}$.

For the overall average, by symmetry,

(10)
$$\bar{d}_1 = \frac{4}{A(D_1)} \iint\limits_{(u,v)\in \Delta_1} \bar{d}_1(u,v)\,du\,dv$$

$$= 2\int_{u=0}^{1}\int_{v=0}^{1-u} (\tfrac{2}{3}+u^2+v^2-\tfrac{1}{3}(u^3+v^3))\,dv\,du$$

$$= \frac{14}{15} = .9333.$$

Next we shall derive $\bar{d}_\infty(u,v)$ and show that $\bar{d}_1=\bar{d}_\infty$. For this purpose, we define a mapping f from D_∞ to D_1 by rotating D_∞ through $-\pi/4$, and then shrinking it around the origin by a factor $1/\sqrt{2}$ (see Fig. 2). It can be shown that every point (x,y) in D_∞ is mapped to $(x'=(y+x)/2,\ y'=(y-x)/2)$ in D_1, creating a one-to-one correspondence between the points of D_∞ and those of D_1.

Let $\zeta,\ \eta$ be any two points in D_∞. Let the line segment joining $\zeta,\ \eta$ have length a and form an angle $0\leqq\theta$ with the x-axis. We shall assume $\theta\leqq\pi/4$. Other cases

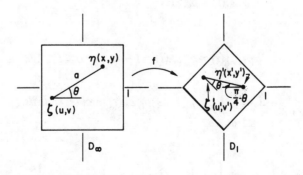

FIG. 2

can be dealt with by symmetry. $d_\infty(\zeta, \eta) = a \cos \theta$. Let $\zeta' = f(\zeta)$, $\zeta' = f(\eta)$. Then,

$$(11) \qquad d_1(\zeta', \eta') = \frac{a}{\sqrt{2}} \left(\cos\left(\frac{\pi}{4} - \theta\right) + \sin\left(\frac{\pi}{4} - \theta\right) \right) = a \cos \theta = d_\infty(\zeta, \eta).$$

Let $\zeta = (u, v)$, $\eta = (x, y)$ be points in D_∞, and let their images be $\zeta' = (u', v')$, $\eta' = (x', y')$; then $u' = (v + u)/2$, $v' = (v - u)/2$, and

$$\bar{d}_\infty(\zeta) = A(D_\infty)^{-1} \int_{\eta \in D_\infty} d_\infty(\zeta, \eta) \, d\eta$$

$$(12)$$

$$= \frac{A(D_1)}{A(D_\infty)} \cdot \frac{1}{A(D_1)} \int_{\eta' \in D_1} d_1(\zeta', \eta') 2 \, d\eta' = \bar{d}_1(\zeta'),$$

or

$$\bar{d}_\infty(u, v) = \bar{d}_1(u', v')$$

$$(13) \qquad = \frac{2}{3} + \left(\frac{v + u}{2}\right)^2 + \left(\frac{v - u}{2}\right)^2 - \frac{|(v + u)/2|^3 + |(v - u)/2|^3}{3}$$

$$= \frac{2}{3} + \frac{u^2}{2} + \frac{v^2}{2} - \frac{1}{24} (|v + u|^3 + |v - u|^3).$$

The minimum is $\bar{d}_\infty(0, 0) = \frac{2}{3}$; the maximum is $\bar{d}_\infty(\pm 1, \pm 1) = \frac{4}{3}$. Further,

$$\bar{d}_\infty = A(D_\infty)^{-2} \iint_{\zeta, \eta \in D_\infty} d_\infty(\zeta, \eta) \, d\zeta \, d\eta$$

$$(14)$$

$$= \frac{1}{4} \cdot \frac{1}{2^2} \iint_{\zeta', \eta' \in D_1} d_1(\zeta', \eta') 4 \, d\zeta' \, d\eta' = \bar{d}_1 = \frac{14}{15}.$$

$\bar{d}_\infty$ can also be computed directly in many different ways. The following is a simple one based on probability arguments. Let (x, y), (u, v) be any points in D_∞. By the assumption of uniform distribution of points in D_∞, we have the density functions $p_x(t) = p_y(t) = p_u(t) = p_v(t) = 1/2$, $-1 \le t \le 1$. It follows that

$$(15) \qquad p_{x-u}(t) = p_{y-v}(t) = \begin{cases} \frac{1}{2} - \frac{1}{4}t, & 0 \le t \le 2, \\ \frac{1}{2} + \frac{1}{4}t, & -2 \le t \le 0. \end{cases}$$

Thus,

$$(16) \qquad p_{|x-u|}(t) = p_{|y-v|}(t) = 1 - \frac{1}{2}t, \qquad 0 \le t \le 2.$$

$$(17) \qquad \bar{d}_\infty = \iint_{0 \le s, t \le 2} p_{|x-u|}(s) p_{|y-v|}(t) \max(s, t) \, ds \, dt.$$

By symmetry,

$$(18) \qquad \bar{d}_\infty = 2 \int_{s=0}^{2} \int_{t=0}^{s} (1 - \frac{1}{2}s)(1 - \frac{1}{2}t)s \, dt \, ds = \frac{14}{15}.$$

As for general $\bar{d}_p$, it is difficult to obtain a closed-form formula for it. However, by the usual continuity arguments, one can show that $\bar{d}_p \to \bar{d}_\infty = \frac{14}{15}$ as $p \to \infty$. Finally, we shall prove that $\bar{d}_p(0) = \frac{2}{3}$ for all $1 \le p \le \infty$. For $0 \le r \le 1$, let $D_p(r) = \{\eta | \eta \in Q \text{ and } d_p(0, \eta) \le r\}$. Thus $D_p(1) = D_p$. Let $A(D_p(r))$ denote the Euclidean area of $D_p(r)$. It can be shown that $A(D_p(r)) = K_p r^2$, where K_p is a constant depending on p only. In particular, $A(D_p) = K_p$. It follows that

(19)
$$\bar{d}_p(0) = A(D_p)^{-1} \int_{\eta \in D_p} d_p(0, \eta)\, d\eta = K_p^{-1} \int_0^1 r\, dA(D_p(r))$$
$$= K_p^{-1} \int_0^1 K_p \cdot 2r^2\, dr = \frac{2}{3}.$$

It should be pointed out that J. S. Lew proved a more general result: in n-dimensional space, the average distance from a point in the unit l_p ball to the center is $n/(n+1)$ for all p.

REFERENCES

[1] H. G. APSIMON, *Mathematical note 2754: A repeated integral*, Math. Gaz., 42 (1958), p. 52.
[2] F. GARWOOD AND J. C. TANNER, *Mathematical note 2800: On note 2754—A repeated integral*, Ibid., 42 (1958), pp. 292–293.
[3] R. M. KARP, A. C. MCKELLAR AND C. K. WONG, *Near-optimal solutions to a 2-dimensional placement problem*, SIAM J. Comput., 4 (1975), pp. 271–286.
[4] J. S. LEW, J. C. FRAUENTHAL AND N. KEYFITZ, *On the average distances in a circular disc*, this Review, to appear.
[5] P. C. YUE AND C. K. WONG, *Near-optimal heuristics for the 2-dimensional storage assignment problem*, Internat. J. Comput. Information Sci., 4 (1975), pp. 281–294.

ON THE AVERAGE DISTANCES IN A CIRCULAR DISC*

JOHN S. LEW†, JAMES C. FRAUENTHAL‡ AND NATHAN KEYFITZ§

Abstract. Using the l^p notion of distance in the Cartesian plane, and assuming a uniform density of locations on a circular disc, we consider the resulting distance to any specified point of this domain, and we determine the first two moments of this random variable for $p = 1, 2, \infty$. We find the maxima and minima of these average distances and their ratios, hence show their almost exact proportionality over the disc. Situations motivating these results include traffic flow on a rectangular street grid in a circular city and physical design of certain computer systems in two dimensions.

1. Introduction. A common model for urban transportation is traffic flow in a circular city. Various authors (Smeed [16], Smeed and Jeffcoate [17], Fairthorne [5], Tan [18], Einhorn [4], Holroyd [11], Pearce [13]) assume a disc city with either a rectangular or a polar street grid, consider travel distances along either grid or

* Received by the editors May 19, 1975, and in revised form January 27, 1977.

† Mathematical Sciences Department, IBM T. J. Watson Research Center, Yorktown Heights, New York 10598.

‡ Department of Applied Mathematics and Statistics, State University of New York at Stony Brook, Stony Brook, New York 11790.

§ Center for Population Studies, Harvard University, Cambridge, Massachusetts 02138.

straight-line paths, obtain double averages over both initial and final points, then essay policy conclusions for both real and hypothetical cities. However, the average distance from a fixed point has received somewhat less attention in this literature (Haight [7], Witzgall [21]). The physical design of a computer generates optimal placement problems in two dimensions which involve similar average distances from a given point (Hanan and Kurtzberg [10], Karp, McKellar and Wong [12], Wong and Chu [22]). Often the appropriate distance for such problems becomes the maximum absolute value of the coordinate differences, while uniform distributions on the relevant domains yield important results for any further analysis, and circular domains, in many cases, provide a close approximation to the optimal shapes.

Certain recent models of disc cities involve some radial dependence for the population density, but cited empirical data on such densities give no clear indication of the functional form (Pearce [13]). Moreover, a probability density, in the contexts of these models, may represent the spatial distribution of other entities, such as fires, workplaces, or accidents. Thus a uniform density offers at least a universal first approximation, and the ultimate issues may demand only broad geometrical assertions (Plattner [14]). Hence we suppose, for concreteness, that a circular city of given radius has a fine rectangular grid of streets; and we require, for simplicity, that the probability distribution of the relevant locations is uniform on the disc: we assume, in other words, that the probability measure of any Borel set is proportional to its area. We choose an arbitrary point in this circular disc, consider its l^p distance to a random location, and calculate the first two distance moments for $p = 1, 2, \infty$. This yields properties of the average distances, and their ratios, which determine maxima and minima on the disc. The almost exact proportionality of these averages is a noteworthy consequence of this work.

We introduce a system (x, y) of rectangular coordinates, and define an associated pair $(\mathbf{i}, \mathbf{j})$ of unit vectors, with the origin located at the disc center and the axes parallel to the street grid. Our unit of length will be the radius of the city, whence the disc city will have a representation

$$(1.1) \qquad\qquad D = \{(x, y) : x^2 + y^2 \leqq 1\},$$

and a location in any area $dx\,dy$ will have a probability of $dx\,dy/\pi$. From an arbitary point (u, v) in these coordinates to a random point (x, y) with uniform distribution, the l^p distance for $1 \leqq p < \infty$ is $\{|u - x|^p + |v - y|^p\}^{1/p}$, so that its nth moment, for $n = 1, 2, \cdots$, is

$$(1.2) \qquad\qquad a_n^p(u, v) = \iint_D \{|u - x|^p + |v - y|^p\}^{n/p}\, dx\,dy/\pi,$$

and the l^∞ distance, as usual, is $\max(|u - x|, |v - y|)$, so that its nth moment, in the same way, is

$$(1.3) \qquad\qquad a_n^\infty(u, v) = \iint_D \{\max(|u - x|, |v - y|)\}^n\, dx\,dy/\pi.$$

The obvious symmetries of (1.2) and (1.3) yield

$$(1.4) \qquad\qquad a_n^p(v, u) = a_n^p(u, v) = a_n^p(|u|, |v|) \quad \text{for all } p \text{ and } n.$$

For any real number ψ we recall the vector function

(1.5) $\mathbf{e}(\psi) = \mathbf{i} \cos \psi + \mathbf{j} \sin \psi,$

and for our generic points we define the further representations

(1.6) $\mathbf{q} = q\mathbf{e}(\phi) = u\mathbf{i} + v\mathbf{j}, \qquad \mathbf{r} = r\mathbf{e}(\theta) = x\mathbf{i} + y\mathbf{j}.$

We use the polar coordinates for (u, v) to express the rotational average of a_n^p:

(1.7) $b_n^p(q) = (2\pi)^{-1} \displaystyle\int_0^{2\pi} a_n^p(q \cos \phi, q \sin \phi) \, d\phi.$

If we rotate the point (u, v) onto the positive x-axis, then we do not change the average $a_n^2(u, v)$ in the Euclidean sense, so that

(1.8) $a_n^2(u, v) = a_n^2(q, 0) = b_n^2(q),$

where $q^2 = u^2 + v^2$. However if $p \neq 2$ then $a_n^p(u, v)$ is not rotation-invariant. The variance of the l^p distance is $a_2^p(u, v) - [a_1^p(u, v)]^2$ by a standard identity. Thus we shall investigate the distance moments for $n = 1, 2$; and we can evaluate the resulting integrals for $p = 1, 2, \infty$. However if, for any real s, t, we recall the identity

(1.9) $|s + t| + |s - t| = 2 \max(|s|, |t|);$

and if, in definition (1.3), we substitute the variables

(1.10) $u' = (u + v)/\sqrt{2}, \qquad v' = (u - v)/\sqrt{2};$

(1.11) $x' = (x + y)/\sqrt{2}, \qquad y' = (x - y)/\sqrt{2};$

then directly, by (1.9), we obtain the reduction

$$a_n^\infty(u, v) = 2^{-n} \iint_D \{|u + v - x - y| + |u - v - x + y|\}^n \, dx \, dy / \pi$$

(1.12)

$$= 2^{-n/2} \iint_D \{|u' - x'| + |v' - y'|\}^n \, dx' \, dy' / \pi = 2^{-n/2} a_n^1(u', v').$$

Hence our study of these moments requires no further mention of $a_n^\infty(u, v)$.

Moreover, our calculated average for the rectangular distance will yield the corresponding result for a nonorthogonal street grid. If $\mathbf{e}_1, \mathbf{e}_2$ are unit vectors parallel to the grid directions, then $\mathbf{k} = (\mathbf{e}_1 \times \mathbf{e}_2)/|\mathbf{e}_1 \times \mathbf{e}_2|$ is a unit vector perpendicular to the disc, and $\mathbf{e}_1^* = \mathbf{e}_2 \times \mathbf{k}$, $\mathbf{e}_2^* = \mathbf{k} \times \mathbf{e}_1$ are unit vectors orthogonal respectively to $\mathbf{e}_2, \mathbf{e}_1$. Moreover, these vectors satisfy

$$\mathbf{e}_1 \cdot \mathbf{e}_1^* = \mathbf{e}_2 \cdot \mathbf{e}_2^* = |\mathbf{e}_1 \times \mathbf{e}_2| = \sin \alpha,$$

where the grid directions define the angle α, and the average distance becomes

(1.14) $\displaystyle\iint_D |e_1 \times e_2|^{-1}\{|e_1^* \cdot (q-r)| + |e_2^* \cdot (q-r)|\}\, d^2 r/\pi = (\csc \alpha) \cdot a_1^1 (e_1^* \cdot q, e_2^* \cdot q).$

Fairthorne [5], for example, considers a triangular grid of streets.

2. Rectangular distance. For $0 \le t \le 1$ we define the function

(2.1) $f(t) = 2t^{1/2} \arcsin t^{1/2} + \tfrac{2}{3}(2+t)(1-t)^{1/2};$

and by direct calculation we obtain its derivative

(2.2) $f'(t) = t^{-1/2} \arcsin t^{1/2} + (1-t)^{1/2}.$

Moreover $t^{1/2} < \arcsin t^{1/2}$ for positive t, whence

(2.3) $f''(t) = \tfrac{1}{2}t^{-1}(1-t)^{1/2} - \tfrac{1}{2}t^{-3/2} \arcsin t^{1/2} < 0 \quad \text{for } 0 < t < 1.$

Thus $f'(t)$ is strictly decreasing on $[0, 1]$, but is positive by (2.2); whereas $f(t)$ is strictly increasing by this remark, and is positive by (2.1). The terms in (2.1) have standard expansions about the origin, which yield a corresponding series for $f(t)$:

(2.4) $\displaystyle f(t) = \sum_{m=0}^{\infty} \Gamma(m - \tfrac{3}{2})t^m / [(1-2m)\pi^{1/2}\Gamma(m+1)] = \tfrac{4}{3}F(-\tfrac{3}{2}, -\tfrac{1}{2}, \tfrac{1}{2}; t).$

The symbol F in this relation is a hypergeometric function of t (Abramowitz and Stegun [1, eq. (15.1.1)]), so that (2.4) offers an analytic continuation to complex t. The coefficient of t^m is $O(m^{-7/2})$ for large m (Abramowitz and Stegun [1, eq. (6.1.47)]), so that (2.4) provides an absolutely convergent series for $|t| \le 1$. Hence $f(t)$, despite its definition (2.1), has no singularity at the origin.

Now we apply relation (1.9) to evaluate an auxiliary integral:

$$\iint_D |u - x|\, dx\, dy = 2 \int_{-1}^{+1} (1-x^2)^{1/2}|u-x|\, dx$$

$$= 4 \int_0^1 (1-x^2)^{1/2} \max(|u|, x)\, dx$$

(2.5)
$$= 4 \int_0^{|u|} (1-x^2)^{1/2}|u|\, dx + 4 \int_{|u|}^1 (1-x^2)^{1/2}x\, dx$$

$$= 4|u| \int_0^{\arcsin|u|} \cos^2 \xi \cdot d\xi - \tfrac{4}{3}[(1-x^2)^{3/2}]_{|u|}^1$$

$$= f(u^2) \quad \text{for } -1 \le u \le +1.$$

Then definition (1.2) and this integral imply

$$\pi a_1^1(u, v) = \iint_D \{|u-x| + |v-y|\}\, dx\, dy$$

(2.6)
$$= f(u^2) + f(v^2) \quad \text{for } -1 \le u, v \le +1.$$

Next we introduce polar coordinates (r, θ) to evaluate a second integral:

(2.7)
$$\iint_D |u - x|^2 \, dx \, dy = \iint_D (u^2 - 2ux + x^2) \, dx \, dy$$

$$= \iint_D u^2 \, dx \, dy - 0 + \frac{1}{2} \iint_D (x^2 + y^2) \, dx \, dy$$

$$= \pi u^2 + \frac{1}{2} \int_0^{2\pi} d\theta \int_0^1 r^3 \, dr = \pi u^2 + \frac{\pi}{4} \quad \text{for real } u.$$

Also we define $Q = D \cap$ first quadrant to abbreviate our notation, and we invoke relation (1.9) to evaluate a third integral:

(2.8)
$$\iint_D |u - x| \, |v - y| \, dx \, dy = \iint_Q \{|u - x| + |u + x|\}\{|v - y| + |v - y|\} \, dx \, dy$$

$$= 4 \iint_Q \max(|u|, x) \max(|v|, y) \, dx \, dy$$

$$= 4|uv| \int_0^{|u|} dx \int_0^{|v|} dy + 2|u| \int_0^{|u|} dx \int_{v^2}^{1-x^2} d(y^2)$$

$$+ 2|v| \int_0^{|v|} dy \int_{u^2}^{1-y^2} d(x^2) + \int_{u^2}^{1-v^2} d(x^2) \int_{v^2}^{1-x^2} d(y^2)$$

$$= \tfrac{1}{2} + u^2 + v^2 + u^2 v^2 - (u^4 + v^4)/6 \quad \text{for } (u, v) \text{ in } D.$$

Thus definition (1.2) and these integrals imply

(2.9)
$$\pi a_2^1(u, v) = \iint_D \{|u - x|^2 + 2|u - x| \, |v - y| + |v - y|^2\} \, dx \, dy$$

$$= (1 + \pi/2)(1 + 2u^2 + 2v^2) + 2u^2 v^2 - (u^4 + v^4)/3 \quad \text{for } (u, v) \text{ in } D.$$

The gradient of $a_1^1(u, v)$ has the form

(2.10)
$$\nabla a_1^1(u, v) = (2u/\pi) f'(u^2) \mathbf{i} + (2v/\pi) f'(v^2) \mathbf{j}$$

inside the disc D. This gradient has an obvious zero at the origin, but has a positive radial component elsewhere in the disc; indeed $f'(u^2), f'(v^2) > 0$, whence

(2.11)
$$(u\mathbf{i} + v\mathbf{j}) \cdot \nabla a_1^1(u, v) = (2/\pi)[u^2 f'(u^2) + v^2 f'(v^2)] > 0.$$

Also the gradient, by the symmetry of relation (2.10), has a precisely radial direction either on the coordinate axes or on the bisectors $u \pm v = 0$. However, the gradient, except on these lines, has an angular component towards the nearest bisector. In proving this assertion, we may recall the symmetries (1.4) and impose the restrictions $0 < v < u < 1$; then

(2.12)
$$(-v\mathbf{i} + u\mathbf{j}) \cdot \nabla a_1^1(u, v) = (2uv/\pi)[f'(v^2) - f'(u^2)]$$

is positive under these assumptions, since $f'(t)$ is decreasing for $0 \leqq t \leqq 1$.

Thus the function $a_1^1(u, v)$, restricted to any circle $u^2 + v^2 = q^2$, assumes its minima on the coordinate axes, and assumes its maxima on the two bisectors. Specifically, the minimum value on the disc boundary is

(2.13) $$a_1^1(1, 0) = 1 + \frac{4}{3\pi} \approx 1.42441$$

while the maximum value on the disc boundary is

(2.14) $$a_1^1(1/\sqrt{2}, 1/\sqrt{2}) = \left[1 + \frac{10}{3\pi}\right] / \sqrt{2} \approx 1.45737.$$

Therefore the absolute maximum on the disc is (2.14), whereas the absolute minimum on the disc is

(2.15) $$a_1^1(0, 0) = \frac{8}{3\pi} \approx 0.84883.$$

We now define s and ψ by

(2.16) $$s\mathbf{e}(\phi + \psi) = r\mathbf{e}(\phi + \theta) - q\mathbf{e}(\phi),$$

given any $u\mathbf{i} + v\mathbf{j} = q\mathbf{e}(\phi)$ and $x\mathbf{i} + y\mathbf{j} = r\mathbf{e}(\phi + \theta)$ in the disc D. If q and r are constants in some calculation then s and ψ are determined by θ; indeed

(2.17) $$s^2 = q^2 + r^2 - 2qr \cos \theta$$

by the law of cosines. We introduce the auxiliary function

(2.18) $$g(q\mathbf{e}(\phi)) = g(u\mathbf{i} + v\mathbf{j}) = (|u| + |v|)/q = |\cos \phi| + |\sin \phi|,$$

and evaluate its rotational average

(2.19) $$(2\pi)^{-1} \int_0^{2\pi} g(q\mathbf{e}(\phi)) \, d\phi = \pi^{-1} \int_0^{2\pi} |\sin \phi| \, d\phi = 4/\pi.$$

We can use identity (2.19) to connect two averages (1.7):

$$b_1^1(q) = \pi^{-1} \int_0^1 r \, dr \int_0^{2\pi} d\theta \cdot (2\pi)^{-1} \int_0^{2\pi} d\phi \cdot s g(s\mathbf{e}(\phi + \psi))$$

(2.20)

$$= \pi^{-1} \int_0^1 r \, dr \int_0^{2\pi} s(\theta, r) \, d\theta \cdot \frac{4}{\pi}$$

$$= \frac{4}{\pi} a_1^2(u, v) = \frac{4}{\pi} b_1^2(q).$$

Hence a byproduct of the later result (3.2) is the boundary average of $a_1^1(u, v)$:

(2.21) $$b_1^1(1) = \frac{4}{\pi} b_1^2(1) = \frac{128}{9\pi^2} \approx 1.44101.$$

Smeed and Jeffcoate [17], via (2.20), evaluate (2.21) in the same way, and Fairthorne [5], via (2.20), calculates the average over (u, v):

(2.22) $$\iint_D a_1^1(u, v) \frac{du \, dv}{\pi} = \frac{512}{45\pi^2} \approx 1.15281.$$

Also Haight [7] obtains (2.13), but these authors do not treat the remaining possibilities.

3. Euclidean distance. We now consider the average of the Euclidean distance: we need only calculate $b_1^2(q)$, by the rotational invariance. We first determine its extrema on the disc; our gradient analysis shows that $a_1^1(u, v)$ is strictly increasing in all radial directions, whence relation (2.20) shows that $b_1^2(q)$ is strictly increasing on $[0, 1]$. Geometric intuition might perhaps suggest using polar coordinates about the disc center, but angular integration will then produce elliptic integrals of the second kind (Fairthorne [5]). Instead we translate the origin to $(q, 0)$ and we take polar coordinates (s, ψ) about this point). Clearly if $q = 0$, so that (u, v) is at the disc center, then

$$(3.1) \qquad b_1^2(0) = a_1^2(0, 0) = \int_0^{2\pi} d\psi \int_0^1 s^2 \, ds / \pi = \tfrac{2}{3} \approx 0.66667;$$

while if $q = 1$, so that (u, v) is on the disc boundary, then

$$b_1^2(1) = a_1^2(1, 0) = \int_{\pi/2}^{3\pi/2} d\psi \int_0^{-2\cos\psi} s^2 \, ds / \pi$$

$$(3.2)$$

$$= -\frac{8}{3\pi} \int_{\pi/2}^{3\pi/2} \cos^3 \psi \, d\psi = \frac{32}{9\pi} \approx 1.13177.$$

Hence (3.1) and (3.2), by these remarks, are the minimum and maximum on the disc.

If q is an arbitrary number in $[0, 1]$ and $(s(\psi), \psi)$ is an arbitrary point on the disc boundary, then the law of cosines asserts

$$(3.3) \qquad 1 = q^2 + s(\psi)^2 + 2qs(\psi) \cos \psi,$$

and the proper choice of signs implies

$$(3.4) \qquad s(\psi) = -q \cos \psi + [1 - q^2 \sin^2 \psi]^{1/2}.$$

However, odd powers of $\cos \psi$ have zero mean, whence the required average on the disc satisfies

$$3\pi b_1^2(q) = 3\pi a_1^2(q, 0)$$

$$= 3 \int_0^{2\pi} d\psi \int_0^{s(\psi)} s^2 \, ds$$

$$(3.5) \qquad = \int_0^{2\pi} s(\psi)^3 \, d\psi$$

$$= \int_0^{2\pi} [1 - q^2 \sin^2 \psi]^{1/2} [1 + 3q^2 - 4q^2 \sin^2 \psi] \, d\psi$$

$$= \int_0^{2\pi} [1 - q^2 \sin^2 \psi]^{-1/2} [1 + 3q^2 - (5q^2 + 3q^4) \sin^2 \psi + 4q^4 \sin^4 \psi] \, d\psi.$$

However we produce no change in the value of (3.5) when we decrease its integrand by the derivative of any 2π-periodic function, while we observe

(3.6)
$$(d/d\psi)[1-q^2\sin^2\psi]^{1/2}\sin\psi\cos\psi$$
$$=[1-q^2\sin^2\psi]^{-1/2}[1-(2+2q^2)\sin^2\psi+3q^2\sin^4\psi]$$

by direct calculation. If we subtract that multiple of (3.6) which eliminates the $\sin^4\psi$ term in (3.5), then we obtain

(3.7)
$$9\pi a_1^2(u,v)=9\pi a_1^2(q,0)=9\pi b_1^2(q)$$
$$=(4q^2-4)\int_0^{2\pi}[1-q^2\sin^2\psi]^{-1/2}\,d\psi+(q^2+7)\int_0^{2\pi}[1-q^2\sin^2\psi]^{1/2}\,d\psi$$
$$=16(q^2-1)K(q^2)+4(q^2+7)E(q^2)\quad\text{for }0\le q\le 1,$$

where respectively $K(m)$ and $E(m)$ are complete elliptic integrals of the first and second kind (Abramowitz and Stegun [1, § 17.3]).

Haight [7] derives (3.2) in the same way, while Witzgall [21] obtains both (3.7) and its analogue for external (u,v). An ingenious argument via elementary functions (Whitworth [20, Exercise 696], ApSimon [2], Garwood and Tanner [6]) yields the additional average over (u,v):

(3.8)
$$\iint_D a_1^2(u,v)\,du\,dv/\pi=\frac{128}{45\pi}\approx 0.90541;$$

and the factor $4/\pi$ from (2.20) then provides the corresponding average in (2.22) (Fairthorne [5]). Various authors (Deltheil [3, pp. 114–120], Hammersley [8], Watson [19], Schweitzer [15], Wyler [23]) calculate higher moments for two random points; moreover the first two consider higher dimensions, while the last two generalize the external average, and Hammersley [9] cites a biological application of such results. We shall not repeat these calculations, but, recalling the integral (2.7), we obtain the desired second moment:

(3.9)
$$\pi b_2^2(q)=\pi a_2^2(u,v)=\iint_D \{|u-x|^2+|v-y|^2\}\,dx\,dy$$
$$=(\pi/2)(1+2u^2+2v^2)=(\pi/2)(1+2q^2)\quad\text{for }0\le q.$$

Some partially numerical results of Karp, McKellar and Wong [12] suggest an almost exact proportionality among the averages $a_1^p(u,v)$. Hence, on the disc D, we consider the ratio $a_1^1(u,v)/a_1^2(u,v)$, or equivalently, by (2.20), we study the ratio $a_1^1(u,v)/b_1^1(q)$. The rotational minimum, average, and maximum of $a_1^1(u,v)$ satisfy respectively

(3.10)
$$\pi a_1^1(q,0)=f(q^2)+f(0)$$
$$=\tfrac{8}{3}+\sum_{m=1}^{\infty}\Gamma(m-\tfrac{3}{2})q^{2m}/[(1-2m)\pi^{1/2}\Gamma(m+1)],$$

$$\pi b_1^1(q)=(2\pi)^{-1}\int_0^{2\pi}[f(q^2\cos^2\phi)+f(q^2\sin^2\phi)\,d\phi$$

$$(3.11) \qquad = \pi^{-1} \sum_{m=0}^{\infty} \Gamma(m - \tfrac{3}{2})q^{2m} \int_0^{2\pi} \sin^{2m}\phi \, d\phi / [(1-2m)\pi^{1/2}\Gamma(m+1)]$$

$$= -\sum_{m=0}^{\infty} \Gamma(m-\tfrac{1}{2})\Gamma(m-\tfrac{3}{2})q^{2m}/[\pi\Gamma(m+1)^2],$$

$$(3.12) \qquad \pi a_1^1(q/\sqrt{2}, q/\sqrt{2}) = 2f(q^2/2)$$

$$= \sum_{m=0}^{\infty} 2\Gamma(m-\tfrac{3}{2})q^{2m}/[(1-2m)2^m \pi^{1/2}\Gamma(m+1)],$$

by (2.4) and (2.6). All three series, for $|q| \leq 1$, are absolutely convergent; the first two terms in each expansion are respectively identical; only these terms in each expansion are ever positive. Thus, in particular,

$$(3.13) \qquad \pi b_1^1(q) = \tfrac{8}{3} + 2q^2 - \cdots;$$

and, by monotonicity,

$$(3.14) \qquad \tfrac{8}{3} \leq \pi b_1^1(q), \qquad (d/dq^2)\pi b_1^1(q) \leq 2.$$

Not only do the three preceding functions, by our gradient analysis, take positive values in the stated order, but also their consecutive differences, in the same order, are power series of positive terms. This follows respectively from the inequalities

$$(3.15) \qquad 2\Gamma(m+\tfrac{1}{2}) < \Gamma(\tfrac{1}{2})\Gamma(m+1) < 2^m\Gamma(m+\tfrac{1}{2})$$

for $m = 2, 3, \cdots$, which follow immediately by induction from the case $m = 1$. Therefore

$$(3.16) \qquad \begin{aligned} &\pi(b_1^1(q) - a_1^1(q, 0)] = q^4 \times \text{strictly increasing function,} \\ &\pi[a_1^1(q/\sqrt{2}, q/\sqrt{2}) - b_1^1(q)] = q^4 \times \text{strictly increasing function.} \end{aligned}$$

However $(d/dq)\log[q^4/b_1^1(q)]$ is strictly positive for $0 \leq q < 1$, since

$$(3.17) \qquad \pi q(d/dq)b_1^1(q) = 2q^2(d/dq^2)\pi b_1^1(q) \leq 4q^2 \leq 4 < \tfrac{32}{3} \leq 4\pi b_1^1(q)$$

by (3.14). Hence the differences (3.16), even multiplied by $1/(\pi b_1^1(q))$, remain strictly increasing functions on $[0, 1]$, and assume their maxima on the disc boundary. However the ratio $a_1^1(u, v)/b_1^1(q)$ is unity at the origin, whereas

$$(3.18) \quad a_1^1(1, 0)/b_1^1(1) = 3\pi(3\pi + 4)/128 \approx 0.98848,$$

$$(3.19) \quad a_1^1(1/\sqrt{2}, 1/\sqrt{2})/b_1^1(1) = 3\pi(3\pi + 10)/(128\sqrt{2}) \approx 1.01135.$$

Clearly (3.18) and (3.19), by these arguments, are the minimum and maximum of $a_1^1(u, v)/b_1^1(q)$, whence

$$(3.20) \quad a_1^1(1, 0)/a_1^2(1, 0) = 3(3\pi + 4)/32 \approx 1.25857,$$

$$(3.21) \quad a_1^1(1/\sqrt{2}, 1/\sqrt{2})/b_1^1(1) = 3(3\pi + 10)/(32\sqrt{2}) \approx 1.28769,$$

by (2.20), are the minimum and maximum of $a_1^1(u, v)/a_1^2(u, v)$. Thus the ratio is indeed very nearly constant.

Acknowledgment. The authors wish to thank the referee for several suggestions which added considerable substance to this discussion.

REFERENCES

[1] M. ABRAMOWITZ AND I. A. STEGUN, *Handbook of Mathematical Functions*, U.S. Government Printing Office, Washington, DC, 1964.

[2] H. G. APSIMON, *Mathematical note 2754: A repeated integral*, Math. Gaz., 42 (1958), p. 52.

[3] R. DELTHEIL, *Probabilités Géométriques, Traité du Calcul des Probabilités et de ses Applications*, Tome II, Fascicule II, Gauthier-Villars, Paris, 1926.

[4] S. J. EINHORN, *Polar vs. rectangular road networks*, Operations Res., 35 (1967), pp. 546–548.

[5] D. FAIRTHORNE, *The distance between pairs of points in towns of simple geometrical shapes*, Proceedings of the Second International Symposium on the Theory of Road Traffic Flow, London 1963, J. Almond, ed., Organisation for Economic Co-operation and Development, Paris, 1965, pp. 391–406.

[6] F. GARWOOD AND J. C. TANNER, *Mathematical note 2800: On note 2754—A repeated integral*, Math. Gaz., 42 (1958), pp. 292–293.

[7] F. A. HAIGHT, *Some probability distributions associated with commuter travel in a homogeneous circular city*, Operations Res., 12 (1964), pp. 964–975.

[8] J. M. HAMMERSLEY, *The distribution of distance in a hypersphere*, Ann. Math. Statist., 21 (1950), pp. 447–452.

[9] ———, *Mathematical note 2936: On note 2871*, Math. Gaz., 44 (1960), pp. 287–288.

[10] M. HANAN AND J. M. KURTZBERG, *A review of the placement and quadratic assignment problems*, this Review, 14 (1972), pp. 324–342.

[11] E. M. HOLROYD, *Polar and rectangular road networks for circular cities*, Transportation Sci., 3 (1969), pp. 86–88.

[12] R. M. KARP, A. C. McKELLAR AND C. K. WONG, *Near-optimal solutions to a 2-dimensional placement problem*, SIAM J. Comput., 4 (1975), pp. 271–286.

[13] C. E. M. PEARCE, *Locating concentric ring roads in a city*, Transportation Sci., 8 (1974), pp. 142–168.

[14] S. PLATTNER, *Rural market networks*, Sci., Amer., 232 (1975), no. 5, pp. 66–79.

[15] P. A. SCHWEITZER, *Problem 5524: Moments of distances of uniformly distributed points*, Amer. Math. Monthly, 74 (1967), p. 1014 and 75 (1968), p. 802.

[16] R. J. SMEED, *The Traffic Problem in Towns*, Manchester Statistical Society, Manchester, U.K., 1961.

[17] R. J. SMEED AND G. O. JEFFCOATE, *Traffic flow during the journey to work in the central area of a town which has a rectangular grid for its road system*, Proceedings of the Second International Symposium on the Theory of Road Traffic Flow, London 1963, J. Almond, ed., Organisation for Economic Co-operation and Development, Paris, 1965, pp. 369–390.

[18] T. TAN, *Road networks in an expanding circular city*, Operations Res., 14 (1966), pp. 607–613.

[19] G. N. WATSON, *Mathematical note 2871: A quadruple integral*, Math. Gaz., 43 (1959), pp. 280–283.

[20] W. A. WHITWORTH, *DCC Exercises in Choice and Chance*, Deighton Bell, Cambridge, England, 1897.

[21] C. WITZGALL, *Optimal location of a central facility: Mathematical models and concepts*, National Bureau of Standards Rep. 8388, Washington, DC, 30 June 1965.

[22] C. K. WONG AND K.-C. CHU, *Average distances in l^p discs*, this Review, 19 (1977), pp. 320–324.

[23] O. WYLER, *Solution of problem 5524*, Amer. Math. Monthly, 75 (1968), pp. 802–804.

Supplementary References
Traffic Flow

[1] D. J. ARMITAGE AND M. McDONALD, *"Traffic flow and control,"* BIMA (1979) pp. 274–278.

[2] W. D. ASHTON, *"Models for road traffic flow,"* BIMA (1972) pp. 318–322.

[3] ———, *The Theory of Road Traffic Flow*, Methuen, London, 1966.

[4] J. E. BAERWALD, ed., *Transportation and Traffic Engineering Handbook*, Prentice-Hall, N.J., 1976.

[5] E. A. BENDER AND L. P. NEUWIRTH, *"Traffic flow: Laplace transforms,"* AMM (1973) pp. 417–422.

[6] R. E. CHANDLER, R. HERMAN AND E. W. MONTROLL, *"Traffic dynamics: Studies in car following,"* Oper. Res. (1958) pp. 165–184.

[7] A. J. H. CLAYTON, *"Road traffic calculations,"* J. Inst. Civil Engrs. (1941) pp. 247–284, 558–594.

[8] D. R. DREW, *Traffic Flow-Theory and Control*, McGraw-Hill, N.Y., 1968.

[9] M. DUNNE AND R. B. POTTS, *"Algorithm for traffic control,"* Oper. Res. (1964) pp. 870–881.

[10] L. C. EDIE, *"Car-following and steady state theory for noncongested traffic,"* Oper. Res. (1961) pp. 66–76.

[11] ———, *"Traffic deals at toll booths,"* J. Oper. Res. Soc. Amer. (1954) pp. 107–138.

[12] D. C. GAZIS, *"Traffic flow and control: Theory and applications,"* Amer. Sci. (1972) pp. 414–424.

[13] ———, *Traffic Science,* Interscience, N.Y., 1974.

[14] D. C. GAZIS, R. HERMAN AND R. B. POTTS, *"Car following theory of steady state traffic flow,"* Oper. Res. (1959) pp. 499–505.

[15] D. C. GAZIS, R. HERMAN AND R. W. ROTHERY, *"Non-linear follow-the-leader models of traffic flow,"* Oper. Res. (1961) pp. 546–567.

[16] D. L. GERLOUGH AND M. J. HUBER, *Traffic Flow Theory, A Monograph,* Special Report 165, Traffic Research Board, National Research Council, Washington, D.C., 1975.

[17] J. D. GRIFFITHS, *"Mathematical models for delays at pedestrian crossings,"* BIMA (1979) pp. 278–282.

[18] F. A. HAIGHT, *Mathematical Theories of Traffic Flow,* Academic Press, N.Y., 1963.

[19] R. HERMAN, E. W. MONTROLL, R. B. POTTS AND R. W. ROTHERY, *"Traffic dynamics: Analysis of stability in car following,"* Oper. Res. (1959) pp. 86–106.

[20] R. HERMAN, ed., *Theory of Traffic Flow,* Elsevier, N.Y., 1961.

[21] S. H. HOLLINGDALE, ed., *Mathematical Aspects of Marine Traffic,* Academic, N.Y., 1979.

[22] A. J. HOWIE, *"Mathematical models for road traffic,"* BIMA (1972) pp. 118–123.

[23] J. F. C. KINGMAN, *"On queues in heavy traffic,"* J. Roy. Statist. Soc. Ser. B. (1962) pp. 383–392.

[24] E. KOMETANI AND T. SASAKI, *"On the stability of traffic flow,"* J. Oper. Res. Japan (1958) pp. 11–26.

[25] B. O. KOOPMAN, *"Air-terminal queues under time-dependent conditions,"* Oper. Res. (1972) pp. 1089–1114.

[26] M. J. LIGHTHILL AND G. B. WHITHAM, *"On kinematic waves II: A theory of traffic flow on long crowded roads,"* Proc. Royal Soc. (1955) pp. 317–345.

[27] A. J. MILLER, *"Settings for fixed-cycle traffic signals,"* Oper. Res. (1963) pp. 373–386.

[28] G. F. NEWELL, *"A theory of platoon formation in tunnel traffic,"* Oper. Res. (1959) pp. 589–598.

[29] ———, *"Approximation methods for queues with application to the fixed-cycle traffic light,"* SIAM Rev. (1965) pp. 223–240.

[30] ———, *"Nonlinear effects in the dynamics of car following,"* Oper. Res. (1961) pp. 209–229.

[31] ———, *"Synchronization of traffic lights for high flow,"* Quart. Appl. Math. (1964) pp. 315–324.

[32] ———, *"Theories of instability in dense highway traffic,"* J. Oper. Res. Japan (1962) pp. 9–54.

[33] P. I. RICHARDS, *"Shock waves on the highway,"* Oper. Res. (1956) pp. 42–51.

[34] H. ROBBINS, *"A theorem on graphs with an application to a problem in traffic control,"* AMM (1939) pp. 281–283.

[35] T. UEMATU, *"On the traffic control at an intersection controlled by a repeated fixed cycle traffic light,"* Ann. Inst. Statist. Math. (1958) pp. 87–107.

[36] J. G. WARDROP, *"Some theoretical aspects of road traffic research,"* Proc. Inst. Civil Engrs. II (1952) pp. 325–362.

[37] M. WOHL AND B. V. MARTIN, *Traffic System Analysis,* McGraw-Hill, N.Y., 1967.

4. Electrical Networks

A Minimum Switching Network

Problem 59-5, by RAPHAEL MILLER (Hermes Electronic Corp.).

In a recent issue of an automation periodical[2], the problem is posed of obtaining a switching network which will actuate a device for a selected range of binary inputs. In particular, it is asked that a valve be opened for binary inputs corresponding to the integers 8, 9, 10, 11, 12, 13, and 14, where possible inputs go up to 7 binary digits. The proposed network is shown in the figure.

The binary significance of a vertical line is read bottom to top from the unencircled crossovers of the vertical line with horizontal lines. Note that such crossovers are *not* electrical connections. If a setting of the switches includes a circled crossover in a vertical line, then the line is shorted directly to ground through the switch. (The short is not indicated on the diagram.) Otherwise, cur-

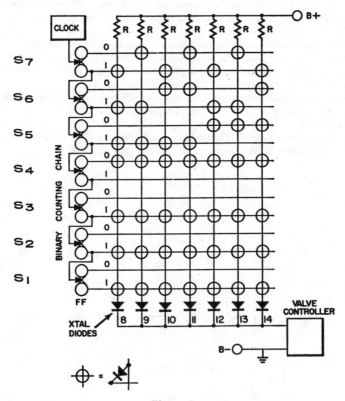

Figure 1.

[1] T. R. Munson and R. J. Spindler, *Transient Thermal Behavior of Decomposing Materials, Part I: General Theory and Application to Convective Heating*, RAD-TR 61–10, April 1961.

[2] Klein, Williams and Morgan, *Digital process control*. Instruments and Automation, October, 1956, pp. 1979–1984.

rent flows through the vertical line from the B+ terminal and a potential drop is produced across the valve controller. A circled crossover represents a diode inserted as shown in Figure 1. Accordingly, current can flow from a vertical line to a horizontal one, but not vice versa. This feature prevents the deliberate shorting of one vertical line from inadvertently shorting another vertical line. For example, the reader should verify that if the diodes were replaced by tie points, then shorting of lines 5, 6, and 7 by switch S_5 would also short lines 1, 2, 3, and 4.

Show that the proposed network can be considerably simplified so as to involve the minimum number of diodes.

Editorial note: In Raphael Miller's solution (July, 1960), the number of diodes needed are 15. In A. H. McMorris' solution, 9 diodes are required. In the solution below, only 8 diodes are required and this is the minimum number.

Solution by Layton E. Butts (Systems Laboratories).

McMorris' simplification of the logical function F is certainly correct, but it seems somewhat simpler to me to start with the logical function $\bar{F}$ on p. 221 (July 1960), and use DeMorgan's theorem to find its negative F. Thus (in the notation of Miller)

$$\bar{F} = p_1 \lor p_2 \lor p_3 \lor \overline{p_4} \lor (p_5 p_6 p_7),$$

and hence

$$F = \overline{p_1 p_2 p_3} p_4 (\overline{p_5 p_6 p_7}) = \overline{p_1 p_2 p_3} p_4 (\overline{p_5} \lor \overline{p_6} \lor \overline{p_7}).$$

Furthermore, the logical diagram given by McMorris involves a redundancy; it should be (in his notation)

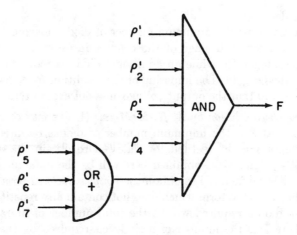

for which the circuit diagram is

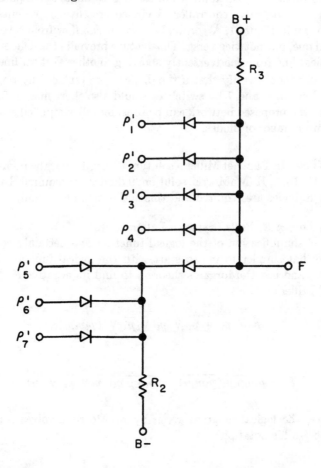

with only 8 diodes.

This is believed to be the minimum number of diodes inasmuch as one diode must be employed for each stage of the counter in order to inhibit the valve response to the integer 15, while an additional diode is required to isolate the And gate from the Or gate. The imposing of the condition $R_2 \leq R_3$ assumes that $|B-| = |B+|$ and that the output voltage does not exceed zero when F is low.

Comment on the above solution by A. H. McMorris (University of Houston):

Butts' belief that 8 is the minimum number of diodes required can be substantiated by applying the methods of Caldwell to obtain the *minimum sum expression* (MSE). The MSE obtained here will be the unfactored form for F. The MSE can then be factored to obtain the *simplest factored form* (SFF) where the SFF is defined as that form which does not contain any repetition of literals. The number of diodes required will be the total number of literals in the SFF plus (L-1) where L is the number of logic levels required by the SFF. In this case, the number of literals is 7 and the number of levels is 2. It should be noted that it is not always possible to obtain the SFF. For example, the function

$$F = WX + WY + XY$$

cannot be factored so that repeated literals do not occur. The three factored forms are:

$$F = W (X+Y) + XY,$$
$$= Y (W+X) + WX,$$
$$= X (W+Y) + WY.$$

In each case, two literals are duplicated.

A Resistor Network Inequality

Problem 60-5, by ALFRED LEHMAN (University of Toronto).

Consider the $m \times n$ series-parallel resistor network of Figure 1. The driving-point resistance between the terminals A and B is given by

$$R_{AB} = \left\{ \sum_{i=1}^{m} \left[\sum_{j=1}^{m} r_{ij} \right]^{-1} \right\}^{-1}.$$

The addition of short-circuits to the network of Figure 1 produces Figure 2. The driving-point resistance between terminals C and D is given by

$$R_{CD} = \sum_{q=1}^{n} \left\{ \sum_{p=1}^{m} [r_{pq}]^{-1} \right\}^{-1}.$$

Since short-circuits in a resistor network cannot increase resistance,

$$R_{AB} \geq R_{CD},$$

or equivalently

$$\sum_{i=1}^{m} \sum_{q=1}^{n} \left\{ \sum_{j=1}^{n} \sum_{p=1}^{m} \frac{r_{ij}}{r_{pq}} \right\}^{-1} \leq 1.$$

Give a direct proof of the latter inequality where $r_{ij} \geq 0$.

Solution by FAZLOLLAH REZA (Syracuse University).

We shall first prove the inequality $R_{AB} \geq R_{CD}$ for $n = m = 2$, and then extend it by induction to the general case. The proof is based on the concavity of the function $\phi(x, y) = (x^{-1} + y^{-1})^{-1}$, $(x, y \geq 0)$. That ϕ is concave follows from

$$Q = \phi_{xx}u^2 + 2\phi_{xy}uv + \phi_{yy}v^2 = -2(uy - vx)^2/(x + y)^3.$$

Whence,

$$2\phi \left(\frac{x_1 + x_2}{2}, \frac{y_1 + y_2}{2} \right) \geq \phi(x_1, y_1) + \phi(x_2, y_2)$$

and

$$2 \left\{ \frac{2}{r_{11} + r_{12}} + \frac{2}{r_{21} + r_{22}} \right\}^{-1} \geq \left\{ \frac{1}{r_{11}} + \frac{1}{r_{21}} \right\}^{-1} + \left\{ \frac{1}{r_{12}} + \frac{1}{r_{22}} \right\}^{-1},$$

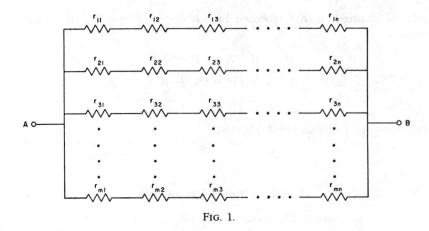

FIG. 1.

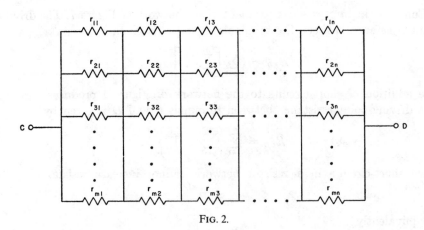

FIG. 2.

(note that $\phi(kx, ky) = k\phi(x, y)$). This establishes the inequality for $n = m = 2$. To extend the proof, it is first shown that the inequality is true for any positive integer $n > 2$. The non-negative numbers r_{ij} can be written in the matrix form

$$\left\| \begin{matrix} r_{11} & r_{12} & \cdots & r_{1n} \\ r_{21} & r_{22} & \cdots & r_{2n} \end{matrix} \right\| .$$

To show that the inequality is valid for $m = 2$, $n = 3$, we note that the inequality is already valid for

$$\left\| \begin{matrix} r_{11} + r_{12} & r_{13} \\ r_{21} + r_{22} & r_{23} \end{matrix} \right\| .$$

On the other hand, it was established that

$$\phi(r_{11} + r_{12}, r_{21} + r_{22}) \geqq \phi(r_{11}, r_{21}) + \phi(r_{12}, r_{22}).$$

Thus the theorem holds for $m = 2$, $n = 3$ and by induction for all $n \geqq 2$.

The theorem holds for the $2 \times n$ array

$$\left\| \begin{matrix} (r_{11}^{-1} + r_{21}^{-1})^{-1} & (r_{12}^{-1} + r_{22}^{-1})^{-1} & \cdots & (r_{1n}^{-1} + r_{2n}^{-1})^{-1} \\ r_{31} & r_{32} & \cdots & r_{3n} \end{matrix} \right\| .$$

Due to the concavity of ϕ, the first row of the latter array can be replaced by the following suitable $2 \times n$ array without violating the inequality. That is, the theorem holds for

$$
\left\| \begin{array}{cccc}
r_{11} & r_{12} & \cdots & r_{1n} \\
r_{21} & r_{22} & \cdots & r_{2n} \\
r_{31} & r_{32} & \cdots & r_{3n}
\end{array} \right\| .
$$

Continuing by induction, the inequality holds for arbitrary positive integers m and n.

Solution by the proposer.

The inequality

$$
\sum_{i=1}^{m} \sum_{q=1}^{n} \left\{ \sum_{j=1}^{n} \sum_{p=1}^{m} \frac{r_{ij}}{r_{pq}} \right\}^{-1} \leq 1, \qquad\qquad (r_{ij} > 0)
$$

is a special case ($\alpha = -1$) of the Minkowski inequality

$$
\left[\sum_{i=1}^{m} \left(\sum_{j=1}^{n} r_{ij} \right)^{\alpha} \right]^{1/\alpha} \overset{\leq}{\geq} \sum_{q=1}^{n} \left[\sum_{p=1}^{m} (r_{pq})^{\alpha} \right]^{1/\alpha}, \qquad \alpha \overset{>}{<} 1 \quad \text{and} \quad r_{ij} \geq 0.
$$

This latter inequality and several proofs are to be found in (3) (see particularly 2.11.4 and 2.11.5 on page 31).

It is the purpose of this solution to discuss the relation between the Minkowski inequality and the concept of a short-circuit in a network of non-linear resistors, each of voltage-current characteristic

$$
E_k = (\text{sign } I_k)(r_k | I_k |^{-1/\alpha}), \qquad\qquad r_k > 0
$$

(E_k denotes the voltage, I_k the current and r_k the resistance associated with the kth resistor. The constant α is to be the same for all resistors of the network.)

In the linear case ($\alpha = -1$) where each resistor obeys Ohm's law the result is essentially given by Jeans (4) on pages 320 to 324. Assume a passive network of linear resistors (i.e. obeying Ohm's law $E_k = r_k I_k$) excited by a single unit-current source. Consider the dissipated power ($\sum_k r_k I_k^2$) resulting from any current distribution obeying Kirchoff's current (continuity) law.* The unique current distribution which also satisfies Kirchoff's voltage law† is the unique distribution which minimizes the dissipated power (Jeans Theorem 357, page 322). Hence consider two such networks, the second being derived from the first by a sequence of short-circuits. The actual current distribution of the first network also satisfies the conditions of Kirchoff's current law for second network. Hence the actual power dissipation in the second network cannot exceed that of the first. Since the networks are excited by a unit current source, the equivalent resistance between the excited terminals is equal to the dissipated power. Hence the terminal-to-terminal resistance of the second network cannot exceed that of the first (essentially Jeans Theorem 359, page 324). The inequality of problem 60–5 then results from a computation of the resistance of the two given series-parallel networks by means of the series and parallel combination rules $a + b$ and $(a^{-1} + b^{-1})^{-1}$.

---

* The algebraic sum of currents entering any vertex is 0.

† The algebraic sum of resistor voltages around any closed circuit is 0.

This result can be generalized to networks of resistors each having a voltage current characteristic of the form $E_k = (\text{sign } I_k)(r_k| I_k |^{-1/\alpha})$ where $r_k > 0$ and $\alpha < 0$, α being the same for all resistors of the network. By Theorems 1 and 3 of Duffin (2) and the remark on page 438 of Birkhoff and Diaz (1) it follows that the unique current distribution which satisfies Kirchoff's current law and minimizes the dissipated power $\sum_k r_k| I_k |^{[1-(1/\alpha)]}$ is the unique current distribution satisfying both of Kirchoff's laws. Since the resulting terminal-to-terminal behavior of the network is easily shown to have the form $E = (\text{sign } I)(r| I |^{-(1/\alpha)})$, E and I being the applied voltage and current, the network has an equivalent resistance r and this resistance cannot be increased by short-circuits. The Minkowski inequality for $\alpha < 0$ then follows, as in the linear case, from the computation of the resistance of the two networks of problem 60–5 by means of the appropriate series and parallel combination rules $a + b$ and $(a^\alpha + b^\alpha)^{1/\alpha}$.

The previous argument holds where E_k is an increasing function of I_k, that is where $\alpha < 0$. For $\alpha > 0$ it is possible to formulate a theory of *series-parallel* resistor networks which yield the Minkowski inequality by power minimization for $0 < \alpha < 1$ and by power maximization for $\alpha > 1$. The electrical significance of such networks is open to question. Consider, for example, the networks a and b of figure 3. For either, the equivalent resistance between the * marked terminals, calculated by the series and parallel formulas, is 8. This corresponds to equality in the 2×2 case of the Minkowski inequality for $\alpha = \frac{1}{2}$. The result can also be obtained from the Kirchoff's law current distribution $I_k = \frac{1}{2}$, $k = 1, 2, 3, 4$. If non-series-parallel networks are considered the Kirchoff's law current distribution need not be unique. For example, in the network c of figure 3, the distribution $I_1 = I_4 = I_5 = \frac{1}{3}$,[a] $I_2 = I_3 = \frac{2}{3}$ satisfies both of

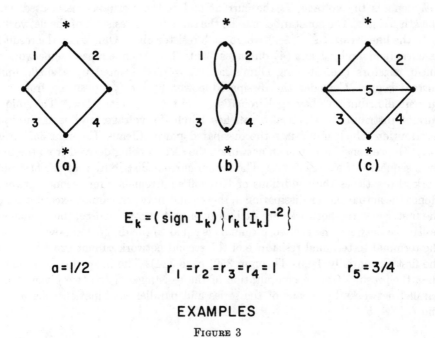

$$E_k = (\text{sign } I_k)\left\{r_k[I_k]^{-2}\right\}$$

$$a = 1/2 \qquad\qquad r_1 = r_2 = r_3 = r_4 = 1 \qquad\qquad r_5 = 3/4$$

EXAMPLES

FIGURE 3

---

[a] With the proper orientation for I_5.

Kirchoff's laws and yields a resistance of 45/4 between the * marked terminals. But by symmetry the same can be said for the current distribution $I_2 = I_3 = I_5 = \frac{1}{3}$,ᵃ $I_1 = I_4 = \frac{2}{3}$. Furthermore the I's may be chosen so as to satisfy Kirchoff's current law and have arbitrarily large magnitudes. Hence the dissipated power may be made arbitrarily small. In the case of series-parallel networks however the restriction of current flows to a single direction insures that a unique current distribution appropriately maximizes or minimizes the dissipated power.

(1) G. BIRKHOFF and J. B. DIAZ, *Non-linear network problems*, Quart. of Appl. Math., vol. 13 (1955) pp. 431–443.
(2) R. J. DUFFIN, *Non-linear networks II A*, Bull. Amer. Math. Soc., vol. 53 (1947) pp. 963–971.
(3) G. H. HARDY, J. E. LITTLEWOOD, and G. PÓLYA, *Inequalities*, Cambridge, 1934.
(4) J. JEANS, *The mathematical theory of electricity and magnetism*, Cambridge, fifth ed. 1925.

It is to be noted that the Reza solution yields a straight-forward easily remembered proof of the Minkowski inequality. Also by starting with the function

$$\phi(x, y) = \{x^\alpha + y^\alpha\}^{1/\alpha}$$

with general α, the inductive proof of Reza (or the similar inductive proof of reference 3, pp. 38–39) then yields the Minkowski inequality for all values of the exponent.

A Network Inequality

Problem 68-1, by J. C. TURNER AND V. CONWAY (Huddersfield College of Technology, England).

If $p + q = 1, 0 < p < 1$, and m, n are positive integers > 1, give a direct analytic proof of the inequality

$$(1 - p^m)^n + (1 - q^n)^m > 1.$$

The problem had arisen in the comparison of the reliabilities R_A and R_B of the two systems in Fig. 1. Here,

$$R_A = 1 - (1 - p^m)^n \quad \text{and} \quad R_B = (1 - q^n)^m,$$

where p denotes the probability of any component operating and it is assumed that all the components operate independently. Reliability is defined as the probability that the system operates successfully under the given conditions. A system is said to operate successfully if at least one path is made from input to output, when all the components are activated simultaneously.

Since the set of all possible paths through the network A is a subset of the set for network B, it is clear that the above inequality must hold.

Editorial note. Particular nice special cases of the above inequality are

$$\left\{1 - \left(\frac{1}{2}\right)^n\right\}^m + \left\{1 - \left(\frac{1}{2}\right)^m\right\}^n > 1, \qquad \frac{1}{2^m} + \frac{1}{2^{1/m}} < 1,$$

the latter inequality being known.

The original inequality can also be shown to hold for nonintegral $m, n > 1$.

Other related inequalities can be obtained by introducing one line of short-circuits at a time, e.g.,

$$1 - (1 - p^m)^n \leq \{1 - (1 - p^r)^n\}\{1 - (1 - p^{m-r})^n\} \leq (1 - q^n)^m.$$

Also, if the probability of each component operating can be different, we obtain

$$\prod_{i=1}^n \left\{1 - \prod_{j=1}^m p_{ij}\right\} + \prod_{j=1}^m \left\{1 - \prod_{i=1}^n q_{ij}\right\} > 1.$$

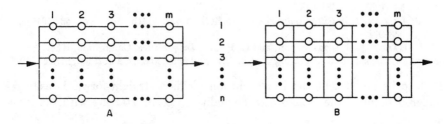

FIG. 1

For a similar derivation of another network inequality, see Problem 60–5, this Review, 4 (1962), pp. 150–155. [M.S.K.]

Solution by F. GÖBEL (Twente Institute of Technology, Enschede, Netherlands).

For $m = 1$, the assertion holds with the equality sign. Assuming the assertion to be true for $m = 1$, we have

$$(1 - q^n)^m = (1 - q^n)^{m-1}(1 - q^n) \geq \{1 - (1 - p^{m-1})^n\}(1 - q^n).$$

It remains to be shown that the last expression is greater than or equal to $1 - (1 - p^m)^n$, or, equivalently

$$(1 - p^{m-1})^n + q^n \leq q^n(1 - p^{m-1})^n + (1 - p^m)^n.$$

This inequality is of the form $a^n + b^n \leq c^n + d^n$ with $a = 1 - p^{m-1}$, $b = q$, $c = q(1 - p^{m-1})$, $d = 1 - p^m$. It is easily seen that $a + b = c + d$ and that $d = \max(a, b, c, d)$. Since x^n is convex for $x \geq 0$, $n \geq 1$, the proof can be completed by invoking the following lemma.

LEMMA. *If* $f(\cdot)$ *is convex,* $d = \max(a, b, c, d)$ *and* $a + b = c + d$, *then* $f(a) + f(b) \leq f(c) + f(d)$.

Proof. Let $\alpha + \beta = 1$, α and $\beta \geq 0$. From the convexity of $f(\cdot)$ we have

$$f(\alpha c + \beta d) \leq \alpha f(c) + \beta f(d)$$

and

$$f(\beta c + \alpha d) \leq \beta f(c) + \alpha f(d).$$

Adding, we obtain

$$f(\alpha c + \beta d) + f(\beta c + \alpha d) \leq f(c) + f(d).$$

Choose α and β such that $\alpha c + \beta d = a$; this implies

$$b = c + d - a = c + d - \alpha c - \beta d = \beta c + \alpha d,$$

and the proof is completed.

Solution by L. CARLITZ (Duke University).

We shall show that, for $n > 1, m > 1$,

(1)
$$\prod_{i=1}^{n} \left\{ 1 - \prod_{j=1}^{m} p_{ij} \right\} + \prod_{j=1}^{m} \left\{ 1 - \prod_{i=1}^{n} q_{ij} \right\} > 1,$$

where

$$p_{ij} + q_{ij} = 1, \qquad 0 < p_{ij} < 1, \qquad i = 1, \cdots, n, \quad j = 1, \cdots, m.$$

To begin with, we show that

(2)
$$(a + b - ab)(c + d - cd) > ac + bd - abcd,$$

where

$$0 < a < 1, \quad 0 < b < 1, \quad 0 < c < 1, \quad 0 < d < 1.$$

Indeed,

$$(a + b - ab)(c + d - cd) - (ac + bd - abcd)$$
$$= ad + bc - ab(c + d) - cd(a + b) + 2abcd$$
$$= ad(1 - b - c + bc) + bc(1 - a - d + ad)$$
$$= ad(1 - b)(1 - c) + bc(1 - a)(1 - d) > 0.$$

This proves (2).

If in (2) we replace a, b, c, d by $1 - a, 1 - b, 1 - c, 1 - d$, respectively, we obtain

$$(1 - ab)(1 - cd) + (1 - a'c')(1 - b'd') > 1,$$

where

$$a' = 1 - a, \qquad b' = 1 - b, \qquad \text{etc.}$$

This evidently proves (2) when $n = m = 2$.

We now show that, for $n \geq 2$,

(3)
$$\prod_{i=1}^{n} (1 - p_{i1}p_{i2}) + \left\{ 1 - \prod_{i=1}^{n} q_{i1} \right\} \left\{ 1 - \prod_{i=1}^{n} q_{i2} \right\} > 1.$$

Since we have already proved (3) when $n = 2$, we assume that it holds up to and including the value n. To carry out the induction, we put

(4)
$$\bar{q}_{ij} = \begin{cases} q_{ij}, & 1 \leq i < n, \\ q_{nj}q_{n+1,j}, & i = n, \end{cases}$$

and $\bar{p}_{ij} = 1 - \bar{q}_{ij}$. By the inductive hypothesis,

$$\prod_{i=1}^{n}(1-\bar{p}_{i1}\bar{p}_{i2}) + \left\{1-\prod_{i=1}^{n}\bar{q}_{i1}\right\}\left\{1-\prod_{i=1}^{n}\bar{q}_{i2}\right\} > 1,$$

that is,

$$\prod_{i=1}^{n-1}(1-p_{i1}p_{i2})\cdot(1-\bar{p}_{n1}\bar{p}_{n2}) + \left\{1-\prod_{i=1}^{n+1}q_{i1}\right\}\left\{1-\prod_{i=1}^{n+1}q_{i2}\right\} > 1.$$

To complete the induction, we need only show that

(5) $$(1-p_{n1}p_{n2})(1-p_{n+1,1}p_{n+1,2}) > 1 - \bar{p}_{n1}\bar{p}_{n2}.$$

Since

(6) $$\bar{p}_{nj} = 1 - \bar{q}_{nj} = 1 - q_{nj}q_{n+1,j},$$

(5) is equivalent to

$$(1-p_{n1}p_{n2})(1-p_{n+1,1}p_{n+1,2}) + (1-q_{n1}q_{n+1,1})(1-q_{n2}q_{n+1,2}) > 1,$$

which has already been proved.

We have therefore proved (1) for $m = 2$ and all n: or equivalently for $n = 2$ and all m. We now assume that (1) holds for all m and for all $i = 1, 2, \cdots, n$. We again use the notation (4). By the inductive hypothesis,

$$\prod_{i=1}^{n}\left\{1-\prod_{j=1}^{m}\bar{p}_{ij}\right\} + \prod_{j=1}^{m}\left\{1-\prod_{i=1}^{n}\bar{q}_{ij}\right\} > 1,$$

that is,

$$\prod_{i=1}^{n-1}\left\{1-\prod_{j=1}^{m}p_{ij}\right\}\cdot\left\{1-\prod_{j=1}^{m}\bar{p}_{nj}\right\} + \prod_{j=1}^{m}\left\{1-\prod_{i=1}^{n+1}q_{ij}\right\} > 1.$$

To complete the induction, we need only show that

$$\left\{1-\prod_{j=1}^{m}p_{nj}\right\}\left\{1-\prod_{j=1}^{m}p_{n+1,j}\right\} > 1 - \prod_{j=1}^{m}\bar{p}_{nj}:$$

in view of (6), this is the same as

$$\left\{1-\prod_{j=1}^{m}p_{nj}\right\}\left\{1-\prod_{j=1}^{m}p_{n+1,j}\right\} + \prod_{j=1}^{m}(1-q_{nj}q_{n+1,j}) > 1.$$

But this has already been proved.

Editorial note. JOEL BRENNER (University of Arizona) in his solution also establishes that for

$$0 < p,q,r < 1, \quad p+q+r=1, \quad m,n,t > 1, \quad s = mnt,$$
$$\{1-p^{s/m}\}^m + \{1-q^{s/n}\}^n + \{1-r^{s/t}\}^t > 2,$$

or in homogeneous form

$$\{(a+b+c)^{s/m} - a^{s/m}\}^m + \{(a+b+c)^{s/n} - b^{s/n}\}^n$$
$$+ \{(a+b+c)^{s/t} - c^{s/t}\}^t > 2(a+b+c)^s$$

with $a,b,c > 0$.

Correspondingly, the proposed inequality in homogeneous form is

$$\{(a + b)^m - a^m\}^n + \{(a + b)^n - b^n\}^m > (a + b)^{mn},$$

$$m, n > 1, \qquad a, b > 0.$$

Also he notes that the inequalities are reversed if $0 < m, n, t < 1$.

Supplementary References
Electrical Networks

[1] W. N. ANDERSON, JR. AND R. J. DUFFIN, *"Series and parallel addition of matrices,"* J. Math. Anal. Appl. (1969) pp. 576–594.

[2] S. L. BASIN, *"The appearance of Fibonacci numbers and the Q matrix in electrical network theory,"* MM (1963) pp. 84–97.

[3] H. W. BODE, *Network Analysis and Feedback Amplifier Design,* Van Nostrand, N.Y., 1949.

[4] F. T. BOESCH, ed., *Large Scale Networks: Theory and Design,* IEEE Press, N.Y., 1976.

[5] R. BOTT AND R. J. DUFFIN, *"On the algebra of networks,"* Trans. AMS (1953) pp. 99–109.

[6] R. K. BRAYTON AND J. K. MOSER, *"Nonlinear networks I, II,"* Quart. Appl. Math. (1964) pp. 1–33, 81–104.

[7] R. L. BROOKS, C. A. B. SMITH, A. H. STONE AND W. T. TUTTE, *"The dissection of rectangles into squares,"* Duke Math. J. (1940) pp. 312–340.

[8] O. BRUNE, *"Synthesis of a finite two-terminal network whose driving-point impedance is a prescribed function of frequency,"* J. Math. Phys. (1931) pp. 191–236.

[9] W. CHEN, *"Boolean matrices and switching nets,"* MM (1966) pp. 1–8.

[10] C. CLOS, *"A study of nonblocking switching networks,"* Bell System Tech. J. (1953) pp. 406–424.

[11] G. C. CORNFIELD, *"The use of mathematics at the electricity council research centre in problems associated with the distribution of electricity,"* BIMA (1977) pp. 13–17.

[12] H. CRAVIS, *Communications Network Analysis,* Heath, Lexington, 1981.

[13] R. J. DUFFIN, *"An analysis of the Wang algebra of networks,"* Trans. AMS (1959) pp 114–131.

[14] ———, *"Distributed and lumped networks,"* J. Math. Mech. (1959) pp. 793–826.

[15] ———, *"The extremal length of a network,"* J. Math. Anal. Appl. (1962) pp. 200–215.

[16] ———, *"Nonlinear networks,"* Bull. AMS (1947) pp. 963–971.

[17] ———, *"Potential theory on a rhombic lattice,"* J. Comb. Th. (1968) pp. 258–272.

[18] ———, *"Topology of series-parallel networks,"* J. Math. Anal. Appl. (1965) pp. 303–318.

[19] R. J. DUFFIN, D. HAZONY AND N. MORRISON, *"Network synthesis through hybrid matrices,"* SIAM J. Appl. Math. (1966) pp. 390–413.

[20] A. M. ERISMAN, K. W. NEVES AND M. H. DWARAKANATH, eds., *Electric Power Problems: The Mathematical Challenge,* SIAM, 1980.

[21] D. GIVONE, *Introduction to Switching Theory,* McGraw-Hill, N.Y., 1970.

[22] E. A. GUILLEMIN, *Communication Networks II: The Classical Theory of Long Lines, Filters, and Related Networks,* Wiley, N.Y., 1935.

[23] F. HOHN, *"Some mathematical aspects of switching,"* AMM (1955) pp. 75–90.

[24] F. E. HOHN AND L. R. SCHISSLER, *"Boolean matrices in the design of combinatorial switching circuits,"* Bell System Tech. J. (1955) pp. 177–202.

[25] Z. KOHVAI, *Switching and Finite Automata Theory,* McGraw-Hill, N.Y., 1970.

[26] R. W. LIU, *"Stabilities of nonlinear networks,"* J. Elec. Eng. (1970) pp. 1–21.

[27] D. E. MULLER, *"Boolean algebras in electric circuit design,"* AMM (1954) pp. 27–29.

[28] P. PENFIELD, JR., R. SPENCE AND S. DUINKER, *Tellegen's Theorem and Electrical Networks,* M.I.T. Press, Cambridge, 1970.

[29] N. PIPPENGER, *"On crossbar switching networks,"* IEEE Trans. Comm. (1975) pp. 646–659.

[30] F. M. REZA, *"A mathematical inequality and the closing of a switch,"* Arch. Elec. Comm. (1981) pp. 349–354.

[31] ———, *Modern Network Analysis,* McGraw-Hill, N.Y., 1959.

[32] ———, *"Schwarz's lemma for n-ports,"* J. Franklin Inst. (1984) pp. 57–71.

[33] J. RIORDAN AND C. E. SHANNON, *"The number of two-terminal series-parallel networks,"* J. Math. Phys. (1942) pp. 83–93.

[34] S. SESHU AND M. B. REED, *Linear Graphs and Electrical Networks,* Addison-Wesley, Reading, 1961.

[35] SIAM-AMS *Proceedings, Mathematical Aspects of Electrical Network Analysis,* AMS, 1971.

[36] M. E. VAN VALKENBERG, *Introduction to Modern Network Synthesis,* Wiley, N.Y., 1960.

[37] L. WEINBERG, *Network Analysis and Synthesis,* McGraw-Hill, N.Y., 1962.

[38] H. S. WILF AND F. HARARY, eds., *Mathematical Aspects of Electrical Network Analysis,* AMS, 1971.

[39] A. N. WILLSON, JR., *Nonlinear Newtorks: Theory and Analysis,* Wiley, Chichester, 1974.

5. Communication and Coding

A Set of Convolution Equations

Problem 60-9, by WALTER WEISSBLUM (AVCO Research and Advanced Development Division).

In simulating the operation of a certain radar detection system on a computer, it was necessary to have a method of producing a stationary sequence of normal random variables with mean zero, and with a prescribed autocorrelation sequence

$$\cdots, 0, 0, 0, C_n, C_{n-1}, \cdots, C_1, C_0, C_1, \cdots, C_{n-1}, C_n, 0, 0, 0, \cdots$$

That is, the sequence X_i of random variables produced must satisfy $E(X_i) = 0$ and

$$E(X_i \cdot X_j) = C_{|i-j|} \quad \text{for} \quad |i - j| \leq n,$$

$$E(X_i \cdot X_j) = 0 \quad \text{for} \quad |i - j| > n.$$

The following method was proposed: First produce a sequence Y_i of independent normal random variables with mean zero and standard deviation 1 and then define

$$X_i = \sum_{j=0}^{n} a_j Y_{i-n+j}$$

with suitable constants $a_0, a_1, \cdots, a_n$. This construction clearly provides stationarity, normality, mean zero, and zero correlation for $|i - j| > n$. The only question is the existence of suitable real $\{a_i\}$ to produce $E(X_i \cdot X_j) = C_{|i-j|}$ for $|i - j| \leq n$. This reduces to determining the existence of a real solution to the following set of convolution equations:

$$
\begin{aligned}
a_0^2 + a_1^2 + \cdots + a_n^2 &= C_0, \\
a_0 a_1 + a_1 a_2 + \cdots + a_{n-1} a_n &= C_1, \\
a_0 a_2 + a_1 a_3 + \cdots + a_{n-2} a_n &= C_2, \\
\vdots \qquad \vdots \qquad\qquad \vdots \qquad \vdots & \\
a_0 a_n &= C_n.
\end{aligned}
$$

(1)

Determine a necessary and sufficient condition on the $\{C_i\}$ such that real $\{a_i\}$ exist, and give a method for finding them.

I. Solution by A. W. MCKINNEY (Sandia Corporation).†

In order that there exist a sequence of real normally distributed random variables $\{X_j\}$ with means equal to zero and with covariance function C_j, where $C_n \neq 0$ and $C_j = 0$ for $j > n$, it is necessary and sufficient that the sequence $\{C_j\}$ be positive definite [1]. The sequence $\{C_j\}$ is positive definite if and only if the function

†For other solutions, see SIAM Rev. 5 (1963) pp. 276–283.

$$G(\lambda)\dot{} = 2C_0 + \frac{2}{\pi}\sum_{h=1}^{n} C_h \frac{\sin 2\pi h\lambda}{h}$$

is monotone nondecreasing for $0 \leq \lambda \leq \frac{1}{2}$ [1], thus if and only if the function

$$F(\lambda) = \frac{1}{2}G'(\lambda) = C_0 + 2\sum_{h=1}^{n} C_h \cos 2\pi h\lambda$$

is nonnegative for $0 \leq \lambda \leq \frac{1}{2}$. Since $\cos 2\pi h \ (\frac{1}{2} + \lambda) = \cos 2\pi h \ (\frac{1}{2} - \lambda)$ for integers h and all real λ, it follows that the sequence $\{C_j\}$ is positive definite if and only if the trigonometric polynomial

(2) $$F(\lambda) = C_0 + 2\sum_{h=1}^{n} C_h \cos 2\pi h\lambda$$

is nonnegative for all real λ. By a theorem of L. Fejér and F. Riesz [2], a trigonometric polynomial $T(\lambda)$ is nonnegative for all real λ if and only if it can be written in the form

$$T(\lambda) = \left| \sum_{h=0}^{n} a_h e^{2\pi ih\lambda} \right|^2,$$

where the coefficients $a_0, \cdots, a_n$ are real or complex numbers. An examination of the proof given by Szego for this theorem [2] yields as an immediate consequence the fact that the coefficients $a_0, \cdots, a_n$ can be taken to be all real if and only if in the trigonometric polynomial

$$T(\lambda) = C_0 + 2\sum_{h=1}^{n} (C_h \cos 2\pi h\lambda + D_h \sin 2\pi h\lambda),$$

the coefficients D_h are equal to zero for $h = 1, \cdots, n$. Since this is the case for $F(\lambda)$ in (2), it follows that the sequence $\{C_j\}$ is positive definite if and only if there exist real numbers $a_0, \cdots, a_n$ such that

(3) $$C_0 + 2\sum_{j=1}^{n} C_j \cos 2\pi j\lambda = \left| \sum_{h=0}^{n} a_h e^{2\pi i\lambda} \right|^2,$$

or

(4) $$C_0 + 2\sum_{j=1}^{n} C_j \cos 2\pi j\lambda = \sum_{h=0}^{n} a_h^2 + 2\sum_{j=1}^{n}\sum_{h=0}^{n-j} (a_h a_k + j \cos 2\pi j\lambda).$$

Multiplying each member of (4) by $\cos 2\pi k\lambda$ and integrating over the range $\lambda = 0$ to $\lambda = \frac{1}{2}$, it follows that

$$C_k = \sum_{h=0}^{n-k} a_h a_{h+k} \text{ for } k = 0, \cdots, n.$$

Therefore, *there exist real numbers* $a_0, \cdots, a_n$ *such that* $\sum_{h=0}^{n-k} a_h a_{h+k} = C_k$ *if and only if the sequence* $\{C_j\}$ *is positive definite*. If a sequence $\{C_j\}$ is to be tested for positive definiteness, the most convenient approach on a digital computer probably is to compute the polynomial $F(\lambda)$ defined by (2) for many values of λ between 0 and $\frac{1}{2}$, and see if $F(\lambda)$ remains positive. If a sequence $\{C_j\}$ is presumed to be positive definite, and if the coefficients $a_0, \cdots, a_n$ are to be found, the only feasible methods are iterative. It should be noted that there are, in general, several real sets of coefficients $a_0, \cdots, a_n$ which satisfy the required

equations—in fact, if a_0, $\cdots$, a_n is one such sequence, then so is $b_0 = a_n$, $b_1 = a_{n-1}$, $\cdots$, $b_n = a_0$—and therefore, the choice of a starting value may well affect the rate of convergence of Newton's method. (Of course, Newton's method always converges if the starting values are close enough to the required solution.)†
One possible choice of starting values is to set $a_h = C_h$ for $h = 0, 1, \cdots, n$. It may prove desirable in an iterative solution to make use of the fact (easily deduced from (3) by putting $\lambda = 0$ or $\frac{1}{2}$, respectively) that

$$a_0 + a_1 + a_2 + \cdots + a_n = (C_0 + 2C_1 + 2C_2 + \cdots + 2C_n)^{1/2},$$

$$a_0 - a_1 + a_2 - \cdots \pm a_n = (C_0 - 2C_1 + 2C_2 - \cdots \pm 2C_n)^{1/2}.$$

REFERENCES

1. J. L. Doob, *Stochastic Processes*, John Wiley and Sons, Inc., New York, 1953, pp. 72, 474.
2. U. Grenander and G. Szego, *Toeplitz Forms and their Applications*, University of California Press, Berkeley, 1958, pp. 20, 21.

Comment by M. J. Levin (RCA Missile Electronics and Controls Division).

A similar problem arises in the design of phased array antennas. A solution for complex $\{C_i\}$ is given by S. Silver, *Microwave Antenna Theory and Design*, Vol. 12, Radiation Laboratory Series, McGraw-Hill, 1949, pp. 280–281.

This problem also arose in connection with a method of filter synthesis suggested by R. Thorton of M.I.T. who has a solution similar to the one above.

Supplementary References
Communication and Coding

[1] N. Abramson, *Information Theory and Coding*, McGraw-Hill, N.Y., 1963.
[2] E. F. Assmus, Jr. and H. F. Mattson, Jr., *"Coding and combinatorics,"* SIAM Rev. (1974) pp. 349–388.
[3] H. Beker and F. Piper, *Cipher Systems: The Protection of Communication*, Northwood, London, 1982.
[4] E. Berlekamp, *Algebraic Coding Theory*, McGraw-Hill, N.Y., 1968.
[5] I. F. Blake, *"Codes and designs,"* MM (1979) pp. 81–95.
[6] P. J. Cameron and J. H. Van Lint, *Graphs, Codes and Designs*, Cambridge University Press, Cambridge, 1980.
[7] D. E. R. Denning, *Cryptography and Data Security*, Addison-Wesley, Reading, 1982.
[8] W. Diffie and M. E. Hellman, *"Privacy and authentication: An introduction to cryptography,"* Proc. IEEE (1979) pp. 397–427.
[9] P. Elias, *"The noisy channel coding theory for erasure channels,"* AMM (1974) pp. 853–862.
[10] R. M. Fano, *Transmission of Information*, Wiley, N.Y., 1961.
[11] H. F. Gaines, *Cryptanalysis*, Dover, N.Y., 1956.
[12] R. G. Gallager, *Information Theory and Reliable Communication*, Wiley, N.Y., 1968.
[13] R. W. Hamming, *Coding and Information Theory*, Prentice-Hall, N.J., 1980.
[14] ———, *Digital Filters*, Prentice-Hall, N.J., 1977.
[15] J. C. Hancock, *Signal Detection Theory*, McGraw-Hill, N.Y., 1966.
[16] H. F. Harmuth, *Transmission of Information by Orthogonal Functions*, Springer-Verlag, Heidelberg, 1972.
[17] W. A. Harman, *Principles of the Statistical Theory of Communication*, McGraw-Hill, N.Y., 1963.
[18] M. E. Hellman, *"The mathematics of public-key chryptography,"* Sci. Amer. (1979) pp. 146–157, 198.
[19] D. Kahn, *The Codebreakers: The Story of Secret Writing*, Macmillan, N.Y., 1968.
[20] A. G. Konheim, *Cryptography: A Primer*, Wiley, N.Y., 1981.
[21] J. Leech and N. J. A. Sloane, *"Sphere packings and error-correcting codes,"* Canad. J. Math. (1971) pp. 718–745.
[22] J. Levine, *"Some elementary cryptanalysis of algebraic cryptography,"* AMM (1961) pp. 411–418.
[23] N. Levinson, *"Coding theory: A counterexample to G. H. Hardy's conception of applied mathematics,"* AMM (1979) pp. 249–258.
[24] S. Lin and D. J. Costello, Jr., *Error Control Coding: Fundamentals and Applications*, Prentice-Hall, N.J., 1983.
[25] F. J. MacWilliams and N. J. A. Sloane, *The Theory of Error Correcting Codes*, North-Holland, Amsterdam, 1977.
[26] R. J. McEliece, *The Theory of Information and Coding: A Framework for Communication*, Addison-Wesley, Reading, 1979.

[27] B. McMILLAN, *"Elementary approach to the theory of information,"* SIAM Rev. (1961) pp. 211–229.

[28] C. H. MEYER AND S. M. MATYAS, *Cryptography: A New Dimension in Computer Data Security,* Interscience, N.Y., 1982.

[29] C. MOLER AND D. MORRISON, *"Singular value analysis of cryptograms,"* AMM (1983) pp. 78–87.

[30] S. J. ORFANIDIS, *Optimum Signal Processing,* Macmillan, N.Y., 1985.

[31] W. W. PETERSON, *Error-Correcting Codes,* Wiley, N.Y., 1961.

[32] V. PLESS, *Introduction to the Theory of Error Correcting Codes,* Interscience, N.Y., 1982.

[33] F. M. REZA, *An Introduction to Information Theory,* McGraw-Hill, N.Y., 1961.

[34] B. R. SCHATZ, *"Automated analysis of cryptograms,"* Cryptologia (1977) pp. 116–142.

[35] M. R. SCHROEDER, *Number Theory in Science and Communication,* Springer-Verlag, Heidelberg, 1984.

[36] C. E. SHANNON AND W. WEAVER, *The Mathematical Theory of Communication,* University of Illinois Press, Urbana, 1964.

[37] G. J. SIMMONS, ed., *Secure Communication and Asymmetric Cryptosystems,* Westview, Boulder, 1982.

[38] N. J. A. SLOANE, *"Error-correcting codes and invariant theory: new applications of a nineteenth-century technique,"* AMM (1977) pp. 82–107.

[39] T. M. THOMPSON, *From Error-Correcting Codes through Sphere Packings to Simple Codes,* MAA, 1983.

[40] J. VAN LINT, *"A survey of perfect codes,"* Rocky Mountain J. Math. (1975) pp. 199–224.

[41] J. H. VAN LINT, *Coding Theory,* Springer-Verlag, Berlin, 1971.

[42] A. J. VITERBI, *Principles of Coherent Communication,* McGraw-Hill, N.Y., 1966.

[43] A. M. YAGLOM AND I. M. YAGLOM, *Probability and Information,* Reidel, Dordrecht, 1983.

6. Graph Theory and Applications

An Inequality for Walks in a Graph

Problem 83-15, by J. C. LAGARIAS, J. E. MAZO, L. A. SHEPP (AT&T Bell Laboratories, Murray Hill, NJ) AND B. McKAY (Vanderbilt University).

Let G be a finite undirected graph with N vertices, which is permitted to have multiple edges and multiple loops. A *directed walk* of length k is a sequence of edges $e_1, \cdots, e_k$ of G together with vertices $v_1, \cdots, v_{k+1}$ such that edge e_i connects vertex v_i and v_{i+1}. Let $w_k = w_k(G)$ denote the number of distinct directed walks of length k in G. For which pairs (r, s) is the inequality

$$w_r w_s \leqq N w_{r+s}$$

true for all graphs G?

Solution by the proposers.

We show that the inequality holds when $r + s$ is even, and exhibit counterexamples whenever $r + s$ is odd.

Let A denote the adjacency matrix of the graph G (for a fixed ordering of its vertices), i.e., $a_{ij} = 1$ if there is an edge between i and j in G and $a_{ij} = 0$ otherwise. Then A is a nonnegative symmetric matrix with integer entries. It turns out that only the symmetry of A is important in proving the result for $r + s$ even.

THEOREM 1. *For any graph G with N vertices, the inequality*

$$w_r w_s \leqq N w_{r+s}$$

holds for all positive integers r, s for which $r + s$ is even. If the adjacency matrix A of G is positive definite, this inequality holds for all positive integers r and s.

Proof. It is well known that the (i, j)th entry of A^k counts the number of directed walks of length k in A which start at vertex i and finish at vertex j. Consequently

$$w_k = 1' A^k 1,$$

where 1 is the $N \times 1$ column vector with all entries 1. The theorem is an immediate consequence of the following result, after noting that $1'1 = N$. $\square$

THEOREM 2. *Let A be a real $N \times N$ symmetric matrix and ψ a real $N \times 1$ column vector. If $z_k = \psi' A^k \psi$ then*

$$z_r z_s \leqq (\psi' \psi) z_{r+s}$$

if $r + s$ is even. It holds for all r, s if A is positive definite.

Proof. Since A is symmetric, A is diagonalizable by an orthogonal matrix, and all its eigenvalues are real. Let $\phi_1, \cdots, \phi_N$ be an orthonormal set of (real) eigenvectors with corresponding eigenvalues $\lambda_1, \cdots, \lambda_N$, i.e., $\phi_i' \phi_j = 1$ if $i = j$, 0 otherwise. Then we have

(1)
$$\psi = \sum_{i=1}^{N} c_i \phi_i$$

where

$$c_i = \langle \psi, \phi_i \rangle = \psi' \phi_i.$$

Then

124

$$A^r \psi = A^r \left(\sum_{i=1}^{N} c_i \phi_i \right) = \sum_{i=1}^{n} c_i \lambda_i^r \phi_i.$$

Hence

$$w_r = \psi'(A^r \psi) = \sum_{i=1}^{n} c_i^2 \lambda_i^r.$$

Now define a discrete random variable X on the set $\{\lambda_1, \cdots, \lambda_n\}$ by

$$\Pr(X = \lambda_i) = \frac{c_i^2}{(\psi, \psi)}.$$

Note that

$$\sum_{i=1}^{N} c_i^2 = (\psi, \psi)$$

using (1). Consequently,

$$E(X^r) = \frac{1}{(\psi, \psi)} \sum_{i=1}^{N} c_i^2 \lambda_i^r = w_r.$$

We also note that if A is positive semidefinite, then all the eigenvalues λ_i are nonnegative so that X is a nonnegative random variable. The proof is completed using the following well-known fact.

FACT. *The inequality*

(2) $$E(X^r) E(X^s) \le E(X^{r+s})$$

holds for all random variables X when $r + s$ is even, and for nonnegative random variables for all positive r and s.

Proof. Let X, Y be independent random variables both having the same distribution. For any realizations X_0, Y_0 we have

(3) $$(X_0^r - Y_0^r)(X_0^s - Y_0^s) \ge 0$$

if $r + s$ is even. If Y_0, X_0 are nonnegative, then (3) holds for all positive integers r and s. Now take expected values of both sides of (3) and note that

$$E((X_0^r - Y_0^r)(X_0^s - Y_0^s)) = 2E(X^{r+s}) - 2E(X^r)E(X^s)$$

using independence. The fact follows. ☐

THEOREM 3. *For any pair r, s with $r + s$ odd there is a graph G with N vertices with*

$$w_r w_s > N w_{r+s}.$$

Proof. We may suppose $r = 2k - 1, s = 2l$ with $k, l \ge 1$. Let K_m denote the complete graph with m vertices and S_p the p-star, the unique tree with p vertices that has a vertex of degree $p - 1$.

We consider the graph G_{m+1,m^2+t+1} which is the disjoint union of K_{m+1} and S_{m^2+t+1}. We use the fact that if G is the disjoint union of the graphs G_1 and G_2 then

(4) $$w_r(G) = w_r(G_1) + w_r(G_2).$$

It is easy to verify that

(5) $$w_r(K_{m+1}) = m^r(m + 1),$$

and that

(6)
$$w_{2r-1}(S_{p+1}) = 2p^r,$$
$$w_{2r}(S_{p+1}) = p^r(p + 1).$$

Now consider G_{m+1,m^2+t+1} where $t \geq 1$ is viewed as fixed and we let $m \to \infty$. A straightforward algebraic calculation using (4)–(6) gives

$$w_{2k-1}w_l = 3m^{2k+2l+2} + 4m^{2k+2l+1} + ((2k + 3l + 3)t + 7)m^{2k+2l} + O(m^{2k+2l-1}),$$

$$Nw_{2k+2l-1} = 3m^{2k+2l+2} + 4m^{2k+2l+1} + ((2k + 2l + 3)t + 7)m^{2k+2l} + O(m^{2k+2l-1}).$$

This shows that for sufficiently large m (depending on t) G_{m+1,m^2+t+1} has

$$v_{2k-1}v_{2l} > Nv_{2k+2l-1},$$

the desired counterexample.

These counterexamples are disconnected graphs. One may obtain connected counterexamples by considering the graph G^*_{m+1,m^2+t+1} obtained by adding an edge connecting K_{m+1} to a vertex of degree of one in S_{m^2+t+1}. More explicitly, for $r = 1$, $s = 2$, we may take the disjoint union of K_3 and S_6. This has 9 vertices, and $w_1 = 16$, $w_2 = 42$, $w_3 = 74$, so that

$$w_1 w_2 = 672 > Nw_3 = 666.$$

We can also take $m = 8$, $t = 7$ to obtain a connected counterexample, i.e. take K_9 and S_{89} with an edge connecting a vertex of K_9 to a vertex of degree 1 of S_{89}. $\square$

Although there are counterexamples to the inequality when $r + s$ is odd, it is possible to show that it is "almost true" in the following sense.

THEOREM 4. *For any graph G there is a constant* $n_0 = n_0(G)$ *such that for all* $r, s \geq n_0$

(7)
$$w_r w_s \leq Nw_{r+s}.$$

We omit the proof.

Supplementary References
Graph Theory and Applications

[1] B. ANDRÁSFAI, *Introductory Graph Theory*, Pergamon, N.Y., 1977

[2] A. T. BALABAN, ed., *Chemical Applications of Graph Theory*, Academic, London, 1976.

[3] A. Battersby, *Network Analysis*, Macmillan, London, 1964.

[4] M. BEHZAD, G. CHARTRAND AND L. LESNIAK-FOSTER, *Graphs and Digraphs*, Wadsworth, Belmont, 1980.

[5] R. BELLMAN, K. L. COOKE AND J. A. LOCKETT, *Algorithms, Graphs, and Computers*, Academic Press, N.Y., 1970.

[6] C. BERGE, *Graphs and Hypergraphs*, North-Holland, Amsterdam, 1973.

[7] ———, *Theory of Graphs and its Applications*, Methuen, London, 1962.

[8] B. Bollobás, *Graph Theory, An Introductory Course*, Springer-Verlag, 1979, N.Y.

[9] ———, *Extremal Graph Theory*, Academic Press, London, 1978.

[10] J. A. BONDY AND U. S. R. MURTY, *Graph Theory with Applications*, Elsevier, Great Britain, 1976.

[11] R. G. BUSACKER AND T. L. SAATY, *Finite Graphs and Networks: An Introduction with Applications*, McGraw-Hill, N.Y., 1965.

[12] B. CARRE, *Graphs and Networks*, Oxford University Press, Oxford, 1979.

[13] K. W. CATTERMODE, "Graph theory and the telecommunications network," BIMA (1975) pp. 94–107.

[14] V. CHACHRA, P. M. GHARE AND J. M. MOORE, *Applications of Graph Theory Algorithms*, Elsevier, N.Y., 1979.

[15] G. CHARTRAND, *Graphs as Mathematical Models*, Prindle, Weber & Schmidt, Boston, 1977.

[16] G. CHARTRAND AND S. F. KAPOOR, *The Many Facets of Graph Theory,* Springer-Verlag, Berlin, 1969.

[17] W. CHEN, *Applied Graph Theory,* Elsevier, N.Y., 1971.

[18] N. CHRISTOFIDES, *Graph Theory and the Algorithmic Approach,* Academic, London, 1977.

[19] N. DEO, *Graph Theory with Applications to Engineering and Computer Science,* Prentice-Hall, N.J., 1974.

[20] C. FLAMENT, *Applications of Graph Theory to Group Structure,* Prentice-Hall, N.J., 1963.

[21] D. R. FULKERSON, ed., *Studies in Graph Theory I, II,* MAA, 1975.

[22] R. E. GOMORY AND T. C. HU, *"Multi-terminal focus in a network,"* SIAM J. Appl. Math. (1961) pp. 551–570.

[23] P. HAGGERT AND R. J. CHORLEY, *Network Analysis in Geography,* St. Martin's Press, N.Y., 1969.

[24] F. HARARY, *Graph Theory,* Addison-Wesley, Reading, 1969.

[25] ———, *Graph Theory and Theoretical Physics,* Academic Press, London, 1967.

[26] F. HARARY, R. Z. NORMAN AND D. CARTWRIGHT, *Structural Models: An Introduction to the Theory of Directed Graphs,* Wiley, N.Y., 1965.

[27] F. HARARY AND L. MOSER, *"The theory of round robin tournaments,"* AMM (1966) pp. 231–246.

[28] E. J. HENLEY AND R. A. WILLIAMS, *Graph Theory in Modern Engineering,* Academic Press, N.Y., 1973.

[29] D. E. JOHNSON AND J. R. JOHNSON, *Graph Theory with Engineering Applications,* Ronald Press, N.Y., 1972.

[30] J. MALKEVITCH AND W. MEYER, *Graphs, Models, and Finite Mathematics,* Prentice-Hall, N.J., 1974.

[31] R. B. MARIMONT, *"Applications of graphs and Boolean matrices to computer programming,"* SIAM Rev. (1960) pp. 259–268.

[32] J. W. MOON, *"A problem of rankings by committees,"* Econometrica (1976) pp. 241–246.

[33] ———, *Topics on Tournaments,* Holt, Rinehart and Winston, N.Y., 1968.

[34] O. ORE, *Graphs and Their Uses,* MAA, 1963.

[35] W. L. PRICE, *Graphs and Networks,* Auerbach, London, 1971.

[36] F. S. ROBERTS, *Graph Theory and its Applications to Problems of Society,* SIAM, 1978.

[37] F. S. ROBERTS AND T. A. BROWN, *"Signed digraphs and the energy crisis,"* AMM (1975) pp. 577–594.

[38] D. F. ROBINSON AND L. R. FOULDS, *Digraphs: Theory and Techniques,* Gordon and Breach, N.Y., 1978.

[39] P. D. STRAFFIN, JR., *"Linear algebra in geography: Eigenvalues of networks,"* MM (1960) pp. 269–276.

[40] K. G. TINKLER, *"Graph theory,"* Prog. Human Geography (1979) pp. 85–116.

[41] ———, *Introduction to Graph Theoretical Methods in Geography,* Geographical Abstracts Ltd., University of East Anglia, Norwich, England, 1977.

[42] R. J. TRUDEAU, *Dots and Lines,* Kent State University Press, Ohio, 1976.

[43] J. TURNER AND W. H. KAUTZ, *"A survey of progress in graph theory in the Soviet Union,"* SIAM Rev. Supplement (1970) pp. 1–68.

7. Applied Geometry

THREE-DIMENSIONAL PIPE JOINING*

ERYK KOSKO†

Abstract. A solution is given to the problem of joining two skew half-lines by a curve composed of two circular arcs of equal radii in such a way that the aggregate curve has a continuous tangent, providing a smooth passage between its component parts. The solution is based on considerations of spherical geometry.

Introduction. The layout of pipelines often requires the joining of two straight runs which end each at a point. If sharp kinks in the line are to be avoided in order to ensure a smooth flow of the fluid, a curved section must be inserted. In the simplest case this may be a circular arc, but that is not always possible. In general, a curve composed of two mutually touching circular arcs of equal radius is always possible. The advantages of using circular arcs are obvious: for a given configuration they minimize the curvature of the line (i.e., provide the largest radius) and are easy to lay out and to manufacture.

When the two straight lines lie in the same plane, the problem is relatively simple. This case has been discussed by the author in [1] where both an algebraic solution and some geometric constructions were indicated. The more complicated three-dimensional case, which arises when the lines are skew to each other, does not seem to have received much attention in the literature. The only exception known to the author is a paper by Fox [2] whose solution involved the consideration of a spatial curve obtained by intersecting two quadric surfaces. We show that a practical solution may be achieved more directly by treating the problem as one of spherical geometry.

Statement of the problem. As noted by Fox, two given skew lines a' and a'', with points A' and A'', respectively, given on them, determine a unique sphere which touches these lines at the given points. The two circular arcs, which smoothly join A' and A'' by touching the lines at these points and touching each other, are parts of two circles Γ' and Γ'' which lie in the surface of the sphere. The center O of the sphere lies on a line z which is the intersection of a plane Π' laid through A' perpendicularly to a' with a plane Π'' laid through A'' perpendicularly to a''. To determine the radius R of the sphere and the position of the center O on the z axis, we proceed as follows (Fig. 1): consider the z axis as being the north-south axis of the sphere and the plane drawn normally to that axis through O as the equatorial plane. The lines a' and a'' lie in planes parallel to the equatorial plane, at distances $z = h'$ and $z = h''$, respectively, from it; these planes intersect the sphere in parallel circles of radii r' and r'', respectively. These radii, being equal to the distances of A' and A'' from the z axis, may be regarded as known quantities. Also known is the distance h between the parallel planes, equal to the smallest distance between a' and a''. The known lengths, r', r'' and h and the unknowns R, h' and h'' are related by the equations

(1) $$h' - h'' = h, \qquad h'^2 + r'^2 = h''^2 + r''^2 = R^2$$

from which the unknowns are readily obtained:

(2) $$h' = \frac{r''^2 - r'^2}{2h} + \frac{h}{2}, \qquad h'' = \frac{r''^2 - r'^2}{2h} - \frac{h}{2},$$

(3) $$R^2 = \left(\frac{r''^2 - r'^2}{2h}\right)^2 + \left(\frac{h}{2}\right)^2 + \frac{r'^2 + r''^2}{2}.$$

---

* Received by the editors July 11, 1978, and in revised form December 7, 1978.

† 2106 Woodcrest Road, Ottawa, Ontario, Canada K1H 6H8.

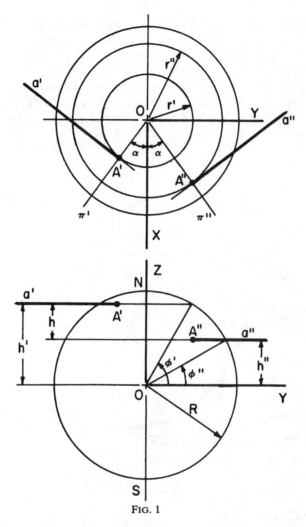

FIG. 1

A convenient cartesian system of coordinates $Oxyz$ may now be introduced by using the two planes which bisect the dihedral angle 2α formed by the planes Π' and Π'', as shown in the figure. The corresponding spherical coordinates are:

$$-\text{longitude} \quad \theta = \tan^{-1}(y/x),$$

$$-\text{latitude} \quad \phi = \tan^{-1}\frac{z}{\sqrt{x^2+y^2}}.$$

For points A' and A'', we thus have

(4a) $\qquad\qquad \theta' = -\alpha, \qquad h' = R\sin\phi, \qquad r' = R\cos\phi',$

(4b) $\qquad\qquad \theta'' = +\alpha, \qquad h'' = R\sin\phi, \qquad r'' = R\cos\phi''.$

The problem now is to determine two small circles Γ' and Γ''' of equal radii on the spherical surface satisfying the following conditions:

(i) circle Γ' must touch the parallel circle ϕ' at A';

(ii) circle Γ''' must touch the parallel circle ϕ'' at A'';

(iii) circles Γ' and Γ''' must touch each other in such way that the passage from A' to A'' along arcs of these circles could be accomplished without reversal of sense.

Outline of solution. Let the (spherical) radius of the circles Γ' and Γ'' be denoted by ρ (see Fig. 2). Then, in order to satisfy condition (i), the center C' of circle Γ' must lie on the meridian passing through A' at a distance ρ from A', i.e., at a latitude $\psi' = \phi' - \rho$. Similarly, condition (ii) will be satisfied if the center C'' of circle Γ'' lies on the meridian passing through A'' at a distance ρ from A'', but here the distance must be taken in the opposite sense to the first one in order to avoid the reversal of sense as postulated in (iii). Therefore the latitude of C'' is $\psi'' = \phi'' + \rho$. The spherical coordinates of the centers are thus $C'(-\alpha, \psi')$, $C''(+\alpha, \psi'')$. For the circles to be touching the spherical distance $C'C''$

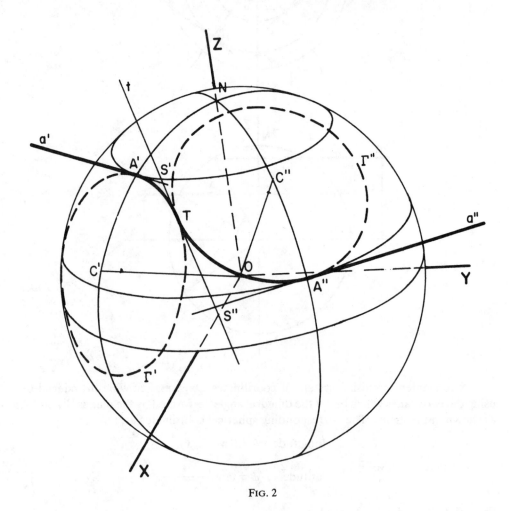

FIG. 2

(i.e., the central angle $C'OC''$) must be equal to 2ρ. Applying the cosine rule to the spherical triangle $C'NC''$ (having sides $C'N = 90° - \psi'$, $C''N = 90° - \psi''$ and angle at N equal to $\theta'' - \theta' = 2\alpha$), the side $C'C''$ is given by

(5) $\cos 2\rho = \sin \psi' \sin \psi'' + \cos \psi' \cos \psi'' \cos 2\alpha.$

Substituting for ψ' and ψ'' and rearranging the terms, this may be written

$$A \sin^2 \rho + 2B \cos \rho \sin \rho - C \cos^2 \rho = 0$$

where

(6)
$$A = 1 - \cos \phi' \cos \phi'' - \sin \phi' \sin \phi'' \cos 2\alpha,$$
$$B = \sin (\phi' - \phi'') \cos^2 \alpha,$$
$$C = 1 - \sin \phi' \sin \phi'' - \cos \phi' \cos \phi'' \cos 2\alpha.$$

Note that both A and C are always positive quantities. Dividing all terms in (5) by $\cos^2 \rho$, we obtain

(7)
$$A \tan^2 \rho + 2B \tan \rho - C = 0,$$

a quadratic equation for $\tan \rho$. Its two roots are

(8)
$$\tan \rho = -\frac{B}{A} \pm \left[\left(\frac{B}{A} \right)^2 + \frac{C}{A} \right]^{1/2}.$$

The square root term is always real and greater in absolute value than the first term. The two roots are therefore always real and have opposite signs; the sign of the root smaller in absolute value is the same as that of the coefficient B, i.e., of $\sin (\phi' - \phi'')$.

Having chosen one of the two values for the root ρ, we next determine the point T where the circles Γ' and Γ'' touch; this is the midpoint of the arc $C'C''$ of the great circle which joins the centers. Its cartesian coordinates are found by obtaining those of the midpoint T^* of the straight line segment $C'C''$ and dividing the values by $\cos \rho$:

(9)
$$x_T = \frac{R}{2}(\cos \psi' + \cos \psi'') \frac{\cos \alpha}{\cos \rho}, \qquad y_T = \frac{R}{2}(\cos \psi'' - \cos \psi') \frac{\sin \alpha}{\cos \rho},$$
$$z_T = \frac{R}{2 \cos \rho}(\sin \psi' + \sin \psi'').$$

The line t, common tangent to the circles Γ' and Γ'' at T (also touching the sphere at this point), is perpendicular to the plane $OC'C''$. The components of its direction vector are obtained by forming the vector product $OC' \times OC''$ and dividing the values by the magnitude of that product, which is $R^2 \sin 2\rho$. After some manipulation, the required direction cosines are found to be

(10)
$$\lambda = -\sin (\psi' + \psi'') \frac{\sin \alpha}{\sin 2\rho}, \qquad \mu = \sin (\psi' - \psi'') \frac{\cos \alpha}{\sin 2\rho},$$
$$\nu = \cos \psi' \cos \psi'' \frac{\sin 2\alpha}{\sin 2\rho}.$$

The coordinates of points S' and S'' at which the tangent t meets the lines a' and a'', respectively, may be found in several ways. The simplest is to write the equation of the plane which touches the sphere at T and which contains the line t:

(11)
$$xx_T + yy_T + zz_T = R^2.$$

This plane intersects the "parallel" plane, $z = h'$, laid through A' (and containing the line a') along a line whose equation in that plane is

$$xx_T + yy_T = R^2 - h'z_T.$$

On the other hand, the equation of line a' in the same plane is

$$x \cos \alpha - y \sin \alpha = r'.$$

Substituting the values of h', r', x_T, y_T, z_T, in terms of the angles α, ϕ', ϕ'', ρ, ψ', ψ'' and solving the foregoing two simultaneous equations for the coordinates of the point of intersection, yields

(12a)

$$x_{S'} = \frac{R}{\cos \alpha \cos \psi''} \cos \frac{\phi'+\phi''}{2} \cos \left(\frac{\phi'+\phi''}{2}+\rho\right),$$

$$y_{S'} = \frac{R}{\sin \alpha \cos \psi''} \sin \frac{\phi'-\phi''}{2} \sin \left(\frac{\phi'-\phi''}{2}-\rho\right).$$

Similar expressions are obtained for the coordinates of point S'' in the plane $z = h''$:

(12b)

$$x_{S''} = \frac{R}{\cos \alpha \cos \psi'} \cos \frac{\phi'+\phi''}{2} \cos \left(\frac{\phi'+\phi''}{2}-\rho\right),$$

$$y_{S''} = \frac{-R}{\sin \alpha \cos \psi'} \sin \frac{\phi'-\phi''}{2} \sin \left(\frac{\phi'-\phi''}{2}-\rho\right).$$

The length s' of the tangents drawn from S' to either A' or T on the circle Γ' is best obtained by setting $s' = (h'-z_T)/\nu$ and substituting for h', z_T, ν their known values, which yields

(13a) $$s' = \frac{2R \sin \rho}{\sin 2\alpha \cos \psi' \cos \psi''} \sin \frac{\phi'-\phi''}{2} \cos \left(\frac{\phi'+\phi''}{2}+\rho\right).$$

Similarly, setting $s'' = S''T = S''A'' = (z_T - h'')/\nu$ and substituting for h'', z_T and ν yields

(13b) $$s'' = \frac{2R \sin \rho}{\sin 2\alpha \cos \psi' \cos \psi''} \sin \frac{\phi'-\phi''}{2} \cos \left(\frac{\phi'+\phi''}{2}-\rho\right).$$

To complete the solution it is necessary to calculate the length of the curve which smoothly joins the points A' and A'' and is composed of the arcs $A'T$ and TA''. The length of the first of these arcs is $l' = 2\delta'r$ where r is the radius of the circle, $r = R \sin \rho$, and δ' is half the angle subtended by the arc. Figure 3 is drawn in the plane of circle Γ' and shows that

(14a) $$\tan \delta' = s'/r.$$

The second arc length, $l'' = 2\delta''r$, is obtained in the same fashion, with

(14b) $$\tan \delta'' = s''/r.$$

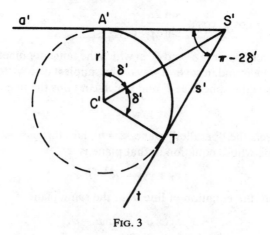

FIG. 3

The somewhat tedious calculation of the tangent lengths s' and s'' may be bypassed by observing that $2\delta'$ is the supplement to the angle formed by the tangents drawn from S', i.e., by the lines a' and t. Therefore its cosine is equal to the scalar product of the direction vectors of these lines and thus

(15a) $\cos 2\delta' = -\lambda \sin \alpha + \mu \cos \alpha,$

which on substitution reduces to

$$\cos 2\delta' = (\cos \psi' \sin \psi'' - \sin \psi' \cos \psi'' \cos 2\alpha)/\sin 2\rho$$

The corresponding expressions for the angle $2\delta''$ are

(15b) $\cos 2\delta'' = -\lambda \sin \alpha - \mu \cos \alpha,$

or, alternatively,

$$\cos 2\delta'' = (-\sin \psi' \cos \psi'' + \cos \psi' \sin \psi'' \cos 2\alpha)/\sin 2\rho.$$

The preceding formulas are all valid for either of the roots (8). Owing to the difference in sign between the two roots ρ the points of tangency tend to lie in opposite hemispheres. In this connection it was found practical for numerical work to extend the range of latitudes to the interval $(-180°, +180°)$ and restrict that of the longitudes to $(-90°, +90°)$. For the sake of clarity only one of the solutions is illustrated in Fig. 2.

Symmetric case. When the points A' and A'' are at equal distance from the z axis, i.e., when $r' = r''$, it is easy to see that we must also have $h' = -h'' = h/2$. The entire configuration then is symmetric with respect to the x axis. All the formulas and their derivations are considerably simplified in this case, which is illustrated in Fig. 4.

The radius of the sphere is now given by

(16) $R^2 = r'^2 + (h/2)^2,$

and the latitudes of points A' and A'' by

$$\phi' = -\phi'' = \tan^{-1} (h/2R).$$

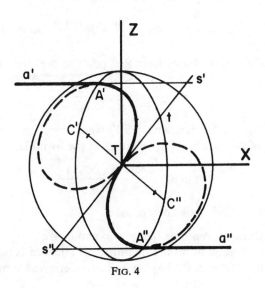

FIG. 4

Equation (5) for the spherical distance between centers C' and C'' takes the form

(17) $$\cos 2\rho = -\sin^2 \psi' + \cos^2 \psi' \cos 2\alpha.$$

Upon substituting $\psi' = \phi' - \rho$, this equation transforms into the quadratic (7) with the coefficients

(18) $$A = 2 \sin^2 \phi' \cos^2 \alpha, \quad B = 2 \sin \phi' \cos \phi' \cos^2 \alpha,$$
$$C = 2(1 - \cos^2 \phi' \cos^2 \alpha).$$

The two roots are

(19) $$\tan \rho = -\cot \phi' \pm \frac{1}{\cos \alpha \sin \phi'}.$$

By virtue of symmetry, the point T where the circles Γ' and Γ'' touch falls on the x axis and its coordinates are

$$x_T = \pm R, \qquad y_T = z_T = 0.$$

The plane tangent to the sphere at that point has the equation $x = \pm R$, the upper and lower signs corresponding to those of the root (19).

A direct application of the addition theorem for the tangent function, yields

$$\tan \psi = -\cot \phi' \pm \frac{\cos \alpha}{\sin \phi'}.$$

On the other hand, by applying the cosine rule to the right-angled spherical triangle having vertices at C', T and at the intersection of the meridian $A'C'$ with the equator, a further useful relation is obtained

$$\cos \rho = \pm \cos \alpha \cos \psi'.$$

The common tangent t to the touching circles at point T has the direction cosines

(20) $$\lambda = 0, \qquad \mu = \frac{\sin \psi'}{\sin \rho}, \qquad \nu = \frac{\tan \alpha}{\tan \rho}.$$

The line t intersects the plane $z = h/2$ at a point S' for which

(21) $$x_{S'} = \pm R, \qquad y_{S'} = \frac{h\mu}{2\nu} = R \frac{\sin \phi'}{\sin \alpha} \tan \psi'$$

and the tangents $S'T = S'A'$ drawn from that point to the circle Γ' have length

(22) $$s' = \frac{h}{2\nu} = R \sin \phi' \frac{\tan \rho}{\tan \alpha}.$$

The length of the arc $A'T$ is again $l' = 2\delta'r$ where $r = R \sin \rho$ and the angle δ' which subtends one half of that arc is given by either of the formulas

$$\tan \delta' = -\frac{s'}{r} = \frac{\sin \phi'}{\sin \alpha} \frac{1}{\cos \psi''}$$

(23)

$$\cos 2\delta' = \mu \cos \alpha.$$

Needless to say, symmetry requires that $s'' = s'$ and $\delta'' = \delta'$.

A graphical evaluation of all the quantities of concern is possible as soon as the position of the point of tangency T has been ascertained. An elementary construction is performed in the plane of each of the circles Γ' and Γ''; it consists in finding the circle which passes through T and touches the line a' (or a'') at point A' (or A''). In the symmetric case, only one of these constructions is necessary and the point T is known as soon as the tangent sphere has been determined.

REFERENCES

[1] E. KOSKO, *A pipelaying problem*, Math. Gaz., 47 (1964), pp. 192–196.
[2] M. D. FOX, *A three-dimensional pipe-joining problem*, Ibid., 53 (1969), pp. 142–147.

ESTIMATING THE LENGTH OF MATERIAL WRAPPED AROUND A CYLINDRICAL CORE*

FRANK H. MATHIS† AND DANNY W. TURNER†

Abstract. We present three methods for estimating the length of material (e.g., carpet, paper) that is wrapped around a cylindrical core. Relationships among the methods are explored.

Introduction. The problem we wish to address originated from a request by our campus Central Receiving Office for "an equation" to compute the number of linear units of carpet remaining on many unlabelled rolls that were in storage. Neither unwinding the rolls nor weighing them was practical. So a relatively simple procedure, using measurements taken from the end of a roll, was desired.

Below we shall develop three methods to approximate the desired length. Each method will calculate the exact length based on specific assumptions about the geometry of the roll. Although the mathematics involved is elementary, we feel that the results are interesting and even somewhat surprising.

For all three methods, we assume that the material is wrapped around a cylindrical core having radius r_1. Consider one circular end of the core and let point P be at the center of this circle. Let point Q, on the circumference of the circle, correspond to the internal starting position of the wrap and let point R, on the inner edge of the material, correspond to the terminal position of the wrap. For computational simplicity we assume that P, Q and R are collinear. Indeed if they are not, our calculations will be off no more than one circumference of the inner core (i.e., $2\pi r_1$), an error which we consider to be acceptable. Fig. 1 illustrates the situation.

We define the outer radius r_2 to be the distance from P to R. We denote by n the number of wraps the material makes from Q to R and by t the thickness of the material which we assume to be constant throughout the wrap.

Method A. In the simplest approach, which requires only elementary geometry, we assume that the material is wrapped in concentric circles for which the difference between consecutive radii is t. We may then either sum up the average circumferences of each layer or equivalently calculate the average circumference overall and multiply by n. We thus obtain the first length formula:

$$L_1 \equiv \sum_{i=1}^{n} 2\pi \left(r_1 + (2i - 1)\frac{t}{2} \right) = n\pi(r_1 + r_2).$$

*Received by the editors August 20, 1982, and in revised form March 3, 1983.
†Department of Mathematics, Baylor University, Waco, Texas 76798.

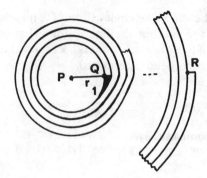

FIG. 1. *End view of material wrapped around a cylindrical core. r_2 is the distance from P to R.*

This method is simple enough; however, it may be challenged since it does not account for the "bump" which occurs when each layer is wrapped above the starting point Q. The next two methods attempt to take care of this.

Method B. If we assume that the material is flexible enough to allow a taut wrapping, then the geometry of the problem will appear as in Fig. 2. Here we are looking at the end of the roll again. The circle with center P and radius r_1 represents an end of the core. Point S is located so that the line through points T and S is tangent to the aforementioned circle at S. Angle QPS is denoted by α and is the same as angle RTV. The length of line segment TS is called l. Now it is an easy exercise in trigonometry and geometry to write down the total length of material, L_2, that is represented in Fig. 2. First, the length corresponding to the circular part of the ith wrap is approximated by

$$\left[r_1 + \frac{t}{2}(2i - 1)\right](2\pi - \alpha), \qquad i = 1, 2, \cdots, n.$$

Note that an average radius is used and that $\alpha = \arccos\left(r_1/(r_1 + t)\right)$. The length associated with each of the n rectangles is $l = \sqrt{t(t + 2r_1)}$. The length associated with the

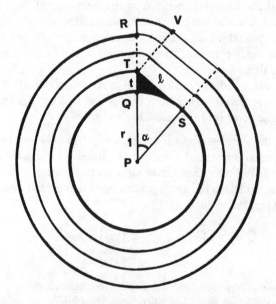

FIG. 2. *Detailed end view for the geometry of Method B ($n = 3$).*

ith sector is $(2i - 1)(t/2)\alpha$. Here we use the average radius again. Adding all of our terms and simplifying yields

$$L_2 \equiv n[\pi(r_1 + r_2) + l - \alpha r_1],$$

where α and l have been previously given in terms of r_1 and t.

Method C. Suppose that the material is not as flexible as the geometry of method B requires, but we may assume that the material forms a spiral from points Q to R. We may define this spiral in polar coordinates by

$$r = r_1 + \left(\frac{t}{2\pi}\right)\theta, \qquad 0 \le \theta \le 2n\pi.$$

Then the length of the spiral is given by the integral

$$L_3 \equiv \int_0^{2n\pi} \sqrt{(r_1 + (t/2\pi)\theta)^2 + (t/2\pi)^2}\, d\theta.$$

Using elementary calculus techniques we obtain a closed form for L_3. Let $g(x) = \sqrt{x^2 + (t/2\pi)^2}$. Then

$$L_3 = \frac{\pi}{t}\left[r_2 g(r_2) - r_1 g(r_1) + \left(\frac{t}{2\pi}\right)^2 \{\ln(r_2 + g(r_2)) - \ln(r_1 + g(r_1))\}\right].$$

Comparison of the methods. At this point we might favor method A because of its simplicity. However, we should ask if the geometries of methods B and C will produce lengths substantially different from that calculated by method A. To answer this we will investigate the differences in the approximations obtained from the three methods. The major tool for this comparison is the following.

THEOREM 1. *Let L_1, L_2, L_3 be the approximations given by methods* A, B, *and* C, *respectively. Then*

(1) $$L_2 - L_1 = \frac{2\sqrt{2}}{3\sqrt{r_1}}(r_2 - r_1)^{3/2} n^{-1/2} + O(n^{-3/2}),$$

and

(2) $$L_3 - L_1 = \frac{(r_2 - r_1)}{4\pi} \ln\left(\frac{r_2}{r_1}\right) n^{-1} + O(n^{-3}).$$

Proof. This is simply an application of Taylor's theorem. We will present the steps to obtain (1). Equation (2) follows in a similar manner.

Since

$$\alpha = \arctan\left(\frac{l}{r_1}\right) = \frac{l}{r_1} - \frac{1}{3}\left(\frac{l}{r_1}\right)^3 + O\left(\left(\frac{l}{r_1}\right)^5\right),$$

$$L_2 = n\pi(r_1 + r_2) + n(l - \alpha r_1)$$

$$= L_1 + nr_1\left[\frac{1}{3}\left(\frac{l}{r_1}\right)^3 + O\left(\left(\frac{l}{r_1}\right)^5\right)\right].$$

But by substituting

$$l = \sqrt{2r_1 t}\sqrt{1 + \frac{t}{2r_1}} = \sqrt{2r_1 t}\left(1 + O\left(\frac{t}{r_1}\right)\right) = \sqrt{\frac{2r_1(r_2 - r_1)}{n}}(1 + O(n^{-1}))$$

into the above, we obtain (1).

The above theorem implies that all three methods yield essentially the same approximation provided n is large relative to $(r_2 - r_1)/r_1$. We may now seek conditions which will assure that the approximations differ by less than some reasonable error, say one percent. For simplicity we will obtain bounds using only the leading terms of the right-hand sides of (1) and (2). Although the results will be meaningful only in an asymptotic sense, i.e., as n tends to infinity, they are sufficient to give a representative comparison. In particular, the next theorem investigates the difference in L_2 and L_1.

THEOREM 2. *Let* $E_1 = (2\sqrt{2}/3\sqrt{r_1})(r_2 - r_1)^{3/2} n^{-1/2}$ *and suppose that there is a positive number* z *such that* $t \leq z^2 r_1$ *and* $n \geq 31\, z$. *Then* E_1/L_1 *is less than* 0.01.

Proof. Note that $t \leq z^2 r_1$ implies that $(r_2 - r_1)/r_1 n \leq z^2$. Then since $r_2 + r_1 > r_2 - r_1$ we have

$$\frac{E_1}{L_1} = \frac{2\sqrt{2}}{3\pi\sqrt{r_1}} \frac{(r_2 - r_1)^{3/2}}{r_2 + r_1} n^{-3/2} \leq \frac{2\sqrt{2}}{3\pi} \left(\frac{r_2 - r_1}{r_1 n}\right)^{1/2} \frac{1}{n}$$

$$\leq \frac{2\sqrt{2}}{3\pi} z \frac{1}{31z} < .01.$$

As an example of the z in the above theorem, consider $z^2 = 0.1$. Then we may expect L_1 and L_2 to differ by less than one percent provided the thickness of the material is less than one tenth of r_1 and there are ten or more wraps.

The next theorem presents a similar result for L_1 and L_3.

THEOREM 3. *Let* $E_2 = ((r_2 - r_1)/4\pi) \ln (r_2/r_1)\, n^{-1}$ *and suppose that* $t \leq r_1$ *and* $n \geq 3$. *Then* E_2/L_1 *is less than* 0.01.

Proof. Since $r_2 > r_1 > 0$, $\ln (r_2/r_1) < (r_2 - r_1)/r_1$. Also since $r_2 + r_1 > r_2 - r_1$ and $(r_2 - r_1)/nr_1 \leq 1$, we have

$$\frac{E_2}{L_1} = \frac{(r_2 - r_1) \ln (r_2/r_1)}{4n^2\pi^2(r_2 + r_1)} \leq \frac{1}{4\pi^2 n} < .01.$$

Conclusion. We have presented three methods for calculating approximations for the length of a material wrapped in a cylindrical roll, each based on different assumptions concerning the geometry of the roll. However, in most practical situations we have shown that all three approximations agree to within an accuracy of roughly one percent.

NAVIGATION ON RIEMANNIAN OCEANS*

YVES NIEVERGELT†

Abstract. This note uses elementary linear algebra and vector products to solve plane or spherical navigation, surveying, and radio goniometry problems. Down-to-water illustrations on San Francisco Bay and the Bermuda Triangle are presented, followed by problems that link navigation to cartography.

Introduction. Spherical goniometry will constitute our main theme, but we shall first treat the goniometric problem in the plane, because its simplicity makes it more accessible to novices in calculus or linear algebra, and prepares students conceptually for direction finding on the sphere, where they will discover an easy solution with cross products. These practical applications will spark enthusiasm in the classroom while illustrating vector

*Received by the editors May 27, 1983, and in revised form November 14, 1983. This work was partially supported by a grant from the Swiss National Research Fund.

†Department of Mathematics, University of Washington, Seattle, Washington 98195.

manipulations. Finally, the problems in the last section will introduce navigation maps, tying together the plane and spherical situations.

1. Plane goniometry. In the "fox hunt" competition, the "hunters" must locate a hidden radio station S (the "fox") by measuring the angle of incidence—called the *azimuth* or *bearing*—between true North and the incoming radio signal from S. Repeating this procedure at two places A and B, they find the "fox" at the intersection of the rays supporting AS and BS, as Fig. 1 shows. A computational method offers greater accuracy and practicality than trying to draw straight lines on a wet map out in the electromagnetic field. Also, it enables one to consider points outside the map, or even to work without a map.

Recalling that freshmen lose their Latin in front of Γρεεκ symbols, we denote the azimuths at A and B by a and b respectively. Then we define the normal unit vectors $\mathbf{U} := (-\cos a, \sin a)$ and $\mathbf{V} := (-\cos b, \sin b)$ to find equations for the intersection $S := (x, y)$ of both lines AS and BS:

(1)
$$\text{line } AS: \quad \mathbf{U} \cdot S = \mathbf{U} \cdot A,$$
$$\text{line } BS: \quad \mathbf{V} \cdot S = \mathbf{V} \cdot B,$$

where the dots represent inner products (taking care of the otherwise infinite slopes due North and South). Setting $A := (A_1, A_2)$ and $B := (B_1, B_2)$, we get the following linear system, which students can solve by their favorite method.

(2)
$$\begin{pmatrix} -\cos a & \sin a \\ -\cos b & \sin b \end{pmatrix} \begin{pmatrix} x \\ y \end{pmatrix} = \begin{pmatrix} -A_1 \cos a + A_2 \sin a \\ -B_1 \cos b + B_2 \sin b \end{pmatrix}.$$

Example 1.1. Suppose that in the Presidio at coordinates $(47.05, 82.95)$, we receive the fox under the azimuth $a = 23°$, whereas on Telegraph Hill, at $(52.34, 83.87)$, we measure $b = 342°$. Substituting these data into system (2), we locate the fox at Mt. Caroline Livermore, $(50.22, 90.41)$, on Angel Island.

In this plane model, the rays AS and BS cross all vertical grid lines under the same azimuths a and b. Therefore, S sees A in the direction $u = a + 180°$ and B in the direction $v = b + 180°$. Replacing a by u and b by v in (2) multiplies both sides by -1, preserving the same solution S. Let us exploit this property to solve the reverse problem: self-location.

Example 1.2. Sailing on the Bay, you see Alcatraz Lighthouse $(50.90, 86.50)$ in the South-West ($u = 225°$) and Blunt Point Lighthouse $(51.22, 89.46)$ in the North-West ($v = 315°$). Solving system (2) with u and v instead of a and b respectively, you determine your own Sail's position at $(52.54, 88.14)$.

This algebraic rather than geometric point-of-view lends itself quite well to least-squares estimates in surveying when fitting together several bearings:

Example 1.3. We see the tip (x, y) of an island under the azimuths $a = 0°$, $b = 84°$, $c = 125°$ respectively from (A) Ghirardelli Square $(50.90, 84.43)$, (B) The Golden Gate Bridge $(46.00, 86.00)$, and (C) Sausalito Point $(46.00, 90.00)$. These measurements give three (generically inconsistent) linear equations $-x + 50.90 = 0$, $-0.1045x + 0.9945y - 80.72 = 0$, $0.5736x + 0.8192y - 100.1 = 0$, derived exactly as (1). Either with a pseudo-inverse or with calculus, we minimize the sum of the squares of the left-hand sides at $(\bar{x}, \bar{y}) = (50.91, 86.53)$.

2. Spherical goniometry. Our plane model yields adequate results at short range, say within 50 miles or so. (In practice, interferences along the wave path create distortions several orders of magnitude worse than our plane approximation.) For longer distances,

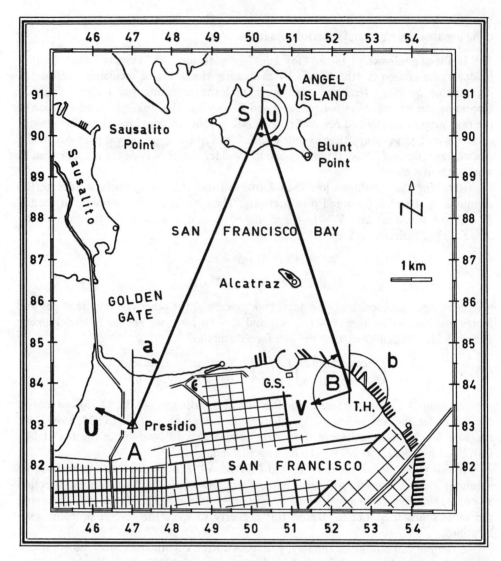

FIG. 1. *The azimuths a and b measure the direction of S clockwise from true North at A and B respectively.* (*Drawing and 1000-meter grid ticks based on the map* SAN FRANCISCO NORTH, *published by the U.S. Geological Survey.*) *G.S. = Ghirardelli Square, T.H. = Telegraph Hill.*

however, we need a spherical model, which we now develop.

A ship S at unknown longitude and latitude[1] (ω, θ) is sending a distress signal received by two coastal stations A at (ϕ, λ) and B at (ψ, γ) under azimuths α and β respectively (see Fig. 2). How do we locate the vessel? We simply determine the intersection of two planes P and Q normal to the surface of the earth and passing through A and B (along great circles) under azimuths α and β. (Radio waves travel from the ship to each station by successive deflections off the ocean and the ionosphere, which is assumed parallel to the surface of the sea. Thus, the direction of propagation remains in a fixed vertical plane, P and Q respectively.)

---

[1]The convention in geographical coordinates is to write the latitude first.

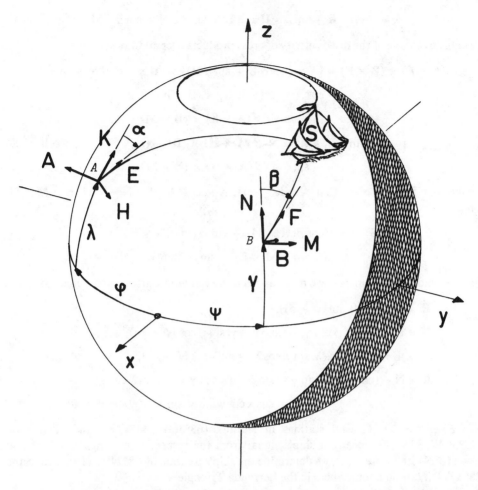

FIG. 2. *Oblique orthographic projection (axonometry) of the earth, showing the ship and both direction-finders at A and B.*

To locate the ship, define a unit vector **E** tangent to the surface at A in the direction α, and a corresponding vector **F** at B. Thus, **E** and **F** point toward the vessel. Let $\mathbf{A} := A/\|A\|$ and $\mathbf{B} := B/\|B\|$; since **A** and **E** span P, then $\mathbf{A} \times \mathbf{E}$ is normal to P, and likewise $\mathbf{B} \times \mathbf{F}$ is normal to Q. Set

$$(3) \qquad\qquad \mathbf{S} := (\mathbf{A} \times \mathbf{E}) \times (\mathbf{B} \times \mathbf{F}).$$

The vector **S** is normal to both $\mathbf{A} \times \mathbf{E}$ and $\mathbf{B} \times \mathbf{F}$, hence it belongs to $P \cap Q$, so we need only convert **S** into spherical coordinates $\mathbf{S} = (\omega, \theta)$, because its length does not matter. However, its sign does: we must choose between (ω, θ) and $(\omega + 180°, -\theta)$, whichever lies closer to A and B. (Select the one for which $\mathbf{S} \cdot \mathbf{E}, \mathbf{S} \cdot \mathbf{F} > 0$. The other one shows the ship's antipode!) Students will appreciate the absence of spherical trigonometry in the concise formula (3).

We can simplify computations by preparing all cross products, and a mere linear combination of four *fixed* vectors will express **S**: endow A and B with orthonormal bases (**A, H, K**) and (**B, M, N**) as in Fig. 2; then

$$\mathbf{E} = \cos \alpha \cdot \mathbf{K} + \sin \alpha \cdot \mathbf{H} \text{ and } \mathbf{F} = \cos \beta \cdot \mathbf{N} + \sin \beta \cdot \mathbf{M}.$$

Taking advantage of the distributivity of cross products over additions, we get

$$\mathbf{S} = (\mathbf{A} \times \mathbf{E}) \times (\mathbf{B} \times \mathbf{F}) = [\mathbf{A} \times (\cos \alpha \, \mathbf{K} + \sin \alpha \, \mathbf{H})] \times [\mathbf{B} \times (\cos \beta \, \mathbf{N} + \sin \beta \, \mathbf{M})]$$

$$= \cos \alpha \cos \beta \, (\mathbf{A} \times \mathbf{K}) \times (\mathbf{B} \times \mathbf{N})$$

$$+ \cos \alpha \sin \beta \, (\mathbf{A} \times \mathbf{K}) \times (\mathbf{B} \times \mathbf{M})$$

$$+ \sin \alpha \cos \beta \, (\mathbf{A} \times \mathbf{H}) \times (\mathbf{B} \times \mathbf{N})$$

$$+ \sin \alpha \sin \beta \, (\mathbf{A} \times \mathbf{H}) \times (\mathbf{B} \times \mathbf{M}).$$

Looking at Fig. 2 we see that $\mathbf{A} \times \mathbf{K} = -\mathbf{H}$, $\mathbf{A} \times \mathbf{H} = \mathbf{K}$, $\mathbf{B} \times \mathbf{N} = -\mathbf{M}$, $\mathbf{B} \times \mathbf{M} = \mathbf{N}$, whence

(4)
$$\mathbf{S} = \cos \alpha \cos \beta (\mathbf{H} \times \mathbf{M}) + \cos \alpha \sin \beta (\mathbf{N} \times \mathbf{H})$$

$$+ \sin \alpha \cos \beta (\mathbf{M} \times \mathbf{K}) + \sin \alpha \sin \beta (\mathbf{K} \times \mathbf{N}).$$

For the reader's convenience, we make all vectors explicit in Cartesian coordinates:

$$\mathbf{H} \times \mathbf{M} = (0, 0, \sin (\psi - \phi)),$$

$$\mathbf{N} \times \mathbf{H} = (-\cos \phi \cos \gamma, -\sin \phi \cos \gamma, \sin \gamma \cos (\psi - \phi)),$$

$$\mathbf{M} \times \mathbf{K} = (\cos \psi \cos \lambda, \sin \psi \cos \lambda, -\sin \lambda \cos (\psi - \phi)),$$

$$\mathbf{K} \times \mathbf{N} = (\cos \lambda \sin \psi \sin \gamma - \sin \phi \sin \lambda \cos \gamma, \cos \phi \sin \lambda \cos \gamma$$

$$- \cos \lambda \cos \psi \sin \gamma, \sin \lambda \sin \gamma \sin (\psi - \phi)).$$

Example 2.1. Coastal stations in Halifax (63°36'W, 44°38'N) and Charleston (79°56'W, 32°47'N) receive a ship's signal from the respective bearings $\alpha = 195°$ and $b = 110°$. With formula (4), we determine the ship's position by 68°24'W of longitude and 28°35'N of latitude: somewhere in the Bermuda Triangle.

3. Relevant problems. Can one reduce the previous spherical computations to plane geometric constructions as in §1? Yes! The following problems will lead students to entertaining discoveries at various levels of difficulty.

Problem 3.1. Prove that no homeomorphic projection $S^2 \to \mathbb{R}^2$ from part of the earth S^2 onto part of the navigation map $\mathbb{R}^2$ can transform *all* great circles into straight lines while preserving *all* angles. (Hint: map a spherical triangle onto a plane one and compare the sums of their angles.) Observe that Mercator projections do preserve all angles, though. (See J. A. Steers, *Introduction to the Study of Map Projections,* 14th ed., University of London Press, London, 1965.)

Problem 3.2. Show that a gnomonic projection—from the center of the earth onto any tangent plane—transforms *all* great circles into straight lines and conserves all angles at the point of tangency. (Hint: great circles are sections through the center of the planet by planes that meet the map along straight lines.)

Problem 3.3. Demonstrate that any projection $p : S^2 \to \mathbb{R}^2$ that maps all great circles onto straight lines must be a gnomonic projection G followed by a linear transformation L and a translation T. (Hint: Set $T := \text{Id} + p \circ G^{-1}(O)$. Then $T^{-1} \circ p \circ G^{-1}$ sends every straight line onto a straight line and fixes the origin. Hence, it must be linear over the rationals $\mathbb{Q}$. Now invoke continuity.)

Problem 3.4. Consider two nonantipodal points A and B on the earth S^2. Find a linear transformation L and a gnomonic projection G such that $p := L \circ G$ preserves all angles at A *and* at B. This Maurer Orthodromic Projection makes goniometry on the sphere look as simple as in the plane.

Problem 3.5. Examine the navigator's problem (Example 1.2) on the sphere: Two beacons A and B send signals received under azimuths μ and ν *at the ship*. Knowing the coordinates of A and B, how can the navigator determine the ship's position? (Caution: the *line of bearing*—locus of all points on S^2 that see A under the same azimuth μ—is neither a geodesic nor a loxodrome.)

Problem 3.6. Are there maps $S^2 \to \mathbb{R}^2$ that project all lines of bearing from A onto straight lines?—and preserve all angles at A?—at A and at B? (See R. Keen, *Wireless Direction Finding*, 3rd ed., Ilife & Sons, London, 1941, p. 264.)

Remark 3.7. Merchant vessels with more sophisticated equipment navigate with the LORAN network, based on distances to four synchronized beacons instead of bearings. (See J. A. Pierce, A. A. McKenzie, and R. H. Woodward, LORAN, McGraw-Hill, New York, 1948.)

Problem 3.8. Study the navigator's problem (3.5) on a toroidal ocean parametrized by the usual covering map $\mathbb{C} \to S^1 \times S^1$. (This flat geometry differs from that on the familiar torus in $\mathbb{R}^3$. Sailing on a doughnut may prove a sticky venture!)

Acknowledgment. The author is very grateful to the U.S. Geological Survey for allowing the use of their maps.

Designing a Three-Edged Reamer

Problem 80-17, by B. C. RENNIE (James Cook University, N.Q., Australia).

A simple closed plane curve and a triangle ABC (regarded as a rigid body movable in the plane) have the property that the triangle can be moved continuously so that each vertex moves monotonically once round the curve. Must the curve be a circle?

The problem arises in the design of a "reamer," which is a tool used by a fitter to finish-cut a hole to an accurate diameter; it may be fixed or adjustable. An adjustable reamer usually has six straight cutting edges equally spaced parallel to the axis of the tool. In reaming a hole it is not enough to enlarge the diameter when it is too small; we also require to ensure that the hole is circular. A two-edged reamer would be of no use since there are noncircular plane curves of constant diameter. Designing a three-edged reamer leads one to the above question in plane geometry.

Solution by M. GOLDBERG (Washington, DC).

The curve does not have to be a circle. It can be a square with rounded corners, as shown in the following example.

The Reuleaux triangle, made by three circular arcs, can be rotated within a square while keeping contact with all the sides of the square. Each of the corners of the Reuleaux triangle traces the square with rounded corners. This is the basis for a commercial drill made by Watts Brothers Tool Works of Wilmerding, PA. It is pictured and described by Martin Gardner in his column in Scientific American, February 1963, p. 150, and it is mentioned in my paper *Rotors in polygons and polyhedra*, Math. Comp., 14 (1960), pp. 229–239.

There are many other possibilities. My papers describe various methods of obtaining noncircular ovals which can rotate within regular polygons,. If a rotor is held fixed while the polygon is rotated about it, then all the vertices of the polygon trace another noncircular curve (see Figures). Therefore, if this new curve is fixed, then the

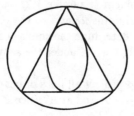 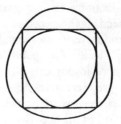

(a) *Triangle rotor in an oval.*   (b) *Square rotor in an oval.*

regular polygon can be rotated within it while all the vertices trace the curve. Any three of the vertices can be the vertices of a triangle rotating within the noncircular oval. Another application of a triangular rotor in an oval is the rotary engine designed by Felix Wankel (presently being used in the sports car, Mazda RX7).

The Shape of Milner's Lamp

Problem 65-9, by ROLAND SILVER (The MITRE Corporation).

Determine the shape of the lamp in the following quotation [1]: "The same gentleman vouches for Milner's lamp: but this had visible *science* in it. . . . A hollow semi-cylinder, but not with a circular curve, revolved on pivots. The curve was calculated on the law that, whatever quantity of oil might be in the lamp, the position of equilibrium just brought the oil up to the edge of the cylinder, at which a bit of wick was placed. As the wick exhausted the oil, the cylinder slowly revolved about the pivots so as to keep the oil always touching the wick." See Fig. 1.

REFERENCE

[1] AUGUSTUS DE MORGAN, *A Budget of Paradoxes*, vol. 1, Dover, New York, 1954, p. 252.

Solution by J. D. LAWSON (University of Waterloo).

We assume for definiteness that the empty lamp has mass m and that the center of gravity is at the lip. We assume further that the distance from pivot to lip is unity, that the lamp is of unit thickness and that the fluid is of unit density.

We consider static equilibrium of the lamp as shown in Fig. 2. The angle θ is measured counterclockwise from the x axis. The equation of the liquid surface is $r = \cos \alpha / \cos \theta$ and the angle β is found from $\rho(\beta + \alpha) = \cos \alpha / \cos \beta$, where $r = \rho(\theta + \alpha)$ is the equation of the lamp. For equilibrium, we have

$$(1) \qquad m \sin \alpha = \int_{\theta=-\alpha}^{\beta} \int_{r=\cos \alpha / \cos \theta}^{\rho(\theta+\alpha)} r \sin \theta \, r dr \, d\theta.$$

Simplifying,

$$(2) \qquad 6m \sin \alpha = 2 \int_{0}^{\alpha+\beta} \rho^3(\theta) \sin (\theta - \alpha) \, d\theta - \cos^3 \alpha / \cos^2 \beta + \cos \alpha.$$

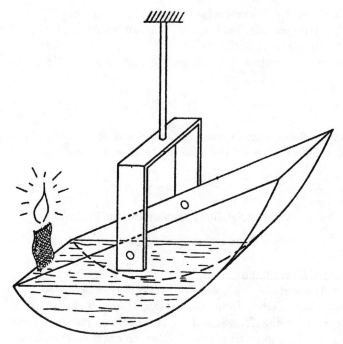

Fig. 1

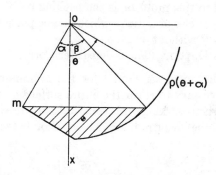

Fig. 2

This equation defines β as a function of α. Differentiating (2) twice with respect to α, noting that $\rho(\beta + \alpha) = \cos \alpha / \cos \beta$, and adding the result to (2) gives the differential equation

(3)
$$\frac{d\beta}{d\alpha} = \tan \alpha / \tan \beta.$$

The solution of (3) is

(4)
$$\cos \beta = c \cos \alpha, \quad c \leqq 1.$$

Thus $\rho(\beta + \alpha) = c^{-1}, \beta + \alpha \geqq \cos^{-1} c$. For $\alpha = 0, \beta = \cos^{-1} c$. The shape of the lamp for $\theta \leq \cos^{-1} c$ is constrained only by $\rho(\theta) \geq [\cos \theta]^{-1}$. Choosing $\rho(\theta)$ $= [\cos\theta]^{-1}$, a straight line, for $\theta \leq \cos^{-1} c$ ensures that the lamp will empty entirely, and is thus a reasonable choice.

Substituting $\rho(\theta) = [\cos \theta]^{-1}$, $\theta \leq \cos^{-1} c$, and $\rho(\theta) = c^{-1}$, $\theta \geq \cos^{-1} c$, in (2) gives

$$m = (1 - c^2)^{3/2}/3c^3.$$

The shape is thus a straight line segment plus a circular arc, the length of the line segment depending upon the mass of the empty lamp.

Another Satellite Communications Problem

Problem 61-7, by ISIDORE SILBERMAN (Raytheon Mfg. Co.).

Determine the minimum number $N(r, R)$ of circles of radius r necessary to cover the entire surface of a sphere of radius R. *Editoral Note:* A lower bound can be gotten immediately from area considerations. If A_r denotes the area of a spherical cap (of radius R) whose base is a circle of radius r, then $N(r, R) >$ $4\pi R^2/A_r$ (since there must be overlap). Since an exact determination of $N(r, R)$ appears to be extremely difficult, reasonably close upper and lower bounds will be acceptable as a solution.

Also, closely allied to this problem, is the packing problem of determining the maximum number of circles of radius r which can be placed on the entire surface of a sphere of radius R without overlap.

Solution by LOUIS D. GREY (The Teleregister Corp.).

We shall determine an upper bound for the minimum number $N(r, R)$ of circles of radius r necessary to cover the entire surface of a sphere of radius R.

It is convenient when working with spherical caps to work with the angular radius A rather than the linear radius r. The relation between these two is given by

(1)
$$\cos A = \frac{\sqrt{R^2 - r^2}}{R}$$

By suitably choosing r the problem can be restricted to the unit sphere without any loss of generality.

To determine an upper bound for the minimum number $N^*(A, 1)$ of spherical caps of angular radius A necessary to cover the entire surface of a unit sphere, we consider a closely allied problem. This latter problem is the determination of the maximum number $K(A)$ of spherical caps of angular radius A which can be placed on a unit sphere without overlapping. The inverse of this latter problem, namely, what is the largest angular radius $A(K)$ such that K spherical caps of angular radius A can be placed on the unit sphere without overlapping, has been treated by L. Fejes Toth [1] who obtained the result.

(2)
$$A(K) \leq \tfrac{1}{2} \cos^{-1} \tfrac{1}{2} \left[\cot^2 \frac{K\pi}{6(K - 2)} - 1 \right].$$

This is an asymptotic upper bound which is exact for $K = 3, 4, 6,$ and 12.

For $K > 2$ which implies $0 < A < \dfrac{\pi}{3}$, we can solve for K to obtain

(3) $$K(A) \le \frac{12 \cot^{-1} \sqrt{4 \cos^2 A - 1}}{6 \cot^{-1} \sqrt{4 \cos^2 A - 1} - \pi}.$$

We shall show that

$$N^*(2A, 1) \le \frac{12 \cot^{-1} [\sqrt{4 \cos^2 A - 1}]}{6 \cot^{-1} [\sqrt{4 \cos^2 A - 1}] - \pi}, \qquad 0 < A < \frac{\pi}{3}.$$

The argument is simple. Imagine that we have placed $K(A)$ nonoverlapping spherical caps of angular radius on the surface of a sphere. If we replace these caps by concentric caps of angular radius $2A$, then we claim the surface of the sphere is completely covered. If it is not, there is a point whose distances from the centers of the caps is greater than $2A$. This implies that we could center a cap of angular radius A at this point and that this cap would not overlap any of the original $K(A)$ caps which contradicts the assumption that $K(A)$ is maximum.

The problem of determining an upper bound for $K(A)$ has been generalized to n dimensions. This bound will provide a bound for $N^*(2A, 1)$ by the argument given above. It is shown [2] that

(4) $$N_n{}^*(2A, 1) \le \frac{\pi^{1/2} \, \Gamma\left(\dfrac{n-1}{2}\right) \sin B \tan B}{2 \, \Gamma\left(\dfrac{n}{2}\right) \displaystyle\int_0^B (\sin\theta)^{n-2}(\cos\theta - \cos B) \, d\theta},$$

$$0 < A < \frac{\pi}{4}, \ B = \sin^{-1}\left(\sqrt{2 \sin A}\right).$$

A summary of results concerning the determination of $K(A)$ and the inverse problem in n-dimensions appears in a paper by the author [3].

REFERENCES

1. L. FEJES TOTH, *On The Densest Packing of Spherical Caps*, American Mathematical Monthly, Vol. 56, pp. 330–331. 1949.
2. R. A. RANKIN, *The Closest Packing Of Spherical Caps in N Dimensions*, Proc. Glasgow Math. Assoc., Vol. 2, p. 139, 195.
3. FLORES AND GREY, *Reference Signals For Character Recognition Systems*, IRE Transactions on Electronic Computers, Vol. EC-9, March 1960, p. 57–60.

N-Dimensional Volume

Problem 59-2, by MAURICE EISENSTEIN AND M. S. KLAMKIN.

Determine the volume in N-space bounded by the region

$$0 \le a_1x_1 + a_2x_2 + \cdots + a_Nx_N < 1, \qquad\qquad a_r \ge 0$$
$$b_r \ge x_r \ge c_r, \qquad\qquad r = 1, 2, \cdots, N.$$

This problem has arisen from the following physical situation: a series-parallel circuit of N resistances is given where each of the resistances R_i are not known exactly but are uniformly distributed in the range $R_i \pm \epsilon_i R_i (\epsilon_i \ll 1)$. We wish to determine the distribution function for the circuit resistance

$$R = F(R_1, R_2, \cdots, R_N).$$

To first order terms

$$\Delta R = \sum_{i=1}^{N} \frac{\partial F}{\partial R_i} dR_i, \qquad\qquad |dR_i| \leq \epsilon_i R_i.$$

The probability that the circuit resistance lies between R and $R + \Delta R$ will be proportional to the volume bounded by the region

$$0 \leq \sum_{i=1}^{N} \frac{\partial F}{\partial R_i} x_i \leq \Delta R, \qquad -\epsilon_i R_i \leq x_i \leq \epsilon_i R_i.$$

Special cases of the problem arise in the two following examples:

(A) A sequence of independent random variables with a uniform distribution is chosen from the interval $(0, 1)$. The process is continued until the sum of the chosen numbers exceeds L. What is the expected number of such choices? The expected number E will be given by

$$E = 1 + F_1 + F_2 + F_3 + \cdots,$$

where F_i is the probability of failure up to and including the ith trial. Geometrically, F_i will be given by the volume enclosed by

$$x_1 + x_2 + \cdots + x_i \leq L,$$

$$0 \leq x_r \leq 1, \qquad\qquad r = 1, 2, \cdots, i.$$

For the case $L = 1$,

$$E = \sum_{m=0}^{\infty} \frac{1}{m!} = e.$$

(D. J. Newman and M. S. Klamkin, *Expectations for Sums of Powers*, American Mathematical Monthly, January, 1959, pp. 50–51.)

(B) What is the probability that N points picked at random in a plane form a convex polygon?

If we denote the interior angles by θ_i, the probability that the polygon will be convex will be proportional to the volume of the region given by

$$\theta_1 + \theta_2 + \theta_3 + \cdots + \theta_N = (N - 2)\pi,$$

$$0 \leq \theta_r \leq \pi$$

The normalizing constant will be given by the volume of the region

$$\theta_1 + \theta_2 + \cdots \theta_N = (N - 2)\pi,$$

$$0 \leq \theta_r < 2\pi,$$

(we are assuming that the angles are uniformly distributed).
Solution by I. J. Schoenberg, University of Wisconsin.

Let B_ω denote the volume of the n-dimensional polyhedron

(1) $0 \leq x_i \leq a_i$ $i = 1, 2, \cdots, n,$

(2) $0 \leq \lambda_1 x_1 + \lambda_2 x_2 + \cdots + \lambda_n x_n \leq \omega,$

where $\lambda_1^2 + \lambda_2^2 + \cdots + \lambda_n^2 = 1$ and a_r, λ_r, $\omega \geq 0$. Also, let $b_i = a_i\lambda_i$. If $F(u)$ is a function of one variable u, we define the operator L_n by

$$(3) \qquad L_nF(u) = \sum_{(\alpha_1, \alpha_2, \cdots, \alpha_i)} (-1)^{n-i} F(b_{\alpha_1} + b_{\alpha_2} + \cdots + b_{\alpha_i})$$

where $(\alpha_1, \alpha_2, \cdots, \alpha_i)$ runs through all the 2^n combinations of the n quantities $b_1, b_2, \cdots, b_n$. For example,

$$L_1F(u) = F(b_1) - F(0),$$

$$L_2F(u) = F(b_1 + b_2) - F(b_1) - F(b_2) + F(0).$$

It follows that

$$L_nF(u) = L_{n-1}F(u + b_n) - L_{n-1}F(u).$$

If $F(u)$ is sufficiently smooth,

$$(4) \qquad \int \cdots \int_B F^{(n)}(\lambda_1 x_1 + \cdots + \lambda_n x_n)\, dx_1 \cdots dx_n = \prod_{r=1}^{n} \lambda_r^{-1} L_n\, F(u)$$

where B denotes the box defined by (1). To establish (4), we assume it holds for $n = 1, 2, \cdots, n - 1$. Then

$$\int_0^{a_n} dx_n \int \cdots \int_{x_n \text{fixed}} F^{(n)}(\lambda_1 x_1 + \cdots + \lambda_n x_n)\, dx_1 \cdots dx_{n-1}$$

$$= \frac{1}{\lambda_1 \cdots \lambda_{n-1}} \int_0^{a_n} L_{n-1}\, F'(u + \lambda_n x_n)\, dx_n$$

$$= \frac{1}{\lambda_1 \cdots \lambda_n} \left\{ L_{n-1}\, F(u + b_n) - L_{n-1}\, F(u) \right\}$$

$$= \prod \lambda_r^{-1} L_n\, F(u).$$

Since (4) is valid for $n = 1$, it is valid for all n by induction.

One consequence of (4) is that $L_nF(u) = 0$ whenever $F(u)$ is a polynomial of degree less than n. By a known theorem of Peano, we can write

$$(5) \qquad L_n\, F(u) = \int_{-\infty}^{\infty} \Phi_n(x) F^{(n)}(x)\, dx,$$

where the kernel Φ_n may be described as follows: If we define the truncated power function x_+^k by

$$(6) \qquad x_+^k = \begin{cases} x^k & \text{if } x \geq 0, \\ 0 & \text{if } x < 0, \end{cases} \qquad\qquad k = 0, 1, 2, \cdots$$

then

$$(7) \qquad \Phi_n(x) = L_n \frac{(u - x)_+^{n-1}}{(n-1)!},$$

where on the right side x is treated as a parameter and L_n operates on the variable u. Since $\Phi_n(x) = 0$ if $x < 0$ or $x > \sum_1^n b_r = b$,

$$(8) \qquad \int \cdots \int_B F^{(n)}(\lambda_1 x_1 + \cdots + \lambda_n x_n)\, dx_1 \cdots dx_n$$

$$= \prod \lambda_r^{-1} \int_0^b \Phi_n(x) F^{(n)}(x)\, dx.$$

Equation (8) shows that $\prod \lambda_r^{-1} \Phi_n(x)$ is the area of the intersection of the box B with the hyperplane $\lambda_1 x_1 + \cdots + \lambda_n x_n = x$ (x-fixed). To see this more clearly, we choose $F(x)$ in (8) such that

$$F^{(n)}(x) = \begin{cases} 1 & \text{if } x \leqq \omega, \\ 0 & \text{if } x > \omega; \end{cases}$$

i.e.

$$F(x) = (-1)^n \frac{(\omega - x)_+^n}{n!}$$

Equation (8) now reduces to

(9) $$B_\omega = \prod \lambda_r^{-1} \int_0^\omega \Phi_n(x)\, dx.$$

Since the operator L_n commutes with the integration,

$$\prod \lambda_r\, B_\omega = L_n \int_0^\omega \frac{(u - x)_+^{n-1}}{(n-1)!}\, dx = -L_n \left\{ \frac{(u - x)_+^n}{n!} \right\}_{x=0}^{x=\omega}$$

$$= \frac{1}{n!} L_n u_+^n - \frac{1}{n!} L_n (u - \omega)_+^n.$$

Writing $B = a_1 a_2 \cdots a_n$ and observing that if $\omega \geqq b$ then $L_n(u - \omega)_+^n = 0$ and $B_\omega = B$. We may now write our final result as

(10) $$B_\omega = B - \frac{\prod \lambda_r^{-1}}{n!} L_n (u - \omega)_+^n.$$

As an example, let us consider the hypercube when $a_r = 1$ and $\lambda_r = n^{-1/2}$, $r = 1, 2, \cdots, n$. Then also $b_r = n^{-1/2}$ and (10) gives

(11) $$B_\omega = 1 - \frac{n^{n/2}}{n!} \Delta^n (u - \omega)_+^n \Big|_{u=0},$$

where Δ^n is the ordinary n^{th} order advancing difference operator of step $h = n^{-1/2}$. Now, if $\omega = 0$ then $B_\omega = 0$ and (11) gives

$$\Delta^n u_+^n \,|_{u=0} = \Delta^n u^n \,|_{u=0} = n^{-n/2} n!$$

which is a known relation. If $\omega = n^{-1/2}$ then again for the ordinary power function

(12) $$\Delta^n (u - \omega)^n \,|_{u=0} = n^{-n/2} n!.$$

Passing to the truncated power function only one term of the left side of (12) drops out so that

$$\Delta^n (u - \omega)_+^n \,|_{u=0} = n^{-n/2} \{n! - 1\}.$$

Finally (11) gives for $\omega = n^{-1/2}$ the value

$$B_{n^{-1/2}} = \frac{1}{n!}$$

which is also known.

The expression (7) shows that $\Phi_n(x)$ is what has been called elsewhere[1] a spline curve of degree $n - 1$, i.e. a composite of different polynomials of degree $n - 1$ having $n - 2$ continuous derivatives while $\Phi_n^{(n-1)}(x)$ has jumps at the "knots" $x = b_{\alpha_i} + \cdots + b_{\alpha_i}$. The Laplace transform of $\Phi_n(x)$, however, has the simple form

$$(13) \qquad \int_{-\infty}^{\infty} e^{-sx}\Phi_n(x)\ dx = \prod_{r=1}^{n} \frac{1 - e^{-sb_r}}{s}.$$

This transform is particularly useful if we wish to discuss the limit properties of the distribution $\Phi_n(x)$ for large n.

Remark: No originality is claimed for the matters presented here. The operator L_n was studied by M. Frechet, T. Popoviciu, and others. Laplace transforms of the kind obtained here were already derived by Laplace himself. Finally, G. Polya's Hungarian doctoral dissertation is devoted to an intensive study of the transforms (13).[2] As a matter of fact, Polya starts from the problem of determining the volume B_ω and also stresses the relations with probability theory which are obtained if n is allowed to tend to infinity.

Also solved by Larry Shepp who shows that the probability that an $n + 1$ sided polygon be convex (the angles of which are assumed uniformly distributed) is

$$P_{n+1} = \cfrac{2^n - n - 1}{(n-1)^n - \binom{n+1}{1}(n-3)^n + \cdots \\ + (-1)^{[n/2]+1}\binom{n+1}{[n/2]-1}(n - 2[n/2] + 1)^n}.$$

This generalizes the result of H. Demir for the case $n = 3$ (Pi Mu Epsilon Journal, Spring 1958).

Editorial note: E. G. Olds in "A Note on the Convolution of Uniform Distributions," Annals of Mathematical Statistics, v. 23, 1952, pp. 282–285, gives a derivation for the probability density function for a sum of independent rectangularly distributed random variables.

[1] Bull. Amer. Math. Soc., vol. 64 (1958), pp. 352–357.
[2] Mathematikai es Physikai Lapok, vol. XXII.

Supplementary References
Applied Geometry

[1] H. ABELSON AND A. DISESSA, *Turtle Geometry: The Computer as a Medium for Exploring Mathematics*, M.I.T. Press, Cambridge, 1981.

[2] H. AHRENS, *Geodesic Mathematics and How to Use it*, University of California Press, Berkeley, 1976.

[3] F. J. ALMGREN, JR., AND J. E. TAYLOR, "The geometry of soap films and soap bubbles," Sci. Amer. (1976) pp. 82–93.

[4] F. L. ALT, "Digital pattern recognition by moments," J.A.C.M. (1962) pp. 240–258.

[5] D. L. ANDERSON AND A. M. DZIEWONSKI, "Seismic tomography," Sci. Amer. (1984) pp. 60–68.

[6] H. G. APSIMON, "Wrapping a parcel," BIMA (1980) pp. 160–164.

[7] I. I. ARTOBOLEVSKII, *Mechanisms for the Generation of Plane Curves*, Pergamon, Oxford, 1964.

[8] E. ASHER AND A. JANNER, "Algebraic aspects of crystallography: Space groups as extensions," Helv. Phys. Acta (1965) pp. 551–572.

[9] J. G. ASPINHALL, "Implementation of a hidden line remover," UMAP J. (1984) pp. 431–461.

[10] F. ATTNEAVE AND M. D. ARNOULT, "The quantitative study of shape and pattern," Psych. Bull. (1956) pp. 452–471.

[11] T. E. AVERY AND G. D. BERLIN, *Interpretation of Aerial Photographs*, Burgess, Minneapolis, 1985.

[12] M. S. BARTLETT, "Stochastic geometry: An introduction and reading list," Int. Statist. Rev. (1982) pp. 179–193.

[13] M. BERMAN, *The Statistical Analysis of Spatial Pattern*, Chapman & Hall, London, 1975.

[14] G. BINNIG AND H. ROHRER, *"The scanning tunneling microscope,"* Sci. Amer. (1985) pp. 50–56.

[15] A. A. BLANK, *"The Luneberg theory of binocular visual space,"* J. Optical Soc. Amer. (1953) pp. 717–727.

[16] F. J. BLOORE AND H. R. MORTON, *"Advice on hanging pictures,"* AMM (1985) pp. 309–321.

[17] E. D. BOLKER AND H. CRAPO, *"Bracing rectangular frameworks I,"* SIAM J. Appl. Math. (1979) pp. 473–490.

[18] ———, *"How to brace a one storey building?"* Environment and Planning (1977) pp. 125–152.

[19] E. D. BOLKER, *"Bracing rectangular frameworks II,"* SIAM J. Appl. Math. (1979) pp. 491–508.

[20] M. BORN AND E. WOLFE, *Principles of Optics*, Pergamon, Oxford, 1980.

[21] O. BOTTEMA AND B. ROTH, *Theoretical Kinematics*, North-Holland, Amsterdam, 1979.

[22] J. B. BOYLING, *"Munro's problem: How to count mountains,"* Math. Scientist (1984) pp. 117–124.

[23] J. W. BOYSE, *"Interference detection among solids and surfaces,"* Comm. Assoc. Comp. Mach. (1979) pp. 3–9.

[24] E. BROOKNER, *"Phased-array radars,"* Sci. Amer. (1985) pp. 94–102.

[25] A. R. BROWN, *Optimum Packing and Depletion*, Elsevier, N.Y., 1971.

[26] C. R. CALLADINE, *"Buckminster Fuller's 'tensegrity' structures and Clerk Maxwell's rules for the construction of stiff frames,"* Int. J. Solids and Structures (1978) pp. 161–172.

[27] T. M. CANNON AND B. R. HUNT, *"Image processing by computer,"* Sci. Amer. (1981) pp. 214–225.

[28] I. CARLBOM AND J. PACIOREK, *"Planar geometric projections and viewing transformations,"* ACM Comp. Surveys (1978) pp. 465–503.

[29] J. L. CHERMANT, ed., *Quantitative Analysis of Microstructures in Biology, Material Science, and Medicine*, Riederer, Stuttgart, 1978.

[30] E. J. COCKAYNE, *"On the Steiner problem,"* Canad. Math. Bull. (1967) pp. 431–450.

[31] D. E. COLE, *"The Wankel engine,"* Sci. Amer. (1972) pp 14–23.

[32] J. J. COLEMAN, *"Kinematics of tape recording,"* Amer. J. Phys. (1982) pp. 184–185.

[33] R. COLEMAN, *An Introduction to Mathematical Stereology*, Memoir Series, Aarhus, 1979.

[34] R. CONNELLY, *"The rigidity of certain cabled frameworks and the second order rigidity of arbitrarily triangulated convex surfaces,"* Adv. in Math. (1980) pp. 272–298.

[35] ———, *"The rigidity of polyhedral surfaces,"* MM (1979) pp. 275–283.

[36] L. A. COOPER AND R. N. SHEPARD, *"Turning something over in the mind,"* Sci. Amer. (1984) pp. 106–114.

[37] H. CRAPO, *"Structural rigidity,"* Struct. Topology (1979) pp. 26–45.

[38] H. CRAPO AND W. WHITELEY, *"Statics of frameworks and motions of panel structures, a projective geometric introduction,"* Struct. Topology (1982) pp. 43–82.

[39] L. CREMONA, *Graphical Statics*, Clarendon Press, Oxford, 1890.

[40] L. M. CRUZ-ORIVE, *"Particle size-shape distributions: The general spheroid problem I, II,"* J. Microscopy (1976) pp. 235–253, (1978) pp. 153–167.

[41] H. M. CUNDY, *"Getting it taped,"* Math. Gaz. (1971) pp. 43–47.

[42] H. M. CUNDY AND A. P. ROLLETT, *Mathematical Models*, Oxford University Press, Oxford, 1961.

[43] R. T. DE HOFF AND F. N. RHINES, eds. *Quantitative Microscopy*, McGraw-Hill, N.Y., 1968.

[44] G. H. DE VISME, *"The length of unexposed film left in a cassette,"* Math. Gaz. (1969) pp. 139–141.

[45] J. C. DAVIS AND M. J. McCULLAGH, *Display and Analysis of Spatial Data*, Wiley, Chichester, 1975.

[46] R. F. DeMAR, *"The problem of the shortest network joining n points,"* MM (1968) pp. 225–231.

[47] A. K. DEWDNEY, *"Computer Recreations" (on fractals)*, Sci. Amer. (1985) pp. 16–24.

[48] J. B. DeVELIS AND G. O. REYNOLDS, *Theory and Applications of Holography*, Addison-Wesley, Reading, 1967.

[49] W. W. DOLAN, *"Early sundials and the discovery of the conic sections,"* MM (1972) pp. 8–12.

[50] ———, *"The ellipse in eighteenth century sundial design,"* MM (1972) pp. 205–209.

[51] R. O. DUDA AND P. E. HART, *Pattern Classification and Scene Analysis*, Wiley, N.Y., 1973.

[52] A. M. DZIEWONSKI AND D. L. ANDERSON, *"Seismic tomography of the earth's interior,"* Amer. Sci. (1984) pp. 483–494.

[53] P. J. ELL AND B. L. HOLMAN, eds., *Computed Emission Tomography*, Oxford University Press, Oxford, 1982.

[54] P. FAZIO, G. HAIDER AND A. BIRIN, *Proceedings of the IASS World Conference on Space Enclosures*, Concordia University, Montreal, 1976.

[55] D. Z. FREEDMAN AND P. VAN NIEUWENHUIZEN, *"Hidden dimensions of spacetime,"* Sci. Amer. (1985) pp. 74–81.

[56] J. G. FREEMAN, *"A method of determining the north and latitude,"* Math. Gaz. (1975) pp. 41–44.

[57] ———, *"How to make a portable sun-dial,"* Math. Gaz. (1975) pp. 261–264.

[58] B. FULLER, *Explorations in the Geometry of Thinking*, Macmillan, N.Y., 1975.

[59] H. FUCHS, Z. M. KEDEM AND S. P. USELTON, *"Optimal surface reconstruction from planar contours,"* Comm. Assoc. Comp. Mach. (1977) pp. 693–702.

[60] P. L. GAILBRAITH AND D. A. PRAEGAR, *"Mathematical sailmanship,"* Math. Gaz. (1975) pp. 94–97.

[61] R. S. GARFINKLE, *"Minimizing wallpaper waste, Part I: A class of travelling salesman problems,"* Oper. Res. (1977) pp. 741–751.

[62] A. GHEORGHIU AND V. DRAGOMIR, *Geometry of Structural Forms*, Applied Science Publ., London, 1978.

[63] E. N. GILBERT AND H. O. POLLAK, *"Steiner minimal trees,"* SIAM Rev. (1968) pp. 1–29.

[64] E. N. GILBERT, *"Distortion in maps,"* SIAM Rev. (1974) pp. 47–62.

[65] B. GILLAM, *"Geometrical illusions,"* Sci. Amer. (1980) pp. 102–11, 162.

[66] H. E. GOHEEN, *"On space frames and plane trusses,"* AMM (1950) pp. 481–516.

[67] M. GOLDBERG, *"A six-plate linkage in three dimensions,"* Math. Gaz. (1974) pp. 287–289.

[68] ———, *"A three-dimensional analogy of a plane Kempe linkage,"* J. Math. Phys. (1946) pp. 96–110.

[69] ———, *"New five-bar and six-bar linkages in three dimensions,"* A.S.M.E. (1943) pp. 649–661.

[70] ———, *"Polyhedral linkages,"* MM (1942) pp. 323–332.

[71] ———, *"Tubular linkages,"* J. Math. Phys. (1947) pp. 10–21.

[72] ———, *"Unstable polyhedral structures,"* MM (1978) pp. 165–170.

[73] R. GREENLER, *Rainbows, Halos, and Glories,* Cambridge University Press, N.Y., 1980.

[74] R. L. GREGORY, *The Intelligent Eye,* McGraw-Hill, N.Y., 1970.

[75] ———, *"Visual illusions,"* Sci. Amer. (1968) pp. 66–76.

[76] U. GRENANDER, *Pattern Synthesis.* Lectures in Pattern Theory I, Springer-Verlag, N.Y., 1976.

[77] W. E. L. GRIMSON, *From Images to Surfaces: A Computational Study of the Human Early Visual System,* M.I.T. Press, Cambridge, 1981.

[78] B. GRUNBAUM AND G. C. SHEPHARD, *"Satin and twills: An introduction to the geometry of fabrics,"* MM (1980) pp. 139–161.

[79] D. HARRIS, *Computer Graphics and Applications,* Chapman and Hall, N.Y., 1984.

[80] J. G. HAYES, *"Numerical methods for curve and surface fitting,"* BIMA (1974) pp. 144–152.

[81] R. M. HAZEN AND L. W. FINGER, *"Crystals at high pressure,"* Sci. Amer. (1985) pp. 110–117.

[82] R. S. HEATH, *Treatise on Geometrical Optics,* Cambridge University Press, Cambridge, 1895.

[83] G. T. HERMAN, ed., *Image Reconstruction from Projections: Implementation and Applications,* Springer-Verlag, N.Y., 1979.

[84] J. HIGGINS, *"Getting it taped,"* Math. Gaz. (1971) pp. 47–48.

[85] T. P. HILL, *"Determining a fair border,"* AMM (1983) pp. 438–442.

[86] J. E. HOCHBERG, *Perception,* Prentice-Hall, N.J., 1978.

[87] D. D. HOFFMAN, *"The interpretation of visual illusions,"* Sci. Amer. (1984) pp. 154–162.

[88] W. C. HOFFMAN, *"Visual illusions of angle as an application of Lie transformation groups,"* SIAM Rev. (1971) pp. 169–184.

[89] A. HOLDEN, *Shapes, Space and Symmetry,* Columbia University Press, N.Y., 1971.

[90] H. HONDA, P. B. TOMLINSON AND J. B. FISHER, *"Two geometrical models of branching of botanical trees,"* Ann. Bot. (1982) pp. 1–11.

[91] B. K. P. HORN AND K. IKEUCHI, *"The mechanical manipulation of randomly oriented parts,"* Sci. Amer. (1984) pp. 100–111.

[92] B. K. P. HORN, R. J. WOODHAM AND W. M. SILVER, *"Determining shape and reflectance using multiple images,"* M.I.T. A.I. Laboratory, Memo 490; August, 1978.

[93] H. S. HORN, *The Adaptive Geometry of Trees,* Princeton University Press, Princeton, 1971.

[94] M. HOTINE, *Mathematical Geodesy,* U.S. Dept. of Commerce, Washington, D.C., 1969.

[95] M. K. HU, *"Visual pattern recognition of moment invariants,"* I. R. E. Trans. Inform. Theory (1962) pp. 179–187.

[96] K. H. HUNT, *Kinematic Geometry of Mechanisms,* Clarendon Press, Oxford, 1978.

[97] P. HUYBERS, *"Polyhedral housing units,"* Int. J. Housing Sci. Appl. (1979) pp. 215–225.

[98] C. ISENBERG, *The Science of Soap Films and Soap Bubbles,* Tieto Ltd., England, 1978.

[99] A. J. JAKEMAN AND R. S. ANDERSSEN, *"Abel type integral equations in stereology. I. General discussion,"* J. Microscopy (1975) pp. 121–134.

[100] M. A. JASWON, *Introduction to Mathematical Crystallography,* Elsevier, N.Y., 1965.

[101] R. V. JEAN, *Mathematical Approach to Pattern and Form in Plant Growth,* Wiley, N.Y., 1984.

[102] E. B. JENSEN, H. J. G. GUNDERSEN AND R. OSTERBY, *"Determination of membrane thickness distribution from orthogonal intercepts,"* J. Microscopy (1979) pp. 19–33.

[103] B. K. JOHNSON, *Optics and Optical Instruments,* Dover, N.Y., 1960.

[104] D. KALMAN, *"A model for playing time,"* MM (1981) pp. 247–250.

[105] W. M. KAULA, *Theory of Satellite Geodesy,* Blaisdell, Waltham, 1967.

[106] J. M. KAPPRAFF, *"A course in the mathematics of design,"* Struct. Topology (1983) pp. 37–52.

[107] J. B. KELLER, *"Parallel reflection of light by plane mirrors,"* Quart. Appl. Math. (1953) pp. 216–219.

[108] E. KHULAR, K. THYAGARAJAN AND A. K. GHATAK, *"A note on mirage formation,"* Amer. J. Phys. (1977) pp. 90–92.

[109] E. KRETSCHMER, ed., *Studies in Theoretical Cartography,* Deuticke, Vienna, 1977.

[110] T. L. KUNII, ed., *Computer Graphics–Theory and Applications,* Springer-Verlag, Tokyo, 1983.

[111] G. LAMAN, *"On graphs and rigidity of plane skeletal structures,"* J. Eng. Math. (1970) pp. 331–340.

[112] C. LANCZOS, *Space Through the Ages,* Academic Press, N.Y., 1970.

[113] A. L. LOEB, *Color and Symmetry,* Wiley, N.Y., 1971.

[114] ———, *Space Structures, Their Harmony and Counterpoint*, Addison-Wesley, Reading, 1976.

[115] M. LUCKIESH, *Visual Illusions: Their Causes, Characteristics, and Applications*, Dover, N.Y., 1965.

[116] R. K. LUNEBERG, *"The metric of binocular visual space,"* J. Optical Soc. Amer. (1950) pp. 627–642.

[117] H. E. MALDE, *"Panoramic photographs,"* Amer. Sci. (1983) pp. 132–140.

[118] B. B. MANDELBROT, *The Fractal Geometry of Nature*, Freeman, San Francisco, 1982.

[119] ———, *"Physical objects with fractional dimension: seacoasts, galaxy clusters, turbulence, and soap,"* BIMA (1977) pp. 189–196.

[120] L. MARCH AND P. STEADMAN, *The Geometry of Environment*, M.I.T. Press, Cambridge, 1971.

[121] J. C. MAXWELL, *"On reciprocal figures, frames, and diagrams of forces,"* Trans. Royal Soc. Edinburgh (1869–1872) pp. 1–40.

[122] D. MARR, *Vision: A Computational Investigation into the Human Representation and Processing of Visual Information*, Freeman, San Francisco, 1982.

[123] J. P. MCKELVEY, *"Dynamics and kinematics of textile machinery,"* Amer. J. Phys. (1982) pp. 1160–1162.

[124] ———, *"Kinematics of tape recording,"* Amer. J. Phys. (1981) pp. 81–83.

[125] R. J. Y. MCLEOD AND E. L. WACHSPRESS, eds., *Frontiers of Applied Geometry*, Math. Modelling (1980) pp. 271–403.

[126] R. K. MELLUISH, *An Introduction to the Mathematics of Map Projection*, Cambridge University Press, Cambridge, 1931.

[127] T. W. MELNYK, O. KNOP AND W. P. SMITH, *"Extremal arrangements of points and unit charges on a sphere: Equilibrium configurations revisited,"* Canad. J. Chem. (1977) pp. 1745–1761.

[128] W. MERZKIRCH, ed., *Flow Visualization* 2, Hemisphere, N.Y., 1982.

[129] H. R. MILLS, *Positional Astronomy and Astro-Navigation Made Easy*, S. Thornes, Halsted, N.Y., 1978.

[130] J. MILNOR, *"A problem in cartography,"* AMM (1969) pp. 1101–1112.

[131] M. MINNAERT, *Light and Colour in the Open Air*, Dover, N.Y., 1954.

[132] P. MOON, *The Scientific Basis of Illuminating Engineering*, Dover, N.Y., 1936.

[133] I. I. MUELLER, *Introduction to Satellite Geodesy*, Ungar, N.Y., 1964.

[134] D. E. NASH, *"Rotary engine geometry,"* MM (1977) pp. 87–89.

[135] L. NEUWIRTH, *"The theory of knots,"* Sci. Amer. (1979) pp. 110–124.

[136] H. M. NUSSENZVIEG, *"The theory of the rainbow,"* Sci. Amer. (1977) pp. 116–127.

[137] T. O'NEILL, *"A mathematical model of a universal joint,"* UMAP J. (1982) pp. 201–219.

[138] F. OTTO, ed., *Tensile Structures: Vol. I – Pneumatic Structures, Vol. II – Cables, Nets, and Membranes*, M.I.T. Press, Cambridge, 1967.

[139] E. W. PARKES, *Braced Frameworks*, Pergamon, Oxford, 1965.

[140] D. PEDOE, *"A geometrical proof of the equivalence of Fermat's principle and Snell's law,"* AMM (1964) pp. 543–544.

[141] ———, *Geometry and the Visual Arts*, Penguin, England, 1976.

[142] P. PEARCE, *Structure in Nature is a Strategy for Design*, M.I.T. Press, Cambridge, 1978.

[143] C. E. PEARSON, *"Optimal mapping from a sphere onto a plane,"* SIAM Rev. (1982) pp. 469–475.

[144] T. POGGIO, *"Vision by man and machine,"* Sci. Amer. (1984) pp. 106–116.

[145] W. K. PRATT, *Digital Image Processing*, Wiley, N.Y., 1978.

[146] A. PUGH, *An Introduction to Tensegrity*, University of California Press, Berkeley, 1976.

[147] D. RAWLINS, *"Doubling your sunsets or how anyone can measure the earth's size with a wristwatch and meter-stick,"* Amer. J. Phys. (1979) pp. 126–128.

[148] A. RECSKI, *"A network theory approach to the rigidity of skeletal structures. Part III. An electric model of planar networks,"* Struct. Topology (1984) pp. 59–71.

[149] W. P. REID, *"Line of flight from shock recordings,"* MM (1968) pp. 59–63.

[150] A. P. ROBERTS AND D. SPREVAK, *"Research on a shoestring,"* BIMA (1980) pp. 215–218.

[151] I. ROSENHOLZ, *"Calculating surface area from a blueprint,"* MM (1979) pp. 252–256.

[152] I. ROSENBERG, *"Structural rigidity: Foundations and rigidity,"* Ann. Discrete Math. (1980) pp. 143–161.

[153] A. ROSENFELD, ed., *Image Modeling*, Academic Press, N.Y., 1981.

[154] B. ROTH, *"Rigid and flexible frameworks,"* AMM (1981) pp. 6–21.

[155] B. ROTH AND W. WHITELEY, *"Rigidity of tensegrity frameworks,"* Trans. AMS (1981) pp. 419–445.

[156] W. H. RUCKLE, *Geometric Games and their Applications*, Pitman, London, 1983.

[157] M. SALVADORI, *Structure in Architecture*, Prentice-Hall, N.J., 1975.

[158] ———, *Why Buildings Stand Up*, Norton, N.Y., 1980.

[159] D. SCHATTSCHNEIDER AND W. WALKER, *M. C. Escher Kaleidocycles*, Ballantine, N.Y., 1977.

[160] B. F. SCHUTZ, *Geometrical Methods of Mathematical Physics*, Cambridge University Press, Cambridge, 1980.

[161] L. A. SHEPP, *Computed Tomography*, AMS, 1983.

[162] L. A. SHEPP AND J. B. KRUSKAL, *"Computerized tomography: The new medical x-ray technology,"* AMM (1978) pp. 420–439.

[163] A. V. SHUBNIKOV AND N. V. BELOV, *Colored Symmetry*, Pergamon, Oxford, 1964.

[164] C. C. SLAMA, C. THEURER AND S. W. HENRIKSEN, eds., *Manual of Photogrammetry*, Amer. Soc. Photogrammetry, Falls Church, 1980.

[165] N. J. A. SLOANE, *"Multiplexing methods in spectroscopy,"* MM (1979) pp. 71–80.

[166] ———, *"The packing of spheres,"* Sci. Amer. (1984) pp. 116–125.

[167] N. J. A. SLOANE AND M. HARWIT, *"Masks for Hadamard transform optics and weighing designs,"* Appl. Optics (1976) pp. 107–114.

[168] W. M. SMART, *Text-Book on Spherical Astronomy,* Cambridge University Press, Cambridge, 1965.

[169] D. A. SMITH, *"A descriptive model for perception of optical illusions,"* J. Math. Psych. (1978) pp. 64–85.

[170] ———, *"Descriptive models for perception of optical illusions: 1,"* UMAP J. (1985) pp. 59–95.

[171] J. V. SMITH, *Geometrical and Structural Crystallography,* John Wiley, N.Y., 1982.

[172] J. P. C. SOUTHALL, *Introduction to Physiological Optics,* Dover, N.Y., 1961.

[173] J. A. STEERS, *Introduction to the Study of Map Projections,* University of London Press, London, 1965.

[174] H. STEINHAUS, *Mathematical Snapshots,* Oxford University Press, N.Y., 1969.

[175] P. STEVENS, *Patterns in Nature,* Little Brown, Boston, 1974.

[176] G. C. STEWARD, *The Symmetrical Optical System,* Cambridge University Press, Cambridge, 1958.

[177] R. S. STICHARTZ, *"What's up moonface?"* UMAP J. (1985) pp. 3–34.

[178] G. STRANG, *"The width of a chair,"* AMM (1982) pp. 529–534.

[179] W. J. SUPPLE, ed., *Proceedings of the Second International Conference on Space Structures,* University of Surrey, Guilford, 1975.

[180] W. SWINDELL AND H. H. BARRETT, *"Computerized tomography: Taking sectional x-rays,"* Phys. Today (1977) pp. 32–41.

[181] J. L. SYNGE, *"Reflection in a corner formed by three plane mirrors,"* Quart. Appl. Math. (1946) pp. 166–176.

[182] P. M. L. TAMMES, *"On the origin of number and arrangement of places of exit on the surface of pollen-grains,"* Recueil des Travaux Botanique Neerlandais (1930) pp. 1–84.

[183] T. TARNEI, *"Simultaneous static and kinematic indeterminacy of space trusses with cyclic symmetry,"* Int. J. Solids and Structures (1980) pp. 347–359.

[184] ———, *"Spherical circle-packing in nature, practice and theory,"* Struct. Topology (1984) pp. 39–58.

[185] D. E. THOMAS, *"Mirror images,"* Sci. Amer. (1980) pp. 206–228.

[186] W. P. THURSTON AND J. R. WEEKS, *"The mathematics of three-dimensional manifolds,"* Sci. Amer. (1984) pp. 108–120.

[187] S. TIMOSHENKO AND D. H. YOUNG, *Theory of Structures,* McGraw-Hill, N.Y., 1965.

[188] S. TOLANSKY, *Optical Illusions,* Pergamon, Oxford, 1964.

[189] P. B. TOMLINSON, *"Tree architecture,"* Amer. Sci. (1983) pp. 141–149.

[190] A. TORMEY AND J. F. TORMEY, *"Renaissance intarsia: The art of geometry,"* Sci. Amer. (1982) pp. 136–143.

[191] L. TORNHEIM, *"Determination of large parallelepipeds,"* SIAM Rev. (1962) pp. 223–226.

[192] G. TOUSSAINT, *"Generalizations of pi: Some applications,"* Math. Gaz. (1974) pp. 289–291.

[193] R. A. R. TRICKER, *Introduction to Meteorological Optics,* Mills and Boon, London, 1970.

[194] E. E. UNDERWOOD, *Quantitative Stereology,* Addison-Wesley, Reading, 1970.

[195] W. R. UTTAL, *Visual Form Detection in 3-Dimensional Space,* Erlbaum Associates, 1983.

[196] H. WALLACH, *"Perceiving a stable environment,"* Sci. Amer. (1985) pp. 118–124.

[197] H. WALLACH AND J. BACON, *"Two kinds of adaptation in the constancy of visual direction and their different effects on the perception of shape and visual direction,"* Perception & Psychophysics (1977) pp. 227–242.

[198] WATTS BROS. TOOL WORKS, *How to Drill Square, Hexagon, Octagon, Pentagon Holes,* Wilmerding, PA, 1966.

[199] A. E. WAUGH, *Sundials, Their Theory and Construction,* Dover, N.Y., 1973.

[200] E. R. WEIBEL, *Stereological Methods I, II,* Academic Press, N.Y., 1980.

[201] A. F. WELLS, *Models in Structural Inorganic Chemistry,* Oxford University Press, Oxford, 1970.

[202] S. J. WILLIAMSON AND H. Z. CUMMINS, *Light and Color,* John Wiley, N.Y., 1983.

[203] D. H. WILSON, *"The geometry of the Wankel engine,"* BIMA (1972) pp. 323–325.

[204] A. T. WINFREE, *The Geometry of Biological Time,* Springer-Verlag, Berlin, 1980.

[205] J. M. WOLFE, *"Hidden visual processes,"* Sci. Amer. (1983) pp. 94–103, 154.

[206] R. W. WOOD, *Physical Optics,* Macmillan, N.Y., 1944.

[207] J. H. WOODHOUSE AND A. M. DZIEWONSKI, *"Mapping the upper mantle: Three dimensional modelling by inversion of seismic waveforms,"* J. Geophys. Res. (1984) pp. 5953–5986.

[208] E. W. WOODLAND AND G. M. CLEMENCE, *Spherical Astronomy,* Academic Press, N.Y., 1966.

[209] L. A. WHYTE, *"Unique arrangements of points on a sphere,"* AMM (1952) pp. 606–612.

[210] T. T. WU, *"Some mathematical problems in fiber x-ray crystallography,"* SIAM Rev. (1972) pp. 420–432.

[211] I. M. YAGLOM, *A Very Simple Non-Euclidean Geometry and its Physical Basis,* Springer-Verlag, Heidelberg, 1979.

[212] J. I. YELLOT, JR., *"Binocular depth inversion,"* Sci. Amer. (1981) pp. 148–159.

[213] W. M. ZAGE, *"The geometry of binocular visual space,"* MM (1980) pp. 289–294.

[214] H. ZIMMER, *Geometrical Optics,* Springer-Verlag, Heidelberg, 1970.

[215] L. ZUSNE, *"Moments of area and of the perimeter of visual forms as predictors of discrimination performance,"* J. Exp. Psych. (1965) pp. 213–220.

7.1 Catastrophe Theory

THE TOY AIRPLANE CATASTROPHE*

JOHN C. SOMMERER†

Abstract. A new demonstration of an elementary catastrophe using a simple machine, in the spirit of the Zeeman machine, is given.

Two of the simplest demonstrations of elementary catastrophes, the Zeeman catastrophe machine and buckling of a solid beam, are mechanical in nature. Complete treatments of both, as well as the seven elementary catastrophes, are to be found in [1] and [2]. Briefly, the catastrophe machine is catastrophic because of the tendency to minimize the potential energy function of an elastic substance under tension. Conversely, the buckling of a beam is catastrophic because of the minimization of the potential energy of a subtance under compression and shear. It therefore seems reasonable to suppose that elastic substances under applied torques can be made to behave catastrophically as well. In fact, experiments with rubber bands and an easily constructed machine show this to be the case. This apparently new demonstration of an elementary catastrophe rounds out a family of similar mechanical examples which are extremely simple and amenable to verification.

The model. The elementary catastrophe of interest is the cusp, and it will be useful to review its salient features. The cusp potential function is

$$\phi = \tfrac{1}{4}x^4 - ax - \tfrac{1}{2}bx^2$$

where x is the behavior variable and a and b are the controls. The behavior of any system governed by the cusp is confined to the three dimensional behavior surface (Fig. 1)

$$\frac{d\phi}{dx} = x^3 - a - bx = 0.$$

The properties of the cusp follow immediately from this behavior surface. They are: 1) bimodal behavior for some values of the controls, 2) discontinuous changes in behavior for small variations in the controls, 3) hysteresis in behavior for a reversal of a change in the controls (processes L and M), 4) divergence in behavior for small variations in initial conditions (processes N and O), and 5) inaccessible portions of the behavior surface.

It is often observed that the twisting of a rubber band, as in the winding of a toy airplane, can proceed only to a limited extent before doubling over occurs. If the band is elastic, or nearly so, the doubling over can occur suddenly or over the course of several twists. The application of a catastrophic model immediately suggests itself.

The first support for such a model was gained in a simple experiment using the length of the band and the number of twists in it as controls, and noting the values at which doubling over occurred. A roughly "cuspish" bifurcation set resulted, indicating that more precise measurements would be worthwhile.

The apparatus. An apparatus was assembled (Figs. 2 and 3) which allowed the quantitative measurement of the torque and tension applied to the elastic. Constant tensions were obtained by hanging small weights from a horizontally constrained

---
* Received by the editors April 1978.

† Department of Systems Science and Mathematics, Washington University, St. Louis, Missouri 63130.

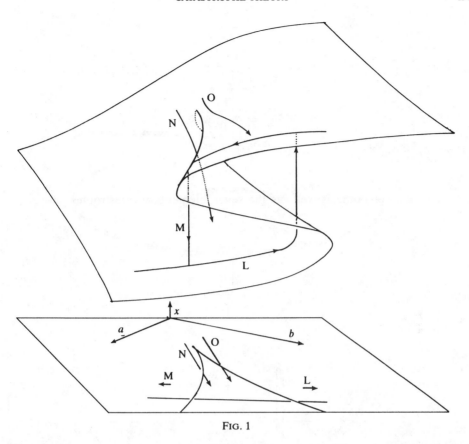

FIG. 1

platform (A) to which one end of a test elastic (B) was attached. The measurement of torque was more difficult, due to the number of rotations required to reach values inducing catastrophic behavior. A relatively simple mechanism was devised to do this. The remaining end of the elastic under test was attached to a torsion shaft (C) which was affixed to a free pointer assembly (D). This assembly was connected through a low friction bearing to a second, controlled, pointer assembly (E). In addition, the torsion shaft was connected by a standard elastic (F) to the controlled pointer assembly. Thus the free and controlled pointers could rotate relative to one another, under the constraint of the standard elastic. This entire assembly was attached to the fixed body of the appartus (G) through a second low friction bearing, and rotated freely relative to it. A protractor fixed to the body of the apparatus allowed the determination of the angular separation of the two pointers.

By rotating the controlled pointer, a torque was applied through the standard elastic to the torsion shaft and the test elastic. Because the torque was not applied through a rigid structure, a separation related to the magnitude of the torque developed between the two pointers. A calibration curve to convert angular separation to torque was easily obtained by applying known torques.

The experiments. The behavior of the test elastic was studied in two types of experiment. In the first type, fixed tensions were applied to single strands of surgical rubber, and the torque necessary to produce doubling over was noted, as was the reduction in torque required for straightening. Properties 1), 2), and 3) of the cusp catastrophe became immediately apparent. In addition, the values of tension and torque producing catastrophe generated a characteristically shaped cusp bifurcation

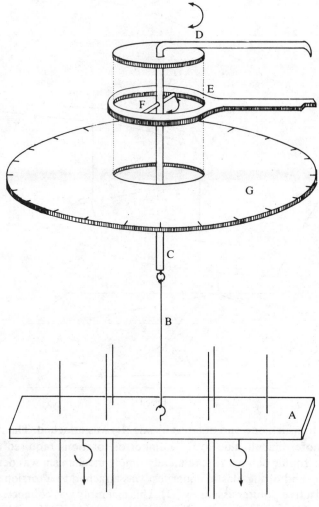

FIG. 2

set graph. The particular shape of such graphs (e.g. Fig. 4) varied considerably from elastic to elastic, which might be expected, in view of the nonuniformity of such materials. A specific bifurcation set graph could be used to predict the subsequent behavior of the same elastic in a second type of experiment, however.

An arbitrary torque was applied to an elastic, and the tension was reduced incrementally by removing weights until discontinuous behavior was observed (process 2). Tension was again increased until straightening occurred. In some cases a particularly good test of the cusp hypothesis was obtained by varying the controls in the vicinity of the cusp point of the bifurcation set. It was possible to produce discontinuous doubling and continuous straightening by setting the initial torque only slightly above the lowest value in the bifurcation set; thus the instantaneous reduction in torque accompanying the doubling over brought the system to a pleatless region of the behavior surface allowing a continuous return to the initial conditions. The success of this test depends on a dynamic relationship between the control and behavior variables which is not required by, but does not exclude, the cusp catastrophe.

Fig. 3

The behavior variable might be identified with the average number of rotations per unit length of the elastic in the single stranded portion. This characterization of the system varies continuously over most of the control plane, yet changes discontinuously across a range of unobserved values when the elastic doubles over. This corresponds to property 5) of the cusp. Four of the five cusp properties have been shown. The remaining one, divergence, is not easily demonstrable since no control splits the bifurcation set, and the simultaneous variation of both controls is not practical with the equipment used.

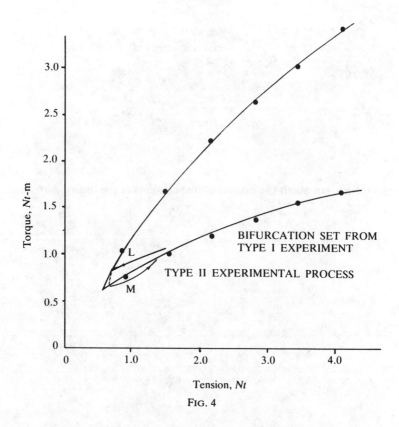

Tension, *Nt*

FIG. 4

The cause of the catastrophic behavior, as in many other mechanical systems, depends on the presence of friction. As torque on the elastic is increased, the elemental contributions to the total are not evenly distributed around the axis of suspension, due to the finite thickness and asymmetry of the band. This asymmetry yields a region where the elastic axis deviates helically from the axis of suspension. When the helical tension becomes greater than the tension in the linear portion, the loops of the helix contract on one another. The total potential energy of the system decreases, since the process is spontaneous.

Because the "toy airplane catastrophe" can be quantitatively demonstrated using simple equipment and because it is a three dimensional cusp, rather than a higher order catastrophe, it is more pedagogically valuable than the buckling beam. It also has the advantage over the Zeeman machine that it exhibits the feedback-like properties resulting from an interdependence of the controls found in many areas of application of catastrophe theory. Yet, unlike many of these applications, the underlying physical laws are well known, and the necessary measurements can be made easily with relative precision.

Acknowledgment. The author is grateful to the referee and Professor E. Y. Rodin of Washington University for their valuable suggestions.

REFERENCES

[1] I. STEWART, *The seven elementary catastrophes*, New Scientist, 68 (1975), pp. 447–450, 452–454.
[2] E. C. ZEEMAN, *Catastrophe theory*, Scientific American, 234 (1976), pp. 65–70, 75–83.
[3] J. GUCKENHEIMER, *The catastrophe controversy*, Mathematical Intelligencer, 1 (1978), pp. 15–20.

CATASTROPHE THEORY AND CAUSTICS*

JENS GRAVESEN†

Abstract. In this paper it is shown by elementary methods that in codimension two and under the assumption that light rays are straight lines, a caustic is the catastrophe set for a timefunction. The general case is also discussed.

1980 AMS Mathematics subject classification 58C28.

Key words. caustic, catastrophe theory, envelope

1. Introduction. Since its appearance, there has been an intense debate about catastrophe theory, not about the mathematical contents of the theory, but about some of the applications of the theory in biology, medicine, sociology, etc. In this paper catastrophe theory is applied to the theory of caustics. This is considered to be one of the more sound applications of catastrophe theory, and it has not been questioned. It has long been known empirically that a caustic has only a finite number of possible shapes. Catastrophe theory has confirmed this result and has even shown that the known lists of stable caustics is complete.

In his book *Stabilité structurelle et morphogénèse,* René Thom explains [7, p. 63] how the elementary catastrophes occur as singularities of propagating wave fronts. Later Klaus Jähnich [8], among others, examined this case more closely and gave a complete proof of Thom's conjecture.

In this paper we present an elementary proof of the special case, where the problem can be considered as two-dimensional and the light rays as straight lines. In addition the general theorem will be made plausible.

The paper is an extension of notes prepared for a seminar on catastrophe theory arranged by the Association of Mathematics Teachers in Denmark, summer, 1979. I wish to thank my teacher Professor Vagn Lundsgaard Hansen who planned the course and encouraged the present work, and my fellow instructors Martin Philip Bendsøe and Henrik Pedersen for valuable discussions.

2. The classification theorem. We start by stating Thom's theorem. A more extensive introduction to catastrophe theory can be found in Callahan [2], [3], and proofs in Zeeman and Trotman [8].

Let $f: \mathbb{R}^n \times \mathbb{R}^r \rightsquigarrow \mathbb{R}$ be a smooth function, i.e. of class C^∞. We let $\mathbf{x} = (x_1, \cdots, x_n)$ denote an element belonging to $\mathbb{R}^n$ and $\mathbf{y} = (y_1, \cdots, y_r)$ denote an element belonging to $\mathbb{R}^r$. Define $M_f \subseteq \mathbb{R}^n \times \mathbb{R}^r$ by

$$M_f = \left\{ (\mathbf{x}, \mathbf{y}) \in \mathbb{R}^n \times \mathbb{R}^r \,\middle|\, \frac{\partial f}{\partial x_i}(\mathbf{x}, \mathbf{y}) = 0, \text{ all } i = 1, \cdots, n \right\}.$$

Generically M_f is an r-manifold (r-dimensional surface in $\mathbb{R}^{n+r}$) because it is the null-space of n equations. Let

$$\chi_f : M_f \to \mathbb{R}^n \times \mathbb{R}^r \to \mathbb{R}^r$$

be the map induced by the projection $\mathbb{R}^n \times \mathbb{R}^r \to \mathbb{R}^r$. The map χ_f is called the *catastrophe map* of f. Let F denote the space of smooth functions on $\mathbb{R}^{n+r}$, with the Whitney

---

* Received by the editors June 25, 1981, and in revised form August 26, 1982.

†Mathematical Institute, The Technical University of Denmark, DK-2800 Lyngby, Denmark.

C^∞-topology (Two functions are close if the values of the functions as well as the values of the derivatives are close, see Zeeman and Trotman [8, p. 316] or Callahan [2, p. 222].

THEOREM 1 (Thom). *If* $r \le 5$, *there exists an open dense set* $F_* \subseteq F$ *called the set of generic functions. If f is generic, then:*

(1) M_f *is an r-manifold.*

(2) *Any singularity of* χ_f *is equivalent (see below) to one of a finite number of types called* <u>*elementary catastrophes.*</u>

(3) χ_f *is* <u>*locally stable with respect to small pertubations of f.*</u>

Two maps $\chi:M \frown \mathbb{R}^r$ and $\chi':M' \frown \mathbb{R}^r$ are equivalent if there exists diffeomorphisms $h:M \frown M'$ and $k:\mathbb{R}^r \frown \mathbb{R}^r$ such that the following diagram commutes:

$$
\begin{array}{ccc}
M & \xrightarrow{\ h\ } & M' \\
\chi \downarrow & & \downarrow \chi' \\
\mathbb{R}^r & \xrightarrow{\ k\ } & \mathbb{R}^r
\end{array}
$$

A diffeomorphism can be considered as a curvilinear change of coordinates. The set of curvilinear singular points $(\mathbf{x}, \mathbf{y}) \in M_f$ of χ_f is denoted Δ_f and is given by

$$
\Delta_f = \left\{ (\mathbf{x}, \mathbf{y}) \in M_f \,\middle|\, \det\left[\frac{\partial^2 f}{\partial x_i \partial x_j}(\mathbf{x}, \mathbf{y}) \right] = 0 \right\}.
$$

The image $\chi_f(\Delta_f)$ is called the *catastrophe set* and is denoted D_f. If χ_f and $\chi_{f'}$ are equivalent and h, k are the associated diffeomorphisms then $M_{f'} = h(M_f)$, $\Delta_{f'} = h(\Delta_f)$ and $D_{f'} = k(D_f)$, so locally D_f and $D_{f'}$ have the same shape.

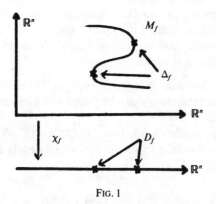

FIG. 1

3. The envelope. We next investigate the possible shape of a caustic formed by a given bundle of light. In order to solve the problem by elementary methods we make two assumptions. Assume the existence of a direction in which the light bundle is translation invariant, and that the speed of light is constant. Then the light rays are straight lines. Thus we consider a family of lines in $\mathbb{R}^2$.

Let a, b and c be real smooth functions defined on an open interval $I \subseteq \mathbb{R}$, such that

$$
\det \begin{bmatrix} a(x) & b(x) \\ a'(x) & b'(x) \end{bmatrix} \ne 0 \quad \text{for all } x \in I.
$$

For every $x \in I$ let $L(x)$ denote the straight line in $\mathbb{R}^2$ given by the equation

$$a(x)u + b(x)v = c(x) \quad \text{(where } (u, v) \text{ are the coordinates in } \mathbb{R}^2\text{)}.$$

Assume that $x \in I$ and $x + \Delta x \in I$, with $\Delta x \neq 0$. We will find a condition that ensures that $L(x)$ and $L(x + \Delta x)$ intersect. By the mean value theorem there exist real numbers ξ, ζ, η between x and $x + \Delta x$ such that

$$a(x + \Delta x) = a(x) + a'(\xi)\Delta x,$$
$$b(x + \Delta x) = b(x) + b'(\zeta)\Delta x,$$
$$c(x + \Delta x) = c(x) + c'(\eta)\Delta x.$$

If (u, v) is a point on both $L(x)$ and $L(x + \Delta x)$ we then have

$$a(x)u + b(x)v = c(x),$$
$$(a(x) + a'(\xi)\Delta x)u + (b(x) + b'(\zeta)\Delta x)v = c(x) + c'(\eta)\Delta x.$$

Since $\Delta x \neq 0$ these equations are equivalent to the equations

$$a(x)u + b(x)v = c(x),$$
$$a'(\xi)u + b'(\zeta)v = c'(\eta).$$

It is well known that the latter pair of equations has a solution if

$$\det \begin{bmatrix} a(x) & b(x) \\ a'(\xi) & b'(\zeta) \end{bmatrix} \neq 0.$$

This is fulfilled when Δx is sufficiently small, because $\det \left(\begin{smallmatrix} a & b \\ a' & b' \end{smallmatrix}\right) \neq 0$ and a, b, a' and b' are continuous. The point of intersection is given by

$$(u, v) = \left(\frac{\det \begin{bmatrix} c(x) & b(x) \\ c'(\eta) & b'(\zeta) \end{bmatrix}}{\det \begin{bmatrix} a(x) & b(x) \\ a'(\xi) & b'(\zeta) \end{bmatrix}}, \frac{\det \begin{bmatrix} a(x) & c(x) \\ a'(\xi) & c'(\eta) \end{bmatrix}}{\det \begin{bmatrix} a(x) & b(x) \\ a'(\xi) & b'(\zeta) \end{bmatrix}} \right).$$

If we let $\Delta x \rightarrow 0$, then ξ, ζ and η converge to x, and by continuity, the point of intersection between $L(x)$ and $L(x + \Delta x)$ converges to

$$(u(x), v(x)) = \left(\frac{\det \begin{bmatrix} c(x) & b(x) \\ c'(x) & b'(x) \end{bmatrix}}{\det \begin{bmatrix} a(x) & b(x) \\ a'(x) & b'(x) \end{bmatrix}}, \frac{\det \begin{bmatrix} a(x) & c(x) \\ a'(x) & c'(x) \end{bmatrix}}{\det \begin{bmatrix} a(x) & b(x) \\ a'(x) & b'(x) \end{bmatrix}} \right).$$

We see that $(u(x), v(x))$ is the unique solution to the equations

$$a(x)u + b(x)v = c(x),$$
$$a'(x)u + b'(x)v = c'(x).$$

The curve $I \frown \mathbb{R}^2 : x \to (u(x), v(x))$ is called the envelope of the family $(L(x))_{x \in I}$ of lines. We have shown that for sufficiently small Δx, $L(x + \Delta x)$ will intersect $L(x)$ near the envelope. If the lines, as in our case, represent light rays, this implies that the light is concentrated on the envelope, i.e. the envelope is the caustic for the light rays.

4. Caustics as catastrophe sets, the special case. Besides describing light as rays, we can describe light as waves. If the waves are known we get the rays as the normals to the wavefronts. Thus the caustic of a wavefront is the envelope for the normals.

Let V be a wavefront in $\mathbb{R}^2$, that is, locally it is nothing but a C^∞-curve $(\tilde{u}, \tilde{v})$: $I \frown \mathbb{R}^2 : x \to (\tilde{u}(x), \tilde{v}(x))$. As we are only interested in local properties, we assume that all of V is given by $(\tilde{u}, \tilde{v})$. We can now define the timefunction T associated to V. T measures the time it takes a light ray to travel from a point belonging to V to a point belonging to $\mathbb{R}^2$. We have assumed that the speed of light is constant so we can use distance as a measure for time. We define

$$T : I \times \mathbb{R}^2 \frown \mathbb{R},$$

$$T : (x, u, v) \to \sqrt{(\tilde{u}(x) - u)^2 + (\tilde{v}(x) - v)^2}.$$

Clearly T is smooth on $I \times (\mathbb{R}^2 \backslash V)$. If we let C denote the caustic of the wavefront V we have

THEOREM 2.

$$C \backslash V = \left\{ (u, v) \in \mathbb{R}^2 \backslash V \;\middle|\; \exists\, x \in I : \frac{\partial T}{\partial x}(x, u, v) = 0 \text{ and } \frac{\partial^2 T}{\partial x^2}(x, u, v) = 0 \right\} = D_T \backslash V.$$

Proof. The last equality is simply the definition of D_T; see §2. The first equality is seen in the following way:

$$\frac{\partial T}{\partial x} = \frac{1}{T}[(\tilde{u} - u)\tilde{u}' + (\tilde{v} - v)\tilde{v}']$$

and

$$\frac{\partial^2 T}{\partial x^2} = -\frac{1}{T^2}\frac{\partial T}{\partial x}[(\tilde{u} - u)\tilde{u}' + (\tilde{v} - v)\tilde{v}'] + \frac{1}{T}[\tilde{u}'^2 + (\tilde{u} - u)\tilde{u}'' + \tilde{v}'^2 + (\tilde{v} - v)\tilde{v}'']$$

$$= \frac{1}{T}\left[-\left(\frac{\partial T}{\partial x}\right)^2 + \tilde{u}'^2 + (\tilde{u} - u)\tilde{u}'' + \tilde{v}'^2 + (\tilde{v} - v)\tilde{v}''\right].$$

For $(u, v) \in \mathbb{R}^2 \backslash V$ we have that

$$\frac{\partial T}{\partial x}(x, u, v) = 0 \quad \text{and} \quad \frac{\partial^2 T}{\partial x^2}(x, u, v) = 0$$

is equivalent to

$$(\tilde{u}(x) - u)\tilde{u}'(x) + (\tilde{v}(x) - v)\tilde{v}'(x) = 0,$$

$$\tilde{u}'(x)^2 + (\tilde{u}(x) - u)\tilde{u}''(x) + \tilde{v}'(x)^2 + (\tilde{v}(x) - v)\tilde{v}''(x) = 0.$$

By rearranging we get

$$\tilde{u}'(x)u + \tilde{v}'(x)v = \tilde{u}(x)\tilde{u}'(x) + \tilde{v}(x)\tilde{v}'(x),$$

$$\tilde{u}''(x)u + \tilde{v}''(x)v = \tilde{u}'(x)^2 + \tilde{u}(x)\tilde{u}''(x) + \tilde{v}'(x)^2 + \tilde{v}(x)\tilde{v}''(x),$$

or equivalently

$$\tilde{u}'(x)u + \tilde{v}'(x)v = (\tilde{u}\tilde{u}' + \tilde{v}\tilde{v}')(x),$$

$$\tilde{u}''(x)u + \tilde{v}''(x)v = (\tilde{u}\tilde{u}' + \tilde{v}\tilde{v}')'(x).$$

The last equations give the envelope for the normals to V, because the normals to the curve $(\tilde{u}, \tilde{v}):x \longrightarrow (\tilde{u}(x), \tilde{v}(x))$ are given by the equation

$$(\tilde{u}(x) - u)\tilde{u}'(x) + (\tilde{v}(x) - v)\tilde{v}'(x) = 0$$

or equivalently

$$\tilde{u}'(x)u + \tilde{v}'(x)v = (\tilde{u}\tilde{u}' + \tilde{v}\tilde{v}')(x).$$

As the caustic of V is the envelope for the normals to V, the theorem follows.   □

5. The general case. Let V be a wavefront in $\mathbb{R}^3$. Locally it is a C^∞-map $\tilde{y}:I^2 \rightsquigarrow \mathbb{R}^3$: $x \longrightarrow \tilde{y}(x)$. ($x = (x_1, x_2)$ denotes a point belonging to $I^2 = I \times I \subseteq \mathbb{R}^2$ and $y = (y_1, y_2, y_3)$ denotes a point belonging to $\mathbb{R}^3$.) As in the preceding paragraphs we are only interested in local properties, so we assume that all of V is parametrized by $\tilde{y}$.

The only assumption made is the existence of a (smooth) timefunction T associated to V, i.e. for a point $\tilde{y} \in V$ and a point $y \in \mathbb{R}$ we can determine the time it takes light to travel from $\tilde{y}$ to y. T can be regarded as a smooth function: $I^2 \times \mathbb{R}^3 \rightsquigarrow \mathbb{R}$.

Let $y \in \mathbb{R}^3$. We wish to determine the points $\tilde{y} \in V$ emitting light passing through y.

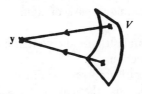

FIG. 2

According to Fermat's principle a light ray travels in such a way that the time taken is the least possible. The points $\tilde{y}(x)$ emitting light hitting y are thus given by the equations

$$\frac{\partial T}{\partial x_i}(x, y) = 0, \qquad i = 1, 2.$$

By definition, this means that (x, y) belongs to M_T. We conclude, $y \in \mathbb{R}^3$ is hit by a light ray from $\tilde{y}(x)$ if and only if $(x, y) \in \chi_T^{-1}(y)$. It is clear that the light is concentrated on the critical values of χ_T, so the caustic is the catastrophe set D_T.

6. Fermat's principle and the timefunction. We will give a brief discussion of the timefunction. In order to do this we state *Fermat's principle* or *the principle of least time*, (see [4, Chap. 26]). The most common variant is: "Out of all possible paths that it might take to get from one point to another, light takes the path which requires the shortest time." To give the precise statement we must look at the space of all paths between the two given points. Then "light takes a path which is a critical point for the timefunction." Similarly we have that light emitted by a wavefront takes a path to a given point which is a

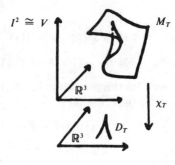

FIG. 3

critical point for the timefunction, this time defined on the space of all paths between the wavefront and the given point.

From Fermat's principle we get the well-known facts that in a medium with constant speed of light the light rays are straight lines and that light rays are orthogonal to the wavefronts. From Fermat's principle we can also deduce the laws of reflection and refraction.

Consider a wavefront V. In the preceding paragraphs we said that V has a timefunction if for a point $\mathbf{x}$ belonging to V and a point $\mathbf{y}$ belonging to $\mathbb{R}^3$ it is possible to determine the time it takes light to travel from $\mathbf{x}$ to $\mathbf{y}$. So in order to define a timefunction for V there must only exist one path from $\mathbf{x}$ to $\mathbf{y}$ which is a critical point for the timefunction. Notice that this unique path or light ray is not necessarily a light ray emitted by the wavefront V.

We will consider two examples that indicate that catastrophe theory also applies to some systems without a timefunction.

Consider an arrangement with a mirror; see Fig. 4. Clearly it is impossible to define a timefunction for this system, because there are two possible light rays from $\tilde{\mathbf{y}}(\mathbf{x})$ to $\mathbf{y}$ so

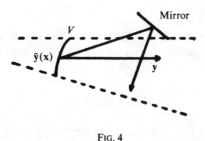

FIG. 4

$T(\mathbf{x}, \mathbf{y})$ would be doublevalued. But if we only consider light rays or paths which do not hit the mirror, it is possible to define a timefunction. The light rays emitted by the wavefront V plus the nearby ones do not hit the mirror, so we can use Fermat's principle and the discussion in the preceding paragraph also applies to this case.

Consider an arrangement with a lens; see Fig. 5. Again it is impossible to define a timefunction, because $T(\mathbf{x}, \mathbf{y})$ would be multivalued. In the first example we disregarded all but one light ray between every pair of points $(\mathbf{x}, \mathbf{y})$ belonging to $V \times \mathbb{R}^3$. If we do this in this example, we would ignore light rays arbitrarily close to the remaining light ray, and

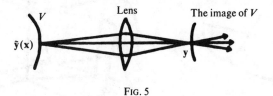

FIG. 5

thus make the use of Fermat's principle impossible. Instead, we observe that we can define a timefunction if we only look at points $\mathbf{y}$ outside the image of V. We conclude that the part of the caustic outside the image of V is the catastrophe set for a timefunction. By choosing another wavefront V' with image disjoint from the image of V, we see that all of the caustic locally is the catastrophe set for a timefunction.

The two examples above do not contradict Fermat's principle. They simply indicate that it is impossible to define a timefunction depending only on the initial points, which is required in our application of catastrophe theory. It is of course possible to define the timefunction on the set of paths and thus make use of Fermat's principle.

7. Conclusion. We have shown that if a given bundle of light has a timefunction, then the caustic is the catastrophe set of this timefunction. Theorem 1 now gives that a stable caustic, locally, only can have a finite number of shapes (stable with respect to small pertubations of the timefunction, i.e., small pertubations of both the wavefront and the media). We have to be careful, because to a given point on the caustic we shall look locally not only on the caustic around the point, but also on the wavefront around the points, from which the light rays come. It is possible that caustics from different locations on the wavefront appear on the same spot, so we can get a caustic, consisting of several elementary catastrophe sets. If we use that stable intersections between surfaces in $\mathbb{R}^3$ only occur as intersections between two or three 2-dimensional surfaces or between a 1-dimensional and a 2-dimensional surface, we get that around a given point the shape of a stable caustic must have one of the basic forms shown in Figs. 6-13; see Callahan [3] or Poston and Stewart [6, Chap. 9].

We have shown that a stable caustic locally at most can have one of the eight shapes mentioned above. In [5] K. Jänich shows that all eight shapes can be realized as caustics. Some of them appear in photos in M.V. Berry [1], which also contains a discussion of the unstable caustics.

FIG. 6. *The fold.*

FIG. 7. *The cusp.*

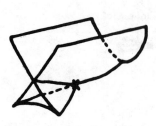

Fig. 8. *The swallow tail.*

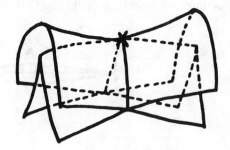

Fig. 9. *The hyperbolic umbilic.*

Fig. 10. *The elliptic umbilic.*

Fig. 11. *Intersection between two folds.*

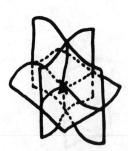

Fig. 12. *Intersection between three folds.*

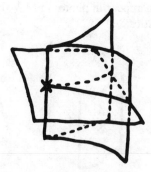

Fig. 13. *Intersection between a fold and a cusp.*

REFERENCES

[1] M. V. BERRY, *Waves and Thom's theorem*. Adv. Phys. 25, (1976), pp. 1–26.
[2] J. CALLAHAN, *Singularities and plane maps*. Amer. Math. Monthly, 81, (1974) pp. 211–240.
[3] J. CALLAHAN, *Singularities and plane maps* II: *Sketching catastrophes*, Amer. Math. Monthly, 84, (1977), pp. 765–803.
[4] R. P. FEYNMAN, *The Feynman Lectures on Physics*, Vol. I, Addison-Wesley, Reading, MA, 1966.
[5] K. JÄHNICH: *Caustics and catastrophes*. Math. Ann., 209, (1974), pp. 166–173.
[6] T. POSTON AND I. STEWART, *Catastrophe Theory and Its Applications*, Pitman, London 1978.
[7] R. THOM, *Stabilité structurelle et morphogénèse*. W.A. Benjamin, New York, 1972.
[8] C. ZEEMAN AND D. TROTMAN: *The classification of elementary catastrophes of codimension ≤ 5*, in Structural Stability, the Theory of Catastrophes and Applications in the Sciences, Battelle Seattle Research Center, 1975, Lecture Notes in Mathematics 525, Springer-Verlag, New York 1976. or C. Zeeman, *Catastrophe Theory. Selected Papers* 1972–1977, Addison-Wesley, Reading, MA, 1977.

Supplementary References
Catastrophe Theory

[1] R. THOM, *"Structural stability, catastrophe theory, and applied mathematics,"* SIAM Rev. (1977) pp. 189–201.
[2] J. W. AUER, *"Mathematical preliminaries to elementary catastrophe theory,"* MM (1980) pp. 13–20.
[3] J. CALLAHAN, *"Singularities and plane maps II: Sketching catastrophes,"* AMM (1977) pp. 765–803.
[4] C. T. J. DODSON AND M. M. DODSON, *"Simple nonlinear systems and the cusp catastrophe,"* BIMA (1977) pp. 77–85.
[5] M. GOLUBITSKY, *"An introduction to catastrophe theory and its application,"* SIAM Rev. (1978) pp. 352–387.
[6] P. HILTON, ed., *Structural Stability, the Theory of Catastrophes, and Applications*, Springer-Verlag, N.Y., 1976.
[7] C. A. ISNARD AND E. C. ZEEMAN, *"Some models from catastrophe theory in the social sciences,"* in L. Collins, ed., *Use of Models in the Social Sciences*, Tavistock, London, 1976, pp. 44–100.
[8] D. A. LAVIS AND G. M. BELL, *"Thermodynamic phase changes and catastrophe theory,"* BIMA (1977) pp. 34–42.
[9] R. S. PECKOVER AND C. G. GIMBLETT, *"A simple electromagnetic catastrophe,"* BIMA (1981) pp. 73–78.
[10] T. POSTON AND I. STEWART, *Catastrophe Theory and its Applications*, Pitman, Great Britain, 1978.
[11] P. T. SAUNDERS, *An Introduction to Catastrophe Theory*, Cambridge University Press, Cambridge, 1980.
[12] M. J. SEWELL, *"A pure catastrophe machine,"* BIMA (1978) pp. 212–214.
[13] ———, *"Some mechanical examples of catastrophe theory,"* BIMA (1976) pp. 163–172.
[14] H. J. SUSSMAN, *"Catastrophe theory,"* Synthèse (1975) pp. 229–270.
[15] I. N. STEWART, *"The seven elementary catastrophes,"* New Sci. (1975) pp. 447–454.
[16] R. THOM, *Structural Stability and Morphogenesis*, Benjamin, N.Y., 1975.
[17] J. M. T. THOMPSON, *Instabilities and Catastrophes in Science and Engineering*, Wiley, Chichester, 1982.
[18] A. WOODCOCK AND M. DAVIS, *Catastrophe Theory*, Dutton, N.Y., 1978.
[19] R. S. ZAHLER AND H. J. SUSSMAN, *Claims and accomplishments of applied catastrophe theory,"* Nature (1977) pp. 759–763.
[20] E. C. ZEEMAN, *"Catastrophe theory,"* Sci. Amer. (1976) pp. 65–83.
[21] ———, *"Duffing's equation in brain modelling,"* BIMA (1976) pp. 207–214.

8. Numerical Analysis

THE NUMERICS OF COMPUTING GEODETIC ELLIPSOIDS*

J. C. ALEXANDER[†]

Abstract. The formulae for calculating ellipsoidal approximations to the earth's shape are analyzed. Numerical convergence and error analyses are done, as well as an investigation of the propagation of observational errors.

Key words. geodetic ellipsoids, reference ellipsoids, numerical error analysis, fixed-point iteration, delta method

Introduction. According to Webster's dictionary, geodesy is "the branch of applied mathematics concerned with measuring, or determining the shape of, the earth..." (Webster [1966]). As a first approximation, the earth is an axially symmetric rotating ellipsoid, and all global geodetic computations are referred to such an ellipsoid. The following equations, given in classic form, are basic to the determination of such an ellipsoid (Heiskanen–Moritz [1967, Chap. 2.9], Moritz [1980]):

$$(1) \qquad 3J_2 = e^2 - \frac{4}{15}\frac{\omega^2 a^3}{GM}\frac{e^3}{2q_0},$$

where

$$(2) \qquad 2q_0 = \left(1 + \frac{3}{e'^2}\right)\arctan e' - \frac{3}{e'}$$

with

$$(3) \qquad e'^2 = \frac{e^2}{1 - e^2}.$$

In current practice, values for a, ω, GM, J_2 are given and a value for e is derived numerically. There are three mathematical points raised. The first is the existence and uniqueness of e. The second is the convergence and accuracy of the numerical procedure used to determine e. The third is the propagation of observational errors in a, ω, GM, J_2 through the computation. Put another way: how many significant digits are there in e given the numbers of significant digits in a, ω, GM, J_2? Such questions arise in the manipulation of any kind of data, and the purpose of this note is to make a case study of this particular example.

Geodetic ellipsoids–discussion of the equations. Since 1930 all global geodetic computations have been referred to a reference ellipsoid (Heiskanen–Moritz [1967, Chap. 2.12]). Computations for an ellipsoid can be done in closed form, and more exact

---

*Received by the editors January 15, 1984, and in revised form September 16, 1984. This work was supported in part by the National Science Foundation.

†Department of Mathematics and Institute for Physical Science and Technology, University of Maryland, College Park, Maryland 20742, and Geodynamics Branch, Code 621, Goddard Space Flight Center, Greenbelt, Maryland 20771.

determinations of the earth's shape (geoidal undulations—up to ± 150 m; see e.g. Lerch et al. [1981]) use an ellipsoid as a zero point. A reference ellipsoid is not, for example, a least squares best fit of the earth's surface, but rather is the solution to a type of free boundary problem.

The shape of an axially symmetric ellipsoid is specified by two parameters, the equatorial and polar radii, a, b, respectively. Usually a and either the (first) eccentricity $e = \sqrt{1 - b^2/a^2}$ or the (linear) flattening $f = 1 - b/a$ are used. Given four parameters: a, e specifying the shape, the rotation rate ω, and the product GM of the universal gravitational constant times mass, there is precisely one gravity field with parameters GM and ω, for which the ellipsoid is an equipotential surface (Heiskanen–Moritz [1967, Chap. 2.7] (a gravity field is the sum of a harmonic gravitational field depending on GM and a rotational "centrifugal field" depending on ω). In particular, the second (unnormalized) coefficient J_2 of the spherical harmonic expansion of the gravitational field is given by (1)–(3). (Remark: e' is called the second eccentricity.)

Thus a geodetic ellipsoid is specified by four parameters. Modern practice is to specify a, ω, GM, J_2 and derive e, and hence the shape. Thus arises the stated problem.

There are two, somewhat conflicting, purposes for such a computation. On the one hand, observed values for a, ω, GM, J_2 are regularly improved. The rotation rate is determined from astronomical observations. The radius a comes from satellite laser altimetry measurements, and GM, J_2 come from analysis of artificial satellite dynamics. Ellipsoids fitting observed data are called mean ellipsoids. Current best values are listed in Table 1. On the other hand, a working geodesist needs a fixed reference ellipsoid. Such ellipsoids are fixed at irregular intervals by international agreement. The values of a, ω, GM, J_2 are set by fiat and all other geodesic parameters are derived. Such ellipsoids are called reference ellipsoids. The current standard, Geodetic Reference System, 1980 (GRS 1980), is listed in Table 2. Reference ellipsoids are also occasionally used in cartography (Snyder [1982]).

To effect computations, (1)–(3) can be solved for e or e^2 as a series in the other parameters and truncated at some point (Heiskanen–Moritz [1967, Chap. 2.10], Chen [1981]) (this last reference also tabulates earlier geodetic reference systems). More standard is to isolate the e^2 term in (1) and iterate to a value for e^2 (Moritz [1980]). A machine or compiler version of arctan is used and the iteration is performed until the desired number of digits stabilizes. Although such a convergence criterion is dangerous in general, we shall see it works in the present case provided enough digits are carried in the computation. In reference to the last point, the reader can verify with a hand calculator that five digits are lost in the subtraction in (2). Thus to obtain, say, 12 significant digits in e^2, a machine with a precision of at least 17 digits must be used. We rearrange the computation so that the calculation is as precise as the machine.

Simplified formula. All the results are based on simplifying (2)–(3). If we substitute (3) in (2) to eliminate e', we find

(4)
$$v = v(e^2) = \frac{2q_0}{e^3} = \frac{(3 - 2e^2)\arcsin e - 3e\sqrt{1 - e^2}}{e^5}$$

$$= \sum_{r=0}^{\infty} \frac{(2r+1)! e^{2r}}{(2r+3)(2r+5)(r!)^2 2^{2r-2}}$$

$$= \frac{4}{15} + \frac{6}{35} e^2 + \frac{5}{42} e^4 + \frac{35}{396} e^6 + \frac{315}{4576} e^8 + \cdots .$$

The analytic expressions comes from routine simplification. The power series can be obtained from the power series for arcsin and $\sqrt{1-z}$ or as follows. Let

$$g(e) = (3 - 2e^2)\arcsin e + 3e\sqrt{1 - e^2}.$$

One differentiation yields

$$g'(e) = 4e^2/\sqrt{1 - e^2} - 4e\arcsin e.$$

Write $eh(e) = g'(e)$ and differentiate again:

$$h'(e) = 4e^2/(1 - e^2)^{3/2}.$$

Now work backwards, starting from the well-known binomial expansion

$$(1 - z)^{-3/2} = \sum_{r=0}^{\infty} \frac{(2r+1)!}{(r!)2^{2r}} z^r.$$

The power series in (4) converges and represents v for $e^2 < 1$.

We can now answer the first question raised in the introduction. All the coefficients in (4) are positive. Thus for $0 \leq e^2 < 1$, (4) is a monotonically increasing function of e^2. Thus so is (1). Moreover, J_2 ranges from $-\omega^2 a^3/3GM$ for $e = 0$ to $\frac{1}{3}$ as $e \to 1$. For this range, as one would hope, a, ω, GM, J_2 uniquely determine an ellipsoid.

Numerical analysis. Let $x = e^2$. The iterative procedure is thus

$$x_{n+1} = 3J_2 + \frac{4}{15} \frac{\omega^2 a^3}{GM} \frac{1}{v}.$$

Let x_∞ denote the (unique) fixed point and let $x_n = x_\infty + \Delta x_n$. Then

$$\Delta x_{n+1} = f'(\bar{x}_n)\Delta x_n = -\frac{4}{15} \frac{\omega^2 a^3}{GM} \frac{v'(\bar{x}_n)}{(v(\bar{x}_n))^2} \Delta x_n$$

for some $\bar{x}_n$ between x_∞ and x_n. A crude calculation yields $x_\infty = e^2 \approx .0067$ and for x_n near x_∞, $\Delta x_{n+1} \approx -2.2 \times 10^{-3}\Delta x_n$. Linear convergence is standard for fixed-point iterations (Householder [1970], Rice [1983, Chap. 8]). Thus near x_∞ the convergence is linear and about $-\log_{10}(2.2 \times 10^{-3}) \approx 2.6$ digits per iteration are gained.

Moreover the Δx_n oscillate between positive and negative values, so x_∞ lies *between* any two successive x_n. Thus digits which have stabilized in the iteration are indeed valid.

If v is truncated to a polynomial, x_∞ is changed (by the truncation error); the convergence of the iteration is unchanged. Moreover we can control the truncation error. The coefficients of (4) are easily seen to be decreasing in magnitude. Thus if the series is truncated after $n - 1$ terms, the truncation error is bounded by the geometric series

$$|\text{error}| < \frac{(2n+1)!}{(2n+3)(2n+5)(n!)^2 2^{2n-1}} \sum_{r=n}^{\infty} x^r$$

$$= \frac{(2n+1)!x^n}{(2n+3)(2n+5)(n!)^2 2^{2n-2}(1-x)} < .25x^n$$

for $x < .01$. Since the value of v at x_∞ is greater than .25, the number of significant digits can be guaranteed a priori. For $x < .01$ a crude, but easy, working rule is the following: if the series is truncated after the term in x^N, the relative truncation error is less than $.5 \times 10^{-(2N+1)}$ (i.e., the number of significant digits is at least $2N + 1$).

The appropriate use of v also minimizes roundoff error, assuming the calculating machine rounds elementary arithmetic operations in an acceptable manner. There is no subtraction to chew up significant digits. In particular, if the truncated v is calculated using the Newton–Horner method, the roundoff error should be within the last digit of the machine numbers (Wilkinson [1964]).

All in all then, the numerical process is highly stable and converges at an acceptable rate. In the next section but one, we investigate the conditioning.

Algorithm. The previous section indicated the following algorithm will accurately compute e^2. For $N = 3$ or 4, the parameters can be stored from (4) and the iterative loop programmed on a small hand-held programmable calculator.

ALGORITHM for computing $X = E**2$ to $2*N + 1$ significant digits, given OMEGA, A, GM, $J2$ (machine precision $\geq 2N + 2$ digits).

Compute and store coefficients $C(K)$, degree 0 through N
    $C(0) = 4/15$
    DO 100, $K = 1, N$
100 $C(K) = C(K-1)*((1+1/(2*K))**2)/(1+5/(2*K))$
Input parameters and initialize
    INPUT OMEGA, A, GM, $J2$
    $Q = 4*(OMEGA**2)*(A**3)/(15*GM)$
    $X = 6*J2$
Iterative step
200 $V = C(N)$
    DO 300, $K = N-1, 0, STEP-1$
300 $V = X*V + C(K)$
    $X1 = Q/V + 3*J2$
    IF $|X1 - X| < .5 * X*(10**(-2*N-1))$ GO TO 400
    $X = X1$
    GO TO 200
Output
400 OUTPUT $X1$
    STOP

In Moritz [1980] geodetic parameters are computed to 12 significant digits. Such an accuracy is more than enough for geodetic computations; 12 digits in e^2 will give the polar circumference to within .00001 cm (exercise). For demonstration purposes, we calculate from Table 2 in 9 iterations using quadruple precision with $N = 14$:

$$e^2 = .006\ 694\ 380\ 022\ 903\ 415\ 749\ 574\ 948\ 586.$$

The concomitant reciprocal flattening

$$f^{-1} = 1/\sqrt{1-(1-e^2)} = (1+\sqrt{1-e^2})/e^2$$

$$= 298.257\ 222\ 100\ 882\ 711\ 243\ 162\ 836\ 6.$$

Thus the polar radius of the earth is 1 part in 298.257... less than the equatorial radius.

Observational data. We consider a mean ellipsoid, which is fit to observed data, and investigate the amount of accuracy available. The data are assumed to be independent and normally distributed about their mean values, as listed in Table 1. The value of J_2 is never available directly from satellite dynamics, but occurs in all formulae in the product $a^2 J_2$ (Kaula [1966]).

For data which have very small relative standard deviations, the "delta method" (Rao [1965]) is useful. If X is a normal random variable with mean $\mu(X)$ and standard deviation $\sigma(X)$, the linear function $mX + b$ is also normal with mean $m\mu(X) + b$ and standard deviation $|m|\sigma(X)$. The standard deviation transforms as the absolute value of the derivative. If $\sigma(X)$ is so small that there is only a negligible difference between $f(X)$ and $f(\mu(X)) + f'(\mu(X))(X - \mu(X))$, then $f(X)$ is essentially a normal variable with mean $f(\mu(X))$ and standard deviation $|f'(\mu(X))|\sigma(X)$. In other words, standard deviations transform as absolute values of differentials. The multivariate version is also valid.

Accordingly, we multiply (4) by a^2, differentiate in differential form, take absolute values of the differentials and solve for $\sigma(e^2)$ (v' denotes the derivative with respect to $x = e^2$). We obtain

$$\sigma(e^2) = \frac{1}{\left(1 - (4\omega^2 a^3/15GM)v'/v\right)} \frac{3\sigma(a^2 J_2)}{a^2} + \frac{8\omega a^3}{15GM} \frac{1}{v} \sigma(\omega)$$

$$+ \frac{\sigma(a)}{a} \left| \frac{4}{3} \frac{\omega^2 a^3}{GM} \frac{1}{v} - 2e^2 \right| + \frac{4\omega^2 a^3}{15(GM)^2} \frac{1}{v} \sigma(GM)$$

$$\approx 7.4 \times 10^{-4} \sigma(a^2 J_2) m^{-2} + 6.1 \times 10^{-10} \sigma(a) m^{-1} + 95\sigma(\omega)\sec$$

$$\approx 4.3 \times 10^{-9}.$$

Accordingly $\sigma(e^2)/e^2$ is about 6×10^{-7}; there are $6+$ observationally significant digits in e^2. There are also $6+$ observationally significant digits in f^{-1}.

Gravity (Heiskanen–Moritz [1967, Chap. 2.8]). The normal gravity is the magnitude of acceleration due to gravity on the ellipsoid. The formula for the normal gravity at geographic lattitude ϕ is

$$y_\phi = \frac{a\gamma_a \cos^2\phi + b\gamma_b \sin^2\phi}{\sqrt{a^2 \cos^2\phi + b^2 \sin^2\phi}}$$

where

$$\gamma_a = \frac{GM}{ab}\left(1 - m - \frac{m}{6}\frac{e'q_0'}{q_0}\right), \qquad \gamma_b = \frac{GM}{a^2}\left(1 + \frac{m}{3}\frac{e'q_0'}{q_0}\right)$$

are the normal gravities at the equator and pole, respectively. Here

$$m = \frac{\omega^2 a^2 b}{GM},$$

$$q_0' = 3\left(1 + \frac{1}{e'^2}\right)\left(1 + \frac{1}{e'}\arctan e'\right) - 1.$$

The interested reader can verify that

$$e_0' q_0' / e^3 = \sum_{r=0}^{\infty} \frac{(2r+1)! e^{2r}}{(r!)^2 (2r+5) 2^{2r-1}}$$

and calculate from Table 1 that $\gamma_a = 9.780\ 322$ m/sec^2 and $\gamma_b = 9.832\ 182$ m/sec^2, both values with a standard deviation of 4×10^{-6}.

TABLE 1

Mean ellipsoid.

Parameter	Value	σ	Units
$a^2 J_2$	$440\ 420.14 \times 10^5$	$.36 \times 10^5$	m^2
GM	$3\ 986\ 004.40 \times 10^8$	$.05 \times 10^8$	m^3/sec^2
a	$6\ 378\ 138$	1	m
ω	$7\ 292\ 116 \times 10^{-11}$	1×10^{-11}	rad/sec

References: Values of $a^2 J_2$ and GM came from Goddard Earth Model L2 (GEM L2) gravitational field solution with speed of light = 299 792 458 m/sec (private communication from F. J. Lerch, see also Lerch et al., [1982], however the values there correspond to an artifically fixed value of $a = 6\ 378\ 145$ m); value of a comes from GEM 10B (Lerch et al. [1981]); value of ω comes from the current UT2 standardization and thus represents a smoothed value—the error is a bound on the smoothing effects.

TABLE 2

Geodetic reference system 1980.

Parameter	Value	Units
a	$6\ 378\ 137$	m
GM	$3\ 986\ 005 \times 10^8$	m^3/sec^2
J_2	$108\ 263 \times 10^{-8}$	—
ω	$7\ 292\ 115 \times 10^{-11}$	rad/sec

Reference: Moritz [1980].

Acknowledgments. The genesis of this note was a class on satellite geodesy and orbit analysis. I am endebted to those students who wondered why their computations of e^2 using equations (1)–(2) differed from published values. Challenged, some of them independently produced the expansion of (4). I am also indebted to Frank Lerch for discussion of GEM models and to R. H. Rapp and E. V. Slud for comments on geodesy and statistics, respectively.

REFERENCES

J. Y. GEN, *Formulae for computing ellipsoidal parameters*, Bull. Géod., 55 (1981), pp. 170–178.

W. A. HEISKANEN AND H. MORITZ, *Physical Geodesy*, W. H. Freeman, San Francisco, 1967.

A. S. HOUSEHOLDER, *The Numerical Treatment of a Single Nonlinear Equation*, McGraw-Hill, New York, 1970.

W. KAULA, *Theory of Satellite Geodesy*, Blaisdell, Waltham, MA, 1966.

F. J. LERCH, B. H. PUTNEY, C. A. WAGNER AND S. M. KLOSKO, *Goddard earth models for oceanographic applications* (GEM 10B and 10C), Marine Geodesy, 5 (1981), pp. 145–187.

F. J. LERCH, S. M. KLOSKO AND G. B. PATEL, *Refined gravity model from Lageos* (GEM L2), Geophys. Res. Let., 9 (1982), pp. 1263–1266.

H. MORITZ, *Geodetic reference system 1980*, Bull. Géod., 54 (1980), pp. 395–405.

C. R. RAO, *Linear Statistical Inference and Its Applications*, John Wiley, New York, 1973.

J. R. RICE, *Numerical Methods, Software and Analysis*, IMSL© Reference Edition, McGraw-Hill, New York, 1983.

J. P. SNYDER, *Map Projections Used by the U. S. Geological Survey*, Geological Survey Bulletin 1532, U. S. Government Printing Office, Washington, DC, 1982.

Webster's New World Dictionary, College Edition, World Publishing Co., New York, 1966.

J. H. WILKINSON, *Rounding Errors in Algebraic Processes*, Prentice-Hall, Englewood Cliffs, NJ, 1964.

CHESS CHAMP'S CHANCES: AN EXERCISE IN ASYMPTOTICS

PETER HENRICI AND CHRISTIAN HOFFMANN†

The model. The following problem was recently proposed by J. G. Wendel [1]: "In one form of chess match $2n$ games are played, wins count 1 point each, draws $\frac{1}{2}$, losses are worth 0. In order to win the match, the defender needs only score at least n, while the challenger must achieve at least $n + \frac{1}{2}$. Suppose that the two players are of equal strength, and that the probability of a draw is a constant δ. Prove or disprove: the defender's chance of keeping his title is an increasing function of δ." The present paper aims at a detailed discussion of this chance as a function of both δ and n, with special regard to its asymptotic behavior as $n \to \infty$.

The solution of Alam and Seo. One published solution of the above problem essentially runs as follows [2]. Let $A_n = A_n(\delta)$ denote the probability that both players score exactly n points. Since the players are of equal strength, the probability that a fixed player scores more than n points is $\frac{1}{2}(1 - A_n)$. Therefore the probability of the defender's keeping his title is

(1) $$P_n(\delta) = A_n + \tfrac{1}{2}(1 - A_n) = \tfrac{1}{2}(1 + A_n).$$

Thus $P_n(\delta)$ is an increasing function of δ if and only if $A_n(\delta)$ is an increasing function of δ.

To compute $A_n(\delta)$, we note that the probability of a win by either player is $\frac{1}{2}(1 - \delta)$. Thus the probability that there will be k wins by the champion, k wins by the challenger, and $2n - 2k$ draws will be

$$\delta^{2n-2k}\left(\frac{1-\delta}{2}\right)^{2k}$$

times the number of ways k wins and k losses of the defender can be distributed among $2n$ games. There are $\dbinom{2n}{k}$ possibilities to distribute the wins, and for each such possibility there exist $\dbinom{2n-k}{k}$ possibilities to distribute the losses. Thus the desired number of ways is

$$\binom{2n}{k}\binom{2n-k}{k} = \frac{(2n)!}{k!(2n-k)!}\frac{(2n-k)!}{k!(2n-2k)!} = \frac{(2n)!}{(k!)^2(2n-2k)!}.$$

* Received by the editors October 22, 1974.

† Seminar für angewandte Mathematik, Eidgenössische Technische Hochschule, Zürich, Switzerland.

The probability $A_n(\delta)$ is the sum of all these probabilities from $k = 0$ to $k = n$; thus

(2) $$A_n(\delta) = \sum_{k=0}^{n} \frac{(2n)!}{(k!)^2(2n - 2k)!} \delta^{2n-2k} \left(\frac{1 - \delta}{2}\right)^{2k}.$$

Differentiating with respect to δ, we see that the derivative $A_n'(\delta)$ satisfies

(3) $$A_n'(0) = -2^{-2n} \binom{2n}{n} 2n < 0,$$

so that $A_n(\delta)$, and hence $P_n(\delta)$, are decreasing functions of δ near 0. Therefore the assertion that $P_n(\delta)$ is increasing in $[0, 1]$ is false for all $n > 0$.

An integral representation. Using the well-known integral

$$\frac{1}{\pi} \int_{-1}^{1} \frac{t^{2k}}{\sqrt{1 - t^2}} dt = \frac{2}{\pi} \int_{0}^{\pi/2} (\cos \phi)^{2k} d\phi = \frac{(2k)!}{2^{2k}(k!)^2},$$

we readily see upon expanding the binomial that

(4) $$A_n(\delta) = \frac{1}{\pi} \int_{-1}^{1} (1 - t^2)^{-1/2} \{\delta + (1 - \delta)t\}^{2n} dt.$$

This could be obtained methodically by noting that, in hypergeometric notation,

$$A_n(\delta) = \delta^{2n} F\left(-n, -n + \frac{1}{2}; 1; \left(\frac{1 - \delta}{\delta}\right)^2\right)$$

and using one of the standard integrals for the hypergeometric function.

The representation (4) is very convenient for discussing the function $A_n(\delta)$, both from an elementary and (because n occurs only as an exponent) from an asymptotic point of view. Differentiating under the integral sign, we get

(5) $$A_n'(\delta) = \frac{2n}{\pi} \int_{-1}^{1} \sqrt{\frac{1 - t}{1 + t}} \{\delta + (1 - \delta)t\}^{2n-1} dt,$$

(6) $$A_n''(\delta) = \frac{2n(2n - 1)}{\pi} \int_{-1}^{1} \frac{(1 - t)^{3/2}}{\sqrt{1 + t}} \{\delta + (1 - \delta)t\}^{2n-2} dt.$$

Because the integrand is positive save at one point at most, we conclude from (6) that $A_n''(\delta) > 0$, $0 \leq \delta \leq 1$, for all $n > 0$. (This was noted without proof in [2].) We have

(7) $$A_n(0) = 2^{-2n} \binom{2n}{n} < 1,$$

and it follows from (2), (4) or directly from the probabilistic interpretation (all games drawn) that $A_n(1) = 1$. We already know that $A_n'(0) < 0$ and thus may draw two simple conclusions:

(i) There exists a unique $\delta > 0$ such that $A_n(\delta) = A_n(0)$. In terms of chess, this is the smallest positive probability of a draw for which the chances of the defender's keeping his title are at least as good as if the probability of a draw were zero. We denote this by δ_n^* and call it the *smallest safe value of* δ.

(ii) The derivative $A_n'(\delta)$ has precisely one zero in $(0, 1)$, at $\delta = \delta_n$, say, which thus is the unique point where $A_n(\delta)$ achieves its smallest value. From the defender's point of view, this is the *most dangerous value of δ*.

Our purpose in the remaining sections will be to study the behavior of δ_n^*, δ_n, and $A_n(\delta_n)$ as $n \to \infty$.

The smallest safe value of δ. By Stirling's formula (see, e.g., [3, p. 188]),

$$(8) \qquad A_n(0) = 2^{-2n}\frac{(2n)!}{(n!)^2} \sim \frac{1}{\sqrt{\pi n}}, \qquad n \to \infty.$$

Thus $A_n(0) \to 0$ for $n \to \infty$, as is probabilistically obvious, although the convergence perhaps is not as fast as expected. For $\delta = \delta_n^*$, the asymptotic behavior of $A_n(\delta)$ should match that of $A_n(0)$. For each fixed $\delta \in (0, 1)$, the function $\phi(t) = |\delta + (1 - \delta)t|$ has the relative extrema $\phi(1) = 1$ and $\phi(-1) = |1 - 2\delta|$. Asymptotically, only the first extremum is relevant. Setting

$$(9) \qquad s = -\log\{\delta + (1 - \delta)t\},$$

we obtain

$$A_n(\delta) = \frac{1}{\pi}\int_0^\infty \frac{e^{-s}}{\sqrt{(1 - e^{-s})(1 + e^{-s} - 2\delta)}}\, e^{-2ns}\, ds + O((1 - 2\delta)^{2n}),$$

$n \to \infty$. The integral now is in a form suitable for applying Watson's lemma (see, e.g., [3, p. 253]), the result being

$$(10) \qquad A_n(\delta) \sim \frac{1}{2\sqrt{\pi n}\sqrt{1 - \delta}}, \qquad n \to \infty.$$

A comparison with (8) yields $2\sqrt{1 - \delta} = 1$ or $\delta = \frac{3}{4}$. We conclude that $\delta_n^* \sim \frac{3}{4}$ as $n \to \infty$. A more elaborate computation, taking into account nondominant terms in the asymptotic expansions, yields

$$(11) \qquad \delta_n^* = \frac{3}{4} - \frac{3}{32n} + O(n^{-2}), \qquad n \to \infty.$$

From the point of view of chess, it is interesting that the smallest safe value of δ is asymptotically independent of n. Independently of the number of games played, the defender should keep the probability of a draw at least equal to $\frac{3}{4}$ in order to make sure that his probability of winning the match at least equals the probability of winning if there were no draws.

The most dangerous value of δ. For fixed $\delta > 0$, an application of Watson's lemma to the integral (5) yields

$$(12) \qquad A_n'(\delta) \sim \frac{1}{4\sqrt{\pi n}\,(1 - \delta)^{3/2}}, \qquad n \to \infty,$$

which we would also get by differentiation (not permissible in general) of the asymptotic formula (10). Thus for every fixed $\delta > 0$, $A_n'(\delta)$ is ultimately positive, showing that $\delta_n \to 0$ for $n \to \infty$. Because for $\delta \to 0$ the two maxima of the function $\phi(t)$ at $t = \pm 1$ are of approximately equal strength, a single application of Watson's lemma does not suffice for a more accurate evaluation of δ_n.

Let $0 < \delta < \frac{1}{2}$, and let $\phi(t_\delta) = 0$. We split the integral (5) into its positive and negative parts,

$$A_n'(\delta) = \frac{2n}{\pi}\{I_1(\delta) - I_2(\delta)\},$$

where

$$I_1(\delta) = \int_{t_\delta}^1 \sqrt{\frac{1-t}{1+t}}\{\delta + (1-\delta)t\}^{2n-1}\,dt,$$

$$I_2(\delta) = \int_{-t_\delta}^1 \sqrt{\frac{1+t}{1-t}}\{(1-\delta)t - \delta\}^{2n-1}\,dt.$$

The integral $I_1(\delta)$ can be dealt with by the substitution (9). If δ is any function of n which tends to zero as $n \to \infty$, Watson's lemma yields

$$I_1(\delta) \sim \frac{\sqrt{\pi}}{8n^{3/2}}, \qquad\qquad n \to \infty.$$

To evaluate $I_2(\delta)$, we set

(13) $$s = -\log\frac{(1-\delta)t - \delta}{1 - 2\delta}.$$

As t runs from 1 to $-t_\delta$, s runs from 0 to ∞, and the integral takes the form

$$I_2(\delta) = \frac{(1-2\delta)^{2n}}{1-\delta}\int_0^\infty \sqrt{\frac{1 + (1-2\delta)e^{-s}}{(1-2\delta)(1-e^{-s})}}e^{-2ns}\,ds.$$

Evaluation by Watson's lemma now yields

$$I_2(\delta) \sim (1 - 2\delta)^{2n}\frac{\sqrt{\pi}}{\sqrt{n}}, \qquad\qquad n \to \infty,$$

for any $\delta = \delta(n)$ such that $\lim_{n\to\infty}\delta(n) = 0$. If $\delta(n) = \delta_n$, then $I_1(\delta_n) = I_2(\delta_n)$, and the two asymptotic expressions must be equal. This requires

$$(1 - 2\delta_n)^{2n} \sim \frac{1}{8n}, \qquad 2n\log(1 - 2\delta_n) \sim -\log 8n,$$

hence

$$\delta_n \sim \frac{\log 8n}{4n}, \qquad\qquad n \to \infty.$$

A more precise computation, taking into account higher terms, yields

(14) $$\delta_n = \frac{\log 8n}{4n} - \left(\frac{\log 8n}{4n}\right)^2 + O(n^{-2}), \qquad\qquad n \to \infty.$$

It remains to evaluate $A_n(\delta_n)$. Again there is the difficulty that (10) is invalid near $\delta = 0$. Splitting the integral (4) into the integrals from -1 to t_δ and from t_δ to 1, and evaluating each integral separately (reversing the sign of t and making the substitution (13) in the first integral), we find for the minimum value of $A_n(\delta)$

(15) $$A_n(\delta_n) \sim \frac{1}{2\sqrt{\pi n}}\left(1 + \frac{\log 8n}{8n}\right),$$ $$n \to \infty.$$

In view of (8) we see that for large n the defender's advantage for $\delta = \delta_n$ drops by nearly 50% below the advantage for $\delta = 0$; moreover, the most dangerous value of δ tends to zero only slowly as $n \to \infty$.

Numerical values. It is of interest to compare the asymptotic values found above to numerical values found by directly solving the equations $A_n(\delta) = A_n(0)$ and $A_n'(\delta) = 0$. The values in the tables below are selected from output generated by a root-finding procedure combining bisection and quadratic inverse interpolation; $A_n(\delta)$ and $A_n'(\delta)$ were computed from their polynomial definition (2). Machine-drawn graphs of $P_n(\delta)$ for selected values of n are shown in Fig. 1. Also shown are the points $(\delta_n^*, P_n(0))$ and $(\delta_n, P_n(\delta_n))$, as well as the curves

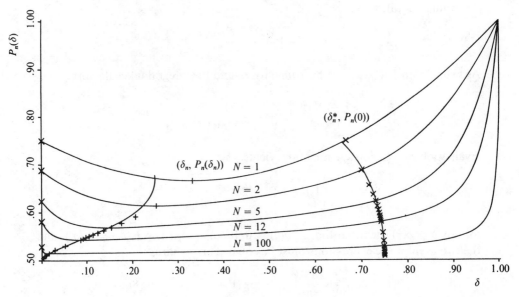

FIG. 1

$$\left(\frac{3}{4} - \frac{3}{32n}, \frac{1}{2} + \frac{1}{2\sqrt{\pi n}}\left(1 - \frac{1}{8n}\right)\right)$$

and

$$\left(\frac{\log 8n}{4n}\left(1 - \frac{\log 8n}{4n}\right), \frac{1}{2} + \frac{1}{4\sqrt{\pi n}}\left(1 + \frac{\log 8n}{8n}\right)\right)$$

($n \geq 1$) describing their asymptotic behavior.

TABLE 1
The smallest safe value of δ

n	δ_n^*	(11)	$A_n(\delta_n^*) = A_n(0)$
2	0.699470	0.703125	0.375000
5	0.730189	0.731250	0.246094
12	0.742050	0.742187	0.161180
100	0.749061	0.749062	0.056348
500	0.749812	0.749812	0.025225

TABLE 2
The most dangerous value of δ

n	δ_n	(14)	$A_n(\delta_n)$
2	0.253122	0.226460	0.228015
5	0.154742	0.150424	0.136865
12	0.086662	0.086048	0.085163
100	0.016436	0.016432	0.028444
500	0.004130	0.004130	0.012642

REFERENCES

[1] J. G. WENDEL, *Problem E 2414*, Amer. Math. Monthly, 80 (1973), p. 559.
[2] *Chess Champ's Chances, Solutions to Problem E 2414*, Ibid., 81 (1974), pp. 519–520.
[3] G. F. CARRIER, M. KROOK AND C. E. PEARSON, *Functions of a Complex Variable*, McGraw-Hill, New York, 1966.

APPROXIMATING THE GENERAL SOLUTION OF A DIFFERENTIAL EQUATION*

JAMES F. McGARVEY†

Abstract. A method for obtaining a series approximation to the general solution of a first order differential equation is presented. An error estimate is included. An example is given which shows how the method may be used in a neighborhood of a branch point.

Introduction. Suppose that one were given the differential equation $dy/dx = f(x, y)$, where f is analytic in x and y, and that one wanted an approximation of known accuracy to the general solution. One might want this for its own sake, or in a variety of situations including the following:

1) A set of integral curves is required valid for the range $0 \leq x \leq x_{max}$ for the set of initial conditions $y(0) = y_0$, where $y_{min} \leq y_0 \leq y_{max}$.

2) A solution is required in the neighborhood of a branch point of $y(x)$.

3) One wants to know if the integral curve through a given point (x_0, y_0) has a branch point for $x_0 \leq x \leq x_{max}$, and if so for what value of x.

A method for obtaining a series approximation to the general solution of 1) is

*Received by the editor August 14, 1981, and in revised form December 8, 1981.
†NASA—Goddard Space Flight Center, Greenbelt, Maryland 20770.

presented herein. Illustrative examples are included which cover all of the above cases. The method consists of generating a sequence of functions of x and y by a recursion relation starting with an arbitrary function of y only. The procedure for getting term $n + 1$ from term n is a three-step operation consisting of differentiating term n with respect to y, multiplying the result by $f(x, y)$, and integrating this result over x. All of the terms required for an error estimate are obtained in the process.

The procedure has the advantage of converging in cases where Picard approximations diverge. Another advantage is that the terms of the sequence tend to be similar to the function $f(x, y)$ itself. For instance, if f is a polynomial in x and y, then each term is also.

The applicability of the method naturally depends on the function $f(x, y)$. For a complicated function, the process may very well get bogged down, in which case one would have to resort to numerical methods.

The general solution of the first order differential equation

$$(1) \qquad \frac{dy}{dx} = f(x, y)$$

is a function of x and y; $z = z(x, y)$. The function z is the solution of the partial differential equation

$$(2) \qquad z_x + f(x, y)z_y = 0.$$

A series approximation to z may be obtained as follows:

A sequence of functions $T[n]$ $(n = 0, 1, 2, 3, \cdots)$, where $T[0]$ is a function of y only and all other $T[n]$ are functions of x and y, is generated by the recursion relation

$$(3) \qquad T[n] = - \int f(x, y)\, T[n - 1]_y\, dx \qquad (\text{for } n \geq 1),$$

or equivalently

$$(4) \qquad T[n]_x + fT[n - 1]_y = 0.$$

In the above, and in the following text, subscripts denote partial differentiation.

Let $S[n]$ be the partial sum of the first $n + 1$ T's, i.e.,

$$(5) \qquad S[n] = \sum_{k=0}^{n} T[k].$$

Applying the property of $T[n]$ given by (4), it may easily be shown that

$$(6) \qquad S[n]_x + fS[n]_y = fT[n]_y.$$

$S[n]$ is an approximation to z. To evaluate how close an approximation it is, let the slopes of the level curves of z and $S[n]$ be compared.

For constant z, we have

$$(7) \qquad \frac{dy}{dx} = - \frac{z_x}{z_y} = f.$$

For constant $S[n]$, we have

$$(8) \qquad \frac{dy}{dx} = - \frac{S[n]_x}{S[n]_y} = f\left(1 - \frac{T[n]_y}{S[n]_y}\right).$$

$S[n]$ converges to z, if

(9) $$\lim_{n \to \infty} \frac{T[n]_y}{S[n]_y} = 0$$

for (x, y) such that $S[n]$ = constant.

For some cases a convergence criterion may be obtained directly as is shown by the following example.

In order to actually carry out the operations indicated in the previous discussion, it is sufficient that each of the $T[n]$ be differentiable and that $f(x, y)$ be analytic in x and y.

An illustrative example. Let the given equation be

(10) $$\frac{dy}{dx} = x + \frac{1}{y},$$

and assume that solutions are required for initial conditions $y(0) = y_0 > 2$.

For $T[0]$, we choose $T[0] = y$, which is consistent with the asymptotic form of (10) for large y. The first four terms of $T[n]$ and $T[n]_y$ then are as follows:

(11) $$T[0] = y, \qquad T[0]_y = 1,$$

(12) $$T[1] = -\int \left(x + \frac{1}{y} \right) dx = -\frac{1}{2} x^2 - \frac{x}{y}, \qquad T[1]_y = \frac{x}{y^2},$$

(13)
$$T[2] = -\int \left(x + \frac{1}{y} \right) \left(\frac{x}{y^2} \right) dx = -\frac{1}{2} \frac{x^2}{y^3} - \frac{1}{3} \frac{x^3}{y^2},$$

$$T[2]_y = \frac{3}{2} \frac{x^2}{y^4} + \frac{2}{3} \frac{x^3}{y^3},$$

(14)
$$T[3] = -\int \left(x + \frac{1}{y} \right) \left(\frac{3}{2} \frac{x^2}{y^4} + \frac{2}{3} \frac{x^3}{y^3} \right) dx = -\frac{1}{2} \frac{x^3}{y^5} - \frac{13}{24} \frac{x^4}{y^4} - \frac{2}{15} \frac{x^5}{y^3},$$

$$T[3]_y = \frac{5}{2} \frac{x^3}{y^6} + \frac{52}{24} \frac{x^4}{y^6} + \frac{6}{15} \frac{x^5}{y^4}.$$

An estimate for the convergence behavior is obtained by finding a condition on x and y that will be sufficient to ensure that both the $T[k]$ and the $T[k]_y$ are monotonically decreasing.

By inspection, it can be seen that for $k \geq 2$

(15) $$T[k] = \sum_{m-k}^{2k-1} A_{m,n} x^m y^{-n}, \quad \text{where } n = 3k - 1 - m.$$

We then have

(16) $$T[k]_y = -\sum n A_{m,n} x^m y^{-(n+1)}$$

and

(17) $$T[k + 1] = -\int \left(x + \frac{1}{y} \right) T[k]_y \, dx = \int \left(x + \frac{1}{y} \right) \sum n A_{m,n} x^m y^{-(n+1)} \, dx.$$

Performing the indicated operations (first interchanging the order of integration and summation), gives the following:

(18) $$T[k + 1] = \sum \left(\frac{n}{m + 2} \frac{x^2}{y} + \frac{n}{m + 1} \frac{x}{y^2} \right) A_{m,n} x^m y^{-n},$$

(19) $$T[k + 1]_y = -\sum \left(\frac{n + 1}{m + 2} \frac{x^2}{y} + \frac{n + 2}{m + 1} \frac{x}{y^2} \right) n A_{m,n} x^m y^{-(n+1)}.$$

But for $n = 3k - 1 - m$ and $k \le m \le 2k - 1$

(20)
$$\max \left(\frac{n}{m + 2} \right) = \frac{2k - 1}{k + 2}, \qquad \max \left(\frac{n + 1}{m + 2} \right) = \frac{2k}{k + 2},$$

$$\max \left(\frac{n}{m + 1} \right) = \frac{2k - 1}{k + 1}, \qquad \max \left(\frac{n + 2}{m + 1} \right) = \frac{2k + 1}{k + 1},$$

and all of the above maximum values are <2 for all $k \ge 2$.

Equations (18) and (19) may therefore be replaced by the inequalities

(21) $$T[k + 1] < 2 \left(\frac{x^2}{y} + \frac{x}{y^2} \right) \sum A_{m,n} x^m y^{-n},$$

(22) $$|T[k + 1]_y| < 2 \left(\frac{x^2}{y} + \frac{x}{y^2} \right) \sum n A_{m,n} x^m y^{-(n+1)},$$

TABLE 1

x	y Num. int.	y $S_3 = 4$	% error	Convergence factor
0.0	4.	4.	.00	.00
0.5	4.24689	4.24688	.00	.09
1.0	4.7338	4.7334	.01	.26
1.5	5.458	5.454	.06	.46
2.0	6.42	6.41	.17	.67
2.5	7.61	7.59	.33	.87
3.0	9.05	9.00	.52	1.04
4.0	12.64	12.53	.87	1.30
5.0	17.21	17.02	1.05	1.49
10.0	54.88	54.38	.91	1.84
20.0	204.97	204.21	.37	1.96

which are equivalent to

(23) $$T[k + 1] < 2 \left(\frac{x^2}{y} + \frac{x}{y^2} \right) T[k],$$

(24) $$|T[k + 1]_y| < 2 \left(\frac{x^2}{y} + \frac{x}{y^2} \right) T[k]_y.$$

The condition on x and y sufficient to ensure monotonically decreasing $T[k]$ and $T[k]_y$ is obviously

(25) $$\frac{x^2}{y} + \frac{x}{y^2} < 2.$$

Numerical results. For initial conditions $x = 0$, $y = 4$ the values of y obtained by setting $S[3] = 4$ and solving by iteration for y for given x are compared with values obtained by numerical integration in Table 1. Values of $S[3]_y$, $T[3]_y$, and error in slope are given in Table 2. Up to $x = 5$ there is a positive correlation between the errors and the convergence factor $(x^2/y + x/y^2)$, but the converse is true thereafter.

TABLE 2

x	y	S_{3y}	T_{3y}	% error in slope
0.0	4.	1.000	.0000	.00
0.5	4.24688	1.030	.000	.02
1.0	4.7334	1.056	.002	.18
1.5	5.454	1.074	.006	.56
2.0	6.41	1.084	.011	1.02
2.5	7.59	1.085	.015	1.41
3.0	9.00	1.082	.018	1.66
4.0	12.53	1.067	.018	1.73
5.0	17.02	1.050	.016	1.51
10.0	54.38	1.012	.005	.46
20.0	204.21	1.002	.001	.07
50.0	1254.09	1.000	.000	.01
100.0	5004.04	1.000	.000	.00

TABLE 3

x	$y:S = .39375$	$dy/dx = -S_x/S_y$	$x + 1/y$	Error in slope
.50	.500	2.63	2.50	.13
.45	.354	3.35	3.28	.07
.40	.114	9.23	9.21	.02
.35	$-.0106 + i.295$	$.222 - i3.349$	$.228 - i3.388$	$-.006 + i.039$
.30	$-.0206 + i.430$	$.179 - i2.274$	$.189 - i2.319$	$-.010 + i.045$

A second solution valid in a neighborhood of the branch point at $y = 0$ may be obtained by interchanging the roles of x and y. The recursion relation (3) is replaced by

(26)
$$T[n] = -\int T[n-1]_x \frac{y}{1+xy}\, dy$$

and $T[0]$ is now a function of x only.

Choosing $T[0] = x$, gives

(27)
$$T[1] = -\int \frac{y}{1+xy}\, dy.$$

Expanding the integrand in a series and integrating term by term, we have

(28)
$$T[1] = -\frac{1}{2}y^2 + \frac{1}{3}xy^3 - \frac{1}{4}x^2y^4 + \cdots.$$

Applying the recursion relation (26) again gives

$$(29) \qquad T[2] = -\frac{1}{15}y^5 + \frac{5}{36}xy^6 - \frac{43}{210}x^2y^7 + \cdots.$$

For an approximation to the general solution, we take $T[0]$ plus the first two terms of $T[1]$ and the first term of $T[2]$:

$$(30) \qquad S = x - \frac{1}{2}y^2 + \frac{1}{3}xy^3 - \frac{1}{15}y^5.$$

This approximation is now used to trace an approximation to the integral curve through the point $(\frac{1}{2}, \frac{1}{2})$ with x decreasing to .30 by steps of .05. Since $S(\frac{1}{2}, \frac{1}{2}) = .39375$, the relation between x and y is

$$(31) \qquad \frac{1}{15}y^5 - \frac{1}{3}xy^3 + \frac{1}{2}y^2 + .39375 - x = 0.$$

For given values of x, (31) is simply a polynomial in y. The values obtained are given in Table 3. The error in the slope was obtained by evaluating S_x and S_y and computing $dy/dx = -S_x/S_y$ which was then compared to $x + 1/y$.

WHICH ROOT DOES THE BISECTION ALGORITHM FIND?*

GEORGE CORLISS†

A student of elementary probability may be amused by an application of probability to numerical analysis.

Let f be a continuous function on the closed interval $[a, b]$ such that $f(a)f(b) < 0$. Then f has at least one root α on (a, b). The well-known bisection algorithm [1, p. 28] generates a sequence $\{[a_k, b_k]\}$ of intervals on which a root is known to lie. Let $a_0 = a, b_0 = b$, and define $c_k = (a_k + b_k)/2$. If $f(c_k) = 0, c_k = \alpha$, and the algorithm terminates. If $f(a_k)f(c_k) < 0$, $\alpha \in (a_k, c_k)$, so let $[a_{k+1}, b_{k+1}] = [a_k, c_k]$. For the other possible case, if $f(a_k)f(c_k) > 0$, $\alpha \in (c_k, b_k)$, so let $[a_{k+1}, b_{k+1}] = [c_k, b_k]$. $|b_k - a_k| = 2^{-k}|b_0 - a_0|$, so that the bisection algorithm is guaranteed to converge to some root of f on $[a, b]$.

If f has more than one root on $[a, b]$, a problem in [1, p. 35] asks which root the bisection algorithm usually locates. If f has n distinct, simple roots $x_1 < x_2 < \cdots < x_n$ on $[a, b]$ ($f(x_i) = 0$ and $f'(x_i) \neq 0$), then it is well-known that the bisection algorithm finds the even numbered roots with probability zero. This paper shows that the probability of finding the odd numbered roots is uniform.

Let C_n denote the class of continuous functions which satisfy $f(a)f(b) < 0$ with exactly n distinct, simple roots on (a, b). Let the roots of $f \in C_n$ be denoted by $x_1 < x_2 < \cdots < x_n$. We assume that the locations of the roots are independent and distributed according to a uniform random distribution on $[a, b]$. Let $x_0 = a$ and $x_{n+1} = b$. Let $P_{i,n}$ denote the probability that the bisection algorithm converges to the ith root of f, given that $f \in C_n$. Let $Q_{i,n}$ denote the probability that

$$x_i < c_0 = (a_0 + b_0)/2 < x_{i+1},$$

* Received by the editors September 8, 1975 and in revised form December 18, 1975.
† Department of Mathematics and Statistics, University of Nebraska, Lincoln, Nebraska 68588.

for $i = 0, 1, \cdots, n$, given that $f \in C_n$.

We first note that n is odd and that $P_{i,n} = 0$ for all even i since at each step, the bisection algorithm discards the subinterval of length $(b_k - a_k)/2$ which contains an even number of roots. Hence for even i, x_i will be found if and only if $c_k = (a_k + b_k)/2 = x_i$ at some step of the algorithm.

THEOREM. *For n odd,*

$$P_{i,n} = \begin{cases} \dfrac{2}{n+1} & \text{for } i \text{ odd,} \\ 0 & \text{for } i \text{ even.} \end{cases}$$

If i is even, $P_{i,n} = 0$ was shown above. The proof for i odd is by induction on n. $P_{1,1} = 1$, for if f has only one root on $[a, b]$, the bisection algorithm is guaranteed to find it.

Assume that

$$P_{i,m} = \begin{cases} \dfrac{2}{m+1} & \text{for } i \text{ odd,} \\ 0 & \text{for } i \text{ even,} \end{cases}$$

for all odd $m < n$.

$Q_{i,n} = \binom{n}{i}/2^n$ since it is equal to the probability that i of n trials (roots) land on the left half of $[a, b]$.

If $x_j < c_0 = (a_0 + b_0)/2 < x_{j+1}$ for some fixed even j, then the probability of finding x_i is given by

$$\begin{cases} P_{i-j,n-j} & \text{for } i > j, \\ 0 & \text{for } i \leq j, \end{cases}$$

since the bisection algorithm will proceed on the interval $[a_1, b_1] = [c_0, b_0]$, which contains $n - j$ roots of f.

Similarly, if $x_j < c_0 < x_{j+1}$ for some fixed odd j, then the probability of finding x_i is given by

$$\begin{cases} 0 & \text{for } i > j, \\ P_{i,j} & \text{for } i \leq j. \end{cases}$$

Then

$$P_{i,n} = \sum_{j=0}^{n} P(\text{finding } x_i \text{ given } x_j < c_0 < x_{j+1})$$

$$= \sum_{\substack{j=0 \\ j \text{ even}}}^{i-1} Q_{j,n} P_{i-j,n-j} + \sum_{\substack{j=i \\ j \text{ odd}}}^{n} Q_{j,n} P_{i,j}$$

$$= Q_{0,n} P_{i,n} + Q_{n,n} P_{i,n} + \sum_{\substack{j=2 \\ j \text{ even}}}^{i-1} Q_{j,n} P_{i-j,n-j}$$

$$+ \sum_{\substack{j=i \\ j \text{ odd}}}^{n-2} Q_{j,n} P_{i,j},$$

where we assume that $\sum_2^0 = \sum_n^{n-2} = 0$. Hence,

$$P_{i,n} = \frac{1}{1 - Q_{0,n} - Q_{n,n}} \left[\sum_{\substack{j=2 \\ j \text{ even}}}^{i-1} Q_{j,n}P_{i-j,n-j} + \sum_{\substack{j=i \\ j \text{ odd}}}^{n-2} Q_{j,n}P_{i,j} \right]$$

$$= \frac{1}{2^n - 2} \left[\sum_{\substack{j=2 \\ j \text{ even}}}^{i-1} \binom{n}{j} \frac{2}{n-j+1} + \sum_{\substack{j=i \\ j \text{ odd}}}^{n-2} \binom{n}{j} \frac{2}{j+1} \right]$$

$$= \frac{1}{2^n - 2} \left[\sum_{\substack{j=2 \\ j \text{ even}}}^{n-1} \binom{n+1}{j} \right] \frac{2}{n+1}$$

$$= \frac{2}{n+1} \frac{1}{2^n - 2} \sum_{\substack{j=2 \\ j \text{ even}}}^{n-1} \left[\binom{n}{j-1} + \binom{n}{j} \right] = \frac{2}{n+1}.$$

We have shown that if a function f has n distinct, simple roots, the bisection algorithm is equally likely to find each odd root.

REFERENCE

S. D. CONTE AND CARL DE BOOR, *Elementary Numerical Analysis: An Algorithmic Approach*, 2nd ed., McGraw-Hill, New York, 1965.

A SIMPLE THEOREM ON RIEMANN INTEGRATION, BASED ON CLASSROOM EXPERIENCE*

LANCE D. DRAGER†

Abstract. At the Georgia Institute of Technology, a computer program is used in freshman calculus which graphically illustrates upper and lower Riemann sums and generates values of their differences. The students often observe that the differences Δ_n seem to be proportional to $1/n$, where n is the number of subdivisions; but this is only approximate.

We make this rigorous by showing that

$$\Delta_n = \frac{V}{n} + O\left(\frac{1}{n^3}\right) \quad \text{as } n \to \infty,$$

for nice functions, where V is the total variation. The proof is simple, and is a nice illustration of the ideas of asymptotic analysis, and several other techniques of analysis.

One of the computer programs used as a classroom demonstration in freshman calculus at the Georgia Institute of Technology is a program called RIEMANN, which graphically illustrates the upper and lower Riemann sums. It also tabulates the difference of the upper and lower Riemann sums to illustrate that the difference becomes small as the number of subdivisions increases.

In fact, the students often spontaneously observe that the difference appears to be proportional to $1/n$ (where n is the number of subdivisions). A few of the more observant will point out that this is not quite correct and the difference is nearly, but not precisely,

---

*Received by the editors March 9, 1982, and in revised form August 10, 1982.
†Department of Mathematics, Texas Tech University Lubbock, Texas 79409.

proportional to $1/n$. We make this observation precise by giving a result on the asymptotic behavior of the difference of the upper and lower Riemann sums.

To fix our notation, we consider a continuous function $f: [0, 1] \rightarrow \mathbb{R}$. If n is a positive integer, we can partition the interval into n equal subintervals, each of length $\Delta x = 1/n$. The endpoints of the subintervals are

$$0 = x_0 < x_1 < \cdots < x_n = 1,$$

where $x_i = i/n$. We refer to $[x_{i-1}, x_i]$ as the ith subinterval. Define

$$M_i = \sup \{f(x) \mid x \in [x_{i-1}, x_i]\},$$
$$m_i = \inf \{f(x) \mid x \in [x_{i-1}, x_i]\}.$$

Then

$$U_n = \sum_{i=1}^{n} M_i \, \Delta x = \frac{1}{n} \sum_{i=1}^{n} M_i$$

is the upper Riemann sum, and

$$L_n = \sum_{i=1}^{n} m_i \, \Delta x = \frac{1}{n} \sum_{i=1}^{n} m_i$$

is the lower Riemann sum.

We set

$$\Delta_n = U_n - L_n = \frac{1}{n} \sum_{i=1}^{n} M_i - m_i,$$

and we wish to study the behavior of Δ_n as $n \rightarrow \infty$. The first column of Table 1 tabulates the values of Δ_n for a few values of n, for the function $f(x) = \sin (4\pi x)$, and one can immediately see that Δ_n is approximately proportional to $1/n$. To check this idea we tabulate $n\Delta_n$, which is not constant but seems to approach something near 8, so Δ_n becomes more nearly proportional to $1/n$ as n increases. (There is an approximation, besides round-off error, involved in computing the values in Table 1. We will discuss this issue later.)

TABLE 1

n	Δ_n	$n\Delta_n$	$\lvert 8/n - \Delta_n \rvert$	$n^2 \lvert 8 - n\Delta_n \rvert$
10	0.77963	7.79633	2.0367×10^2	20.367
20	0.39021	7.80422	9.7888×10^{-3}	78.310
30	0.26590	7.9772	7.5972×10^{-4}	20.513
40	0.20000	7.9999	1.6391×10^{-7}	0.0010681
50	0.15984	7.9918	1.6397×10^{-4}	20.497
60	0.13297	7.9781	3.6533×10^{-4}	78.911
70	0.11423	7.9958	5.9895×10^{-5}	20.545
80	0.09999	7.9999	1.6391×10^{-7}	0.085449
90	0.08886	7.9975	2.8118×10^{-5}	20.498
100	0.079921	7.99213	7.8670×10^{-5}	78.673

To analyze this situation, we will impose a restriction on our function f, which is satisfied for most functions encountered in freshman calculus, and for any function for which the computer program could draw a reasonable picture. This condition says that the montonicity of f changes only finitely many times.

Condition (FMC). There are points

$$0 = t_0 < t_1 < \cdots < t_K < t_{K+1} = 1$$

so that f is (weakly) monotone on $[t_{j-1}, t_j], j = 1, \cdots, K + 1$.

We may also assume that no t_j may be removed without destroying this property. We set $V = \sum_{j=1}^{K+1} |f(t_{j-1}) - f(t_j)|$. This is the total variation of f on $[0, 1]$, in the sense of the theory of functions of bounded variation. (For $\sin(4\pi x)$, $V = 8$). If f' is continuous, $V = \int_0^1 |f'(x)| \, dx$.

If f is monotone on $[0, 1]$, the values M_i and m_i are taken on at the endpoints of $[x_{i-1}, x_i]$ and the sum $\sum_{i=1}^{n} M_i - m_i = n\Delta_n$ telescopes to $|f(0) - f(1)| = V$. So for a monotone function, $\Delta_n = V/n$. We will see below that in general $\Delta_n \leq V/n$, and we wish to measure how much Δ_n deviates from V/n. (Column 3 of Table 1 tabulates $|\Delta_n - V/n|$.)

THEOREM. *Let f satisfy condition* (FMC). *If f is Hölder continuous of order α, $0 < \alpha \leq 1$. (i.e., there exists $C > 0$ so that $|f(x) - f(y)| \leq C|x - y|^\alpha$), then*

$$\Delta_n = \frac{V}{n} + O\left(\frac{1}{n^{1+\alpha}}\right).$$

If f has a continuous second derivative, this can be improved to

$$\Delta_n = \frac{V}{n} + O\left(\frac{1}{n^3}\right).$$

The last statement, for example, means there is a constant $C > 0$ so that

$$\left|\Delta_n - \frac{V}{n}\right| \leq \frac{C}{n^3},$$

so Δ_n lies inside an envelope closing down on V/n. This is the same as saying $n^3 |\Delta_n - (1/n)V| = n^2 |n\Delta_n - V|$ is bounded. (This quantity is displayed for the function $f(x) = \sin(4\pi x)$ in column 4 of Table 1.)

To prove the theorem, first note that if $Y = \{t_0, \cdots, t_{K+1}\} \cup \{z\}$ and Y is numbered as $0 = y_0 < y_1 < \cdots < y_q = 1$, we have $V = \sum_{i=1}^{q} |f(y_{i-1}) - f(y_i)|$. Indeed, if $z = t_j$ this result is trivial, while if $t_{j-1} < z < t_j$ we have replaced the term $|f(t_{j-1}) - f(t_j)|$ in the sum defining V by $|f(t_{j-1}) - f(z)| + |f(z) - f(t_j)|$. But since f is monotone on $[t_{j-1}, t_j]$, this telescopes back to $|f(t_{j-1}) - f(t_j)|$. Applying this argument repeatedly, we can add any number of points to $\{t_0, \cdots, t_{K+1}\}$ to get $\{y_0, \cdots, y_q\}$ and still have $\sum_{i=1}^{q} |f(y_{i-1}) - f(y_i)| = V$.

To simplify the notation, assume $K = 1$, i.e., there is only one point t_1 where the monotonicity of f changes. Suppose $t_1 \in [x_{k-1}, x_k]$. We want to estimate $|V - n\Delta_n|$. We have, by our previous observation,

$$V = |f(0) - f(x_1)| + \cdots + |f(x_{k-1}) - f(t_1)| + |f(t_1) - f(x_k)|$$
$$+ \cdots + |f(x_{n-1}) - f(1)|,$$

while $n\Delta_n = \sum_{i=1}^{n} M_i - m_i$. If f is monotone on $[x_{i-1}, x_i]$, M_i and m_i occur at the endpoints and $M_i - m_i = |f(x_{i-1}) - f(x_i)|$. So, if we compute $V - n\Delta_n$, the terms in each sum corresponding to the intervals $[x_{i-1}, x_i]$, $i \neq k$, cancel out. If t_1 is an endpoint of $[x_{k-1}, x_k]$ all the terms cancel, while otherwise we have

$$V - n\Delta_n = |f(x_{k-1}) - f(t_1)| + |f(t_1) - f(x_k)| - (M_k - m_k).$$

Since the monotonicity changes at t_1, one of M_k or m_k occurs at t_1 while the other occurs at one or both of the endpoints x_{k-1}, x_k. Thus $(M_k - m_k)$ cancels with one of the two other terms. Let $\tilde{x}(1, n)$ denote the endpoint in the term which does not cancel. (If t_1 is an endpoint, set $\tilde{x}(1, n) = t_1$.) We thus have

$$V - n\Delta_n = |f(t_1) - f(\tilde{x}(1, n))|.$$

If $K > 1$, the same analysis may be applied as soon as n is large enough that each (x_{i-1}, x_i) contains at most one t_j. Then,

$$(1) \qquad V - n\Delta_n = \sum_{j=1}^{K} |f(t_j) - f(\tilde{x}(j, n))|,$$

where $\tilde{x}(j, n)$ is an endpoint of the interval $[x_{i-1}, x_i]$ containing t_j (with $\tilde{x}(j, n) = t_j$, if t_j is one of the endpoints). (1) implies, in particular, that $V - n\Delta_n \geq 0$ so $\Delta_n \leq V/n$. But if we can estimate $|f(t_j) - f(\tilde{x}(j, n))|$, we will get more precise information. Since $|t_j - \tilde{x}(j, n)| \leq \Delta x = 1/n$, if f is Hölder continuous of order α, we get

$$|f(t_j) - f(\tilde{x}(j, n))| \leq \frac{C}{n^\alpha}.$$

Thus,

$$|V - n\Delta_n| = V - n\Delta_n \leq \frac{KC}{n^\alpha}$$

for large n, and for all n if we increase C. In this case,

$$\left| \Delta_n - \frac{V}{n} \right| \leq \frac{CK}{n^{1+\alpha}}$$

or

$$\Delta_n = \frac{V}{n} + O(1/n^{1+\alpha}).$$

If f' exists, each t_j $(j = 1, \cdots, K)$ must be a critical point of f, since the monotonicity changes at t_j. If f'' is continuous, we can use Taylor's formula with remainder of order 2 to show $|f(t_j) - f(x)| \leq C|t_j - x|^2$.

In this case

$$|V - n\Delta_n| \leq \frac{CK}{n^2}$$

for large n, so

$$(2) \qquad \Delta_n = \frac{V}{n} + O\left(\frac{1}{n^3}\right).$$

Indeed, it is easy to see that if f' is Hölder continuous of order α, $\Delta_n = V/n + O(1/n^{2+\alpha})$.

If f is infinitely differentiable, we can use Taylor's formula in (1) to develop a complete asymptotic expansion of Δ_n involving the rather messy functions $t_j - \tilde{x}(j, n)$ of n. This does not seem interesting enough to write out, except to note that it does show that (2) cannot, in general, be improved to an error of $O(1/n^4)$ unless all $f''(t_j) = 0$, and this argument can be extended to higher derivatives.

Using our theorem above, and a somewhat more delicate analysis using Taylor's theorem on each subinterval, we can find the asymptotic behavior of U_n and L_n. In fact, if f satisfies condition (FMC) and has a continuous third derivative,

$$(3) \qquad U_n = \int_0^1 f(x)\,dx + \frac{V}{2n} + \frac{1}{12n^2} \int_0^1 f''(x)\,dx + O\left(\frac{1}{n^3}\right),$$

$$(4) \qquad L_n = \int_0^1 f(x)\,dx - \frac{V}{2n} + \frac{1}{12n^2} \int_0^1 f''(x)\,dx + O\left(\frac{1}{n^3}\right).$$

To write a computer program to calculate Δ_n, V_n, L_n, for fairly general functions, one has to approximate the values of M_i, m_i. An obvious approximation scheme (used by RIEMANN and in computing Table 1) is the following: for some fixed integer p, divide each of the subintervals $[x_{i-1}, x_i]$ into p subintervals, of length $1/pn$, which we refer to for convenience as the "sub-subintervals," and take the maximum and minimum of the values of the function at the endpoints of these sub-subintervals as the approximate values for M_i, m_i. Let us denote these approximate values by $\tilde{M}_{i,p}$, $\tilde{m}_{i,p}$ and the corresponding approximations for U_n, L_n, Δ_n by $\tilde{U}_{n,p}$, $\tilde{L}_{n,p}$, $\tilde{\Delta}_{n,p}$. To analyze the error in these approximations, suppose f has a continuous third derivative and satisfies condition (FMC), and that n is large enough that each subinterval $[x_{i-1}, x_i]$ contains at most one t_j. If f is monotone on $[x_{i-1}, x_i]$ the approximations $\tilde{M}_{i,p}$, $\tilde{m}_{i,p}$ will be exact (since we are checking the values of f at x_{i-1} and x_i), while if $t_j \in [x_{k-1}, x_k]$ one of the approximations $\tilde{M}_{i,p}$, $\tilde{m}_{i,p}$ will be exact, and the other will be $f(\xi)$, where ξ is one of the endpoints of the sub-subinterval which contains t_j. Thus $|\xi - t_j| \le 1/pn$ and, by Taylor's theorem, $|f(\xi) - f(t_j)| \le c/p^2n^2$. Since at most K of the approximations are not exact, we have $|U_n - \tilde{U}_{n,p}| \le Kc/p^2n^3$, or $\tilde{U}_{n,p} = U_n + O(1/p^2n^3)$. Similarly $\tilde{L}_{n,p} = L_n + O(1/p^2n^3)$ and $\tilde{\Delta}_{n,p} = \Delta_n + O(1/p^2n^3)$. As expected, for n fixed, the approximations become more accurate as $p \to \infty$, while for a fixed p, $\tilde{L}_{n,p} = L_n + O(1/n^3)$ and $\tilde{\Delta}_{n,p} = \Delta_n + O(1/n^3)$. Thus the approximations satisfy the same asymptotic formulas (2), (3), (4) as Δ_n, U_n and L_n, and it is no accident that the values in Table 1 show the correct behavior, even though they are only approximate.

The asymptotic formulas (3), (4) show that $U_n = \int_0^1 f(x)\,dx + O(1/n)$, and $L_n = \int_0^1 f(x)\,dx + O(1/n)$ and the approximations $\tilde{U}_{n,p}$, $\tilde{L}_{n,p}$ satisfy the same formulas. Thus these are not very attractive numerical integration schemes. If we let $A_n = 1/2(L_n + U_n)$, $\tilde{A}_{n,p} = 1/2(\tilde{L}_{n,p} + \tilde{U}_{n,p})$ (3) and (4) give us

$$(5) \qquad A_n = \int_0^1 f(x)\,dx + \frac{1}{12n^2} \int_0^1 f''(x)\,dx + O\left(\frac{1}{n^3}\right),$$

$$(6) \qquad \tilde{A}_{n,p} = \int_0^1 f(x)\,dx + \frac{1}{12n^2} \int_0^1 f''(x)\,dx + O\left(\frac{1}{n^3}\right),$$

so $\tilde{A}_{n,p}$ looks more attractive as a numerical integration scheme, since $\tilde{A}_{n,p} = \int_0^1 f(x)\,dx + O(1/n^2)$. If we take $p = 1$,

$$\tilde{A}_{n,1} = \frac{1}{n} \sum_{i=1}^{n} \frac{f(x_{i-1}) + f(x_i)}{2}$$

is just the familiar trapezoidal approximation. For purposes of practical numerical integration, there would not be much point in taking $p > 1$, since computing $\tilde{A}_{n,p}$ will involve much more computation, with no improvement in the order $O(1/n^2)$ of the error.

A NOTE ON THE ASYMPTOTIC STABILITY OF PERIODIC SOLUTIONS OF AUTONOMOUS DIFFERENTIAL EQUATIONS*

H. ARTHUR DE KLEINE

Abstract. In this note we provide two examples to illustrate that the asymptotic stability of a periodic solution of an autonomous differential equation is not necessarily determined by the sign of the real part of the eigenvalues of the linear variational equation.

1980 AMS mathematics classifications. 34C25, 34D05

Key words. asymptotic stability, periodic solutions, linear variational equation, Floquet theory, characteristic exponents

1. Introduction. In this note, two examples of relatively simple ordinary differential equations are given to illustrate that the asymptotic stability of a periodic solution of an autonomous differential equation is not necessarily determined by the sign of the real part of the eigenvalues of the linear variational equation. These examples were obtained during a summer research project at the Lawrence Livermore Laboratory MFE Computer Center involving the study of numerical algorithms for highly oscillatory differential equations [3], [12]. Many numerical schemes for solving ordinary differential equations use the eigenvalue structure of the linear variational equation to estimate the stability of the solution in question. The examples presented here were used to estimate how long the algorithms would continue before the instability was detected. These examples may be of use to numerical analysts because both the characteristic roots of the linear variational equation and the characteristic exponents are easily calculated.

Motivation for these examples comes from a study of diffusively coupled chemical oscillators, where the stability question has received considerable interest. See for example, [4], [6], [7], [9], [11], [14]. Such a system can be represented by a pair of equations,

$$x' = F(x) + K_1(y - x),$$

$$y' = G(y) + K_2(x - y).$$

The first example given represents a pair of stable limit cycles coupled by a nonstabilizing coupling and the second example represents a stable oscillator and an unstable oscillator coupled by a stabilizing coupling.

2. Preliminary considerations. Let $\tilde{x}(t)$ be a nonconstant periodic solution of period τ of the n-dimensional autonomous differential equation

$$(1) \qquad\qquad x' = F(x).$$

We assume that F is continuously differentiable. It is well known [2], [5], [15] that the asymptotic stability of $\tilde{x}(t)$ is determined in part by the stability properties of the null solution of the corresponding linear variational equation

$$(2) \qquad\qquad y' = F_x[\tilde{x}(t)]y.$$

Let $Y(t)$ be the fundamental matrix solution of the variational equation satisfying

---

*Received by the editors April 10, 1983, and in final form September 30, 1983. This research was supported by the Northwest College and University Association for Science (University of Washington) under Contract DE-AM06-76-RL02225 with the U.S. Department of Energy.

†Mathematics Department, California Polytechnic State University, San Luis Obispo, California 93407.

the initial condition $Y(0) = I$. By Floquet's theorem, $Y(t) = P(t)e^{tB}$ where $P(t)$ has period τ in t, $P(\tau) = I$, and B is a constant $n \times n$ matrix. The characteristic roots of B, called the *characteristic exponents* of (2) and denoted by μ_i ($i = 1, 2, \cdots, n$), are uniquely determined mod $(2\pi i/\tau)$. The real part of the characteristic exponents determines the asymptotic stability of $Y(t)$. Note that $\tilde{x}'(t)$ satisfies the variational equation (2) and thus 0 is a characteristic exponent. We let $\mu_1 = 0$.

THEOREM. *If* 0 *is a simple characteristic exponent of the variational equation and all other characteristic exponents have their real part less than* $-\gamma < 0$, *then* $\tilde{x}(t)$ *is orbitally asymptotically stable with asymptotic phase. If the real part of one of the characteristic exponents is greater than zero,* $\tilde{x}(t)$ *is unstable.*

As the Floquet theorem is constructive, the characteristic exponents are determined only after the fundamental matrix $Y(t)$ is known and there is no obvious relation between the characteristic exponents of (2) and the eigenvalues of $F_x[\tilde{x}(t)]$, henceforth denoted by $\lambda_i(t)$ ($i = 1, 2, \cdots, n$). Hale [5] gives an example by Markus and Yamabe [10] of a continuous 2×2 periodic matrix $A(t)$ with constant eigenvalues, $\lambda_i(t) = [-1 \pm i\sqrt{7}]/4$, such that the zero solution of the linear equation $y' = A(t)y$ is unstable. We remark in passing that Coppel [2] provides a direct approach to the analysis of 2×2 linear systems. Laloy [8], [13] has provided an extension of the Markus and Yamabe example.

These observations prompted a search for an example of an autonomous differential equation (1) with an unstable nonconstant periodic solution (where zero is a simple characteristic exponent of the linear variational equation (2)) such that the real parts of all eigenvalues of the variational equation (2) are less than $-\gamma < 0$.

The dimension of such an example will necessarily require $n \geq 3$.

$$\det Y(\tau) = e^{\tau\,\mathrm{tr}\,B}.$$

By Jacobi's formula

$$\det Y(\tau) = \exp \int_0^\tau \mathrm{div}\, F\, dt,$$

where

$$\mathrm{div}\, F = \frac{\partial F_1}{\partial x_1} + \frac{\partial F_2}{\partial x_2} + \cdots + \frac{\partial F_n}{\partial x_n}$$

and the partial derivatives are evaluated at $\tilde{x}(t)$. Thus, mod $(2\pi i/\tau)$,

$$\mathrm{tr}\, B = \tau^{-1} \int_0^\tau \mathrm{div}\, F\, dt = \tau^{-1} \int_0^\tau \sum_{i}^{n} \lambda_i(t)\, dt.$$

If $n = 2$,

$$\mathrm{Re}\,(\mu_2) = \mathrm{Re}\,(\mathrm{tr}\, B) < 0.$$

An example, extending both the Markus and Yamabe example and the Laloy example, will be used in the construction of our model. The matrix

$$R(t) = \begin{pmatrix} e^{\alpha t}\cos \gamma t & -e^{\beta t}\sin \gamma t \\ e^{\alpha t}\sin \gamma t & e^{\beta t}\cos \gamma t \end{pmatrix}$$

is a fundamental matrix solution of the linear equation

(3) $r' = A(t)r,$

where

$$A(t) = \begin{pmatrix} \alpha \cos^2 \gamma t + \beta \sin^2 \gamma t & (\alpha - \beta) \sin \gamma t \cos \gamma t - \gamma \\ (\alpha - \beta) \sin \gamma t \cos \gamma t + \gamma & \alpha \sin^2 \gamma t + \beta \cos^2 \gamma t \end{pmatrix}.$$

The characteristic exponents of (3) are α and β while the eigenvalues of $A(t)$ are

$$\lambda_i(t) = \frac{(\alpha + \beta) \pm \sqrt{(\alpha - \beta)^2 - 4\gamma^2}}{2}.$$

These eigenvalues have a negative real part if and only if $\alpha + \beta < 0$ and $\gamma^2 > -\alpha\beta$. Furthermore, they are complex if $(\alpha - \beta)^2 < 4\gamma^2$.

The stable and unstable sinusoidal oscillators,

$$
\begin{aligned}
x' &= -y + x(1 - x^2 - y^2), \\
y' &= x + y(1 - x^2 - y^2),
\end{aligned}
\tag{4}
$$

and

$$
\begin{aligned}
\xi' &= -\eta - \xi(1 - \xi^2 - \eta^2), \\
\eta' &= \xi - \eta(1 - \xi^2 - \eta^2),
\end{aligned}
\tag{5}
$$

are pedagogical examples [1] of stable and unstable limit cycles, respectively. For both of these examples the limit cycle can be represented by $(\bar{x}(t), \bar{y}(t)) = (\cos t, \sin t)$. A transformation to polar coordinates provides an analytical solution of these equations and a direct analysis of the stability properties. The corresponding linear variational equations are given by

$$Y' = \begin{vmatrix} -2 \cos^2 t & -1 - 2 \cos t \sin t \\ 1 - 2 \cos t \sin t & -2 \sin^2 t \end{vmatrix} Y$$

and

$$Y' = \begin{vmatrix} 2 \cos^2 t & -1 + 2 \cos t \sin t \\ 1 + 2 \cos t \sin t & +2 \sin^2 t \end{vmatrix} Y.$$

From our previous observations, $(-\sin t, \cos t)$ and $(e^{-2t} \cos t, e^{-2t} \sin t)$ are solutions of the first variational equation; $\lambda_1(t) = \lambda_2(t) = -1$ are the corresponding eigenvalues. Similarly, $(-\sin t, \cos t)$ and $(e^{2t} \cos t, e^{2t} \sin t)$ are solutions of the second variational equation while $\lambda_1(t) = \lambda_2(t) = +1$ are the eigenvalues.

3. First example. Consider two stable sinusoidal oscillators coupled by a nonstabilizing linear difference coupling,

$$
\begin{aligned}
x' &= -y + x(1 - x^2 - y^2), \\
y' &= x + y(1 - x^2 - y^2), \\
\xi' &= -\eta + \xi(1 - \xi^2 - \eta^2) - K(x - \xi), \\
\eta' &= \xi + \eta(1 - \xi^2 - \eta^2) - K(y - \eta).
\end{aligned}
$$

The function $(\bar{x}, \bar{y}, \bar{\xi}, \bar{\eta}) = (\cos t, \sin t, \cos t, \sin t)$ is a periodic solution. The corresponding linear variational equation is given by

$$Y' = \begin{bmatrix} -2\cos^2 t & -1 - 2\cos t \sin t & 0 & 0 \\ 1 - 2\cos t \sin t & -2\sin^2 t & 0 & 0 \\ -K & 0 & K - 2\cos^2 t & -1 - 2\cos t \sin t \\ 0 & -K & 1 - 2\cos t \sin t & K - 2\sin^2 t \end{bmatrix} Y.$$

Again making use of our previous observations, the matrix

$$Y(t) = \begin{bmatrix} -\sin t & e^{-2}t \cos t & 0 & 0 \\ \cos t & e^{-2}t \sin t & 0 & 0 \\ -\sin t & e^{-2}t \cos t & -e^{Kt} \sin t & e^{Kt} e^{-2t} \cos t \\ \cos t & e^{-2}t \sin t & e^{Kt} \cos t & e^{Kt} e^{-2t} \sin t \end{bmatrix}$$

is a fundamental solution of the variational equation and the eigenvalues are -1, -1, $K - 1$, $K - 1$. For $0 < K < 1$, all eigenvalues are negative and yet the periodic solution $(\bar{x}, \bar{y}, \bar{\xi}, \bar{\eta})$ is unstable.

4. Second example. Consider a stable oscillator and an unstable oscillator coupled by a stabilizing difference coupling,

$$x' = -y + x(1 - x^2 - y^2),$$

$$y' = x + y(1 - x^2 - y^2),$$

$$\xi' = -\eta - \xi(1 - \xi^2 - \eta^2) + K(x - \xi),$$

$$\eta' = \xi - \eta(1 - \xi^2 - \eta^2) + K(y - \eta).$$

The function $(\bar{x}, \bar{y}, \bar{\xi}, \bar{\eta}) = (\cos t, \sin t, \cos t, \sin t)$ is a periodic solution. The corresponding linear variational equation is given by

$$Y' = \begin{bmatrix} -2\cos^2 t & -1 - 2\cos t \sin t & 0 & 0 \\ 1 - 2\cos t \sin t & -2\sin^2 t & 0 & 0 \\ K & 0 & -K + 2\cos^2 t & -1 + 2\cos t \sin t \\ 0 & K & 1 + 2\cos t \sin t & -K + 2\sin^2 t \end{bmatrix} Y.$$

A fundamental solution for this variational equation is given by

$$Y(t) = \begin{bmatrix} -\sin t & e^{-2t} \cos t & 0 & 0 \\ \cos t & e^{-2t} \sin t & 0 & 0 \\ -\sin t & \phi_1 & -e^{-Kt} \sin t & e^{2t} e^{-Kt} \cos t \\ \cos t & \phi_2 & e^{-Kt} \cos t & e^{2t} e^{-Kt} \sin t \end{bmatrix}$$

where $\phi = [K/(K - 4)]e^{-2t}(\cos t, \sin t)$ if $K \neq 4$ and $te^{-2t}(\cos t, \sin t)$ if $K = 4$. The periodic solution in question is stable if and only if $K > 2$. The eigenvalues of the linear variational equation are given by -1, -1, $1 - K$, $1 - K$. Thus for $1 < K < 2$ all eigenvalues are negative, but the periodic solution is unstable.

Acknowledgment. The author wishes to express his thanks to Dr. Kirby Fong and the staff of the Magnetic Fusion Exchange Computer Center at Lawrence Livermore Laboratory for their support.

REFERENCES

[1] W. E. BOYCE AND R. C. DIPRIMA, *Elementary Differential Equations and Boundary Value Problems,* 3rd ed., John Wiley, New York, 1977.

[2] W. A. COPPEL, *Stability and Asymptotic Behavior of Differential Equations,* D. C. Heath, Boston, 1965.

[3] C. W. GEAR AND K. A. GALLIVAN, *Automatic methods for highly oscillatory ordinary differential equations,* Numerical Analysis (Dundee, 1981), Lecture Notes in Mathematics 912, Springer-Verlag, Berlin, 1982.

[4] M. GRATTAROLA AND V. TORRE, *Necessary and sufficient conditions for synchronization of nonlinear oscillators with a given class of coupling,* IEEE Trans. Circuit. Syst., 24 (1977), pp. 209–215.

[5] J. K. HALE, *Ordinary Differential Equations,* Wiley-Interscience, New York, 1969.

[6] T. KAWAHARA, *Coupled van der Pol oscillators—a model of excitatory and inhibitory neural interactions,* Biol. Cybern., 39 (1980), pp. 37–43.

[7] M. KAWATO AND R. SUZUKI, *Two coupled neural oscillators as a model of the circadian pacemaker,* J. Theoret. Biol., 86 (1980), pp. 547–575.

[8] M. LALOY, *Problèmes d'instabilité pour des équations différentielles ordinaires et fonctionnelles,* Thése de doctorat, Louvain-La-Neuve.

[9] D. A. LINKENS, *The stability of entrainment conditions for RLC coupled van der Pol oscillators used as a model for intestinal electrical rhythms,* Bull. Math. Biol, 39 (1977), pp. 359–372.

[10] L. MARKUS AND H. YAMABE, *Global stability criteria for differential systems,* Osaka Math. J., 12 (1960), pp. 305–317.

[11] H. G. OTHMER AND L. E. SCRIVEN, *Instability and dynamic pattern in cellular networks,* J. Theoret. Biol., 32 (1971), pp. 507–537.

[12] L. R. PETZOLD, *An efficient method for highly oscillatory ordinary differential equations,* SIAM J. Numer. Anal., 18 (1981), pp. 455–479.

[13] N. ROUCHE, P. HABETS AND M. LALOY, *Stability Theory by Liapunov's Direct Method,* Applied Math. Sciences 22, Springer-Verlag, New York, 1977.

[14] S. SMALE, *A mathematical model of two cells via Turing's equations,* Lectures on Mathematics in the Life Sciences 6, J. Cowan ed., American Mathematical Society, Providence, RI, 1974.

[15] M. URABE, *Nonlinear autonomous oscillations,* Analytical Theory, Academic Press, New York, 1967.

APPROXIMATING FACTORIALS:
A DIFFERENCE EQUATION APPROACH*

A. C. KING[†] AND M. E. MORTIMER[†]

Abstract. An algebraic derivation of Stirling's formula for $n!$ is developed based on the asymptotic solution of a difference equation.

Key words. factorial, asymptotic expansion, difference equation

Elementary methods of approximating factorials are of interest since they not only provide an indication of how quickly the factorials grow, but also usually contain some

---

* Received by the editors February 12, 1985, and in final form April 27, 1985.

† Department of Mathematics, City of Birmingham Polytechnic, Birmingham, B42 2SU, United Kingdom.

interesting mathematical ideas. The following novel method, based upon a difference equation, provides the first few terms in the asymptotic (Stirling) approximation to the factorial $n! = \sqrt{2\pi}\, n^{n+1/2} e^{-n} \cdots$. By definition,

$$n! = n(n-1)(n-2) \cdots (n-(n-1))$$

and this can be rewritten as

$$\frac{n!}{n^n} = \left(1-\frac{1}{n}\right)\left(1-\frac{2}{n}\right) \cdots \left(1-\left(\frac{n-1}{n}\right)\right).$$

The R.H.S. is now a product of factors all less than 1 and thus converges to zero; to discover how quickly or slowly this product converges some algebraic rearrangements are necessary. Define

$$\Lambda(n) = \left(1-\frac{1}{n}\right)\left(1-\frac{2}{n}\right) \cdots \left(1-\left(\frac{n-1}{n}\right)\right)$$

$$= \left(1-\frac{1}{n}\right)^{n-1}\left(1-\frac{1}{n-1}\right)\left(1-\frac{2}{n-1}\right) \cdots \left(1-\left(\frac{n-2}{n-1}\right)\right).$$

Thus,

$$\Lambda(n) = \left(1-\frac{1}{n}\right)^{n-1}\Lambda(n-1).$$

From this equation it is possible to conclude

$$\lim_{n \to \infty}\left\{\frac{\Lambda(n)}{\Lambda(n-1)}\right\} = e^{-1}.$$

To obtain a more precise estimate of $\Lambda(n)$ it is convenient to work with the logarithm of the difference equation. Writing $\overline{\Lambda}(n) = \log \Lambda(n)$, we have

$$\overline{\Lambda}(n) = (n-1)\log\left(1-\frac{1}{n}\right) + \overline{\Lambda}(n-1).$$

Expanding the logarithm for large n using the standard $\log(1+x)$ result,

(1) $$\overline{\Lambda}(n) - \overline{\Lambda}(n-1) = -1 + \frac{1}{2n} + \frac{1}{6n^2} + O\left(\frac{1}{n^3}\right).$$

This asymptotic difference equation is now solved subject to $\overline{\Lambda}(n) \to -\infty$ as $n \to \infty$.
    First a solution to the homogeneous problem

$$\overline{\Lambda}(n) - \overline{\Lambda}(n-1) = 0$$

is found. By inspection this has the solution $\overline{\Lambda}(n) = \text{Constant} = C$.
    The construction of a particular solution for the equation is more difficult. The standard finite difference method is algebraically complicated so the following method is used. Let

$$\overline{\Lambda}(n) = a_1 b_1(n) + a_2 b_2(n) + \cdots + a_m b_m(n) + \cdots$$

where the a's are constants and the $b(n)$'s are a sequence of functions which, to give an asymptotic series with $\overline{\Lambda}(n) \to -\infty$ as $n \to \infty$, must satisfy $b_1(n) > b_2(n) > \cdots > b_m(n) > \cdots$ and at least $b_1(n) \to \infty$ as $n \to \infty$. Equation (1) now takes the form

$$a_1(b_1(n) - b_1(n-1)) + a_2(b_2(n) - b_2(n-1)) + \cdots = -1 + \frac{1}{2n} + \frac{1}{6n^2} + O\left(\frac{1}{n^3}\right).$$

The solutions to this set of equations are again found by inspection, subject to the conditions above.

Order 1.

$$a_1(b_1(n) - b_1(n-1)) = -1: \quad \text{choose } b_1(n) = n, \quad a_1 = -1.$$

Order $1/n$.

$$a_2(b_2(n) - b_2(n-1)) = \frac{1}{2n}: \quad \text{choose } b_2(n) = \log n, \quad a_2 = \frac{1}{2}.$$

Note that after expansion of this logarithm for large n this choice of $b_2(n)$ introduces an error of magnitude $1/4n^2 + 1/6n^3$ which must be accounted for when solving at order $1/n^2$.

Order $1/n^2$.

$$a_3(b_3(n) - b_3(n-1)) = \left(\frac{1}{6} - \frac{1}{4}\right) \cdot \frac{1}{n^2}: \quad \text{choose } b_3(n) = \log\left(1 + \frac{1}{n}\right), \quad a_3 = \frac{1}{12}.$$

This process can be continued to a higher order, in fact $b_4(n) = \log(1 + 1/n^2)$, but sufficient terms have now been obtained for the purposes of this article. We have

$$\overline{\Lambda}(n) = C - n + \frac{1}{2}\log n + \frac{1}{12}\log\left(1 + \frac{1}{n}\right) + O\left(\frac{1}{n^2}\right),$$

$$\Lambda(n) = \exp\left(C - n + \frac{1}{2}\log n + \frac{1}{12}\log\left(1 + \frac{1}{n}\right) + O\left(\frac{1}{n^2}\right)\right).$$

Writing C for e^C and using

$$\left(1 + \frac{1}{n}\right)^{1/12} = 1 + \frac{1}{12n} + O\left(\frac{1}{n^2}\right),$$

we finally obtain

$$\Lambda(n) = Cn^{1/2}e^{-n}\left(1 + \frac{1}{12n} + O\left(\frac{1}{n^2}\right)\right),$$

or

$$n! = Cn^{n+1/2}e^{-n}\left(1 + \frac{1}{12n} + O\left(\frac{1}{n^2}\right)\right).$$

Having obtained this approximation for $n!$, the standard method for evaluating C is to use the result

$$\sqrt{\pi} = \lim_{n \to \infty} \left\{ \frac{1 \cdot 2 \cdot 4 \cdot 6 \cdot \ldots \cdot 2n}{n^{1/2} 1 \cdot 3 \cdot 5 \cdot \ldots \cdot 2n-1} \right\} = \lim_{n \to \infty} \left\{ \frac{2^{2n}(n!)^2}{(2n)! n^{1/2}} \right\} = \frac{C}{\sqrt{2}},$$

giving $C = \sqrt{2\pi}$ and

$$n! = \sqrt{2\pi}\, n^{n+1/2} e^{-n} \left(1 + \frac{1}{12n} + O\left(\frac{1}{n^2} \right) \right).$$

Supplementary References
Numerical Analysis

[1] F. S. Acton, *Analysis of Straight-Line Data*, Dover, N.Y., 1959.

[2] ———, *Numerical Methods that Work*, Harper and Row, N.Y., 1970.

[3] J. H. Ahlberg, E. N. Nilson and J. L. Walsh, *The Theory of Splines and their Applications*, Academic, N.Y., 1967.

[4] W. F. Ames, *Nonlinear Partial Differential Equations in Engineering, II*, Academic, N.Y., 1972.

[5] ———, *Numerical Methods for Partial Differential Equations*, Barnes & Noble, N.Y., 1969.

[6] L. Auslander and R. Tolimieri, *"Is computing with the finite Fourier transform pure or applied mathematics?"* Bull. AMS (1979) pp. 847-897.

[7] P. B. Bailey, L. F. Shampine and D. E. Waltman, *Nonlinear Two Point Boundary Value Problems*, Academic, N.Y., 1968.

[8] R. Bellman, K. L. Cooke and J. A. Lockette, *Algorithms, Graphs, and Computers*, Academic, N.Y., 1970.

[9] I. S. Berezin and N. P. Zhidkov, *Computing Methods, I, II*, Pergamon, Oxford, 1965.

[10] G. H. Brown, Jr., *"On Halley's variation of Newton's method,"* AMM (1977) pp. 726-728.

[11] R. L. Burden, J. D. Faires and A. C. Reynolds, *Numerical Analysis*, Prindle, Weber & Schmidt, Boston, 1978.

[12] E. W. Cheney, *Introduction to Approximation Theory*, McGraw-Hill, N.Y., 1966.

[13] A. M. Cohen, ed., *Numerical Analysis*, McGraw-Hill, N.Y., 1973.

[14] L. Collatz, *Numerical Treatment of Differential Equations*, Springer-Verlag, Berlin, 1960.

[15] E. T. Copson, *Asymptotic Expansions*, Cambridge University Press, Cambridge, 1965.

[16] G. Dahlquist and A. Bjorck, *Numerical Methods*, Prentice-Hall, N.J., 1974.

[17] P. J. Davis, *Interpolation and Approximation*, Blaisdell, Waltham, 1963.

[18] P. J. Davis and P. Rabinowitz, *Numerical Integration*, Blaisdell, Waltham, 1967.

[19] N. G. De Bruijn, *Asymptotic Methods in Analysis*, Interscience, N.Y., 1961.

[20] F. De Kok, *"On the method of stationary phase for multiple integrals,"* SIAM J. Math. Anal. (1971) pp. 76-104.

[21] R. B. Dingle, *Asymptotic Expansions: Their Derivation and Interpretation*, Academic, N.Y., 1973.

[22] A. Erdelyi, *Asymptotic Expansions*, Dover, N.Y., 1956.

[23] A. Finbow, *"The bisection method: A best case analysis,"* AMM (1985) pp. 285-286.

[24] G. E. Forsythe, *"Generation and use of orthogonal polynomials for data fitting with a digital computer,"* SIAM J. Appl. Math. (1957) pp. 74-88.

[25] G. E. Forsythe, M. A. Malcolm and C. B. Moler, *Computer Methods for Mathematical Problems*, Prentice-Hall, N.J., 1977.

[26] G. E. Forsythe and W. R. Wasow, *Finite Difference Methods for Partial Differential Equations*, Wiley, N.Y., 1960.

[27] M. R. Garey and D. S. Johnson, *"Approximation algorithms for combinatorial problems: An annotated bibliography,"* in J. F. Traub, ed., *Algorithms and Complexity: New Directions and Recent Results*, Academic, N.Y., 1976, pp. 41-52.

[28] ———, *Computers and Intractability: A Guide to the Theory of NP-Completeness*, Freeman, San Francisco, 1979.

[29] J. M. Hammersley, *"Monte Carlo methods for solving multivariable problems,"* Ann. N.Y. Acad. Sci. (1960) pp. 844-874.

[30] R. W. Hamming, *Numerical Methods for Scientists and Engineers*, McGraw-Hill, N.Y., 1973.

[31] D. R. Hartree, *Numerical Analysis*, Clarendon Press, Oxford, 1958.

[32] P. Henrici, *Elements of Numerical Analysis*, Wiley, N.Y., 1964.

[33] ———, *Essentials of Numerical Analysis with Pocket Calculator Demonstration*, Wiley, N.Y., 1982.

[34] F. B. Hildebrand, *Introduction to Numerical Analysis*, McGraw-Hill, N.Y., 1974.

[35] A. S. Householder, *Principles of Numerical Analysis*, McGraw-Hill, N.Y., 1953.

[36] E. Isaacson and H. B. Keller, *Analysis of Numerical Methods*, Wiley, N.Y., 1966.

[37] L. V. Kantorovich and V. I. Krylov, *Approximate Methods of Higher Analysis*, Noordhoff, Groningen, 1958.

[38] S. KARLIN AND W. J. STUDDEN, *Tchebycheff Systems: With Applications in Analysis and Statistics*, Interscience, N.Y., 1966.

[39] D. KATZ, *"Optimal quadrature points for approximating integrals when function values are observed with error,"* MM (1984) pp. 284–290.

[40] H. B. KELLER, *Numerical Methods for Two-Point Boundary Value Problems*, Blaisdell, Waltham, 1968.

[41] M. S. KLAMKIN, *"Transformation of boundary value problems into initial value problems,"* J. Math. Anal. Appl. (1970) pp. 308–330.

[42] D. E. KNUTH, *"Algorithms,"* Sci. Amer. (1977) pp. 63–80, 148.

[43] ———, *The Art of Computer Programming I, II, III*, Addison-Wesley, Reading, 1973.

[44] Z. KOPAL, *Numerical Analysis*, Chapman & Hall, London, 1961.

[45] C. LANCZOS, *Applied Analysis*, Prentice-Hall, N.J., 1956.

[46] G. G. LORENTZ, *Approximation of Functions*, Holt, Rinehart and Winston, N.Y., 1966.

[47] J. E. MCKENNA, *"Computers and experimentation in mathematics,"* AMM (1972) pp. 294–295.

[48] G. MEINHARDUS, *Approximations of Functions: Theory and Numerical Methods*, Springer-Verlag, N.Y., 1967.

[49] G. H. MEYER, *Initial Value Methods for Boundary Value Problems*, Academic, N.Y., 1973.

[50] G. MIEL, *"Calculator calculus and roundoff error,"* AMM (1980) pp. 243–351.

[51] J. J. H. MILLER, ed., *Topics in Numerical Analysis I, II, III*, Academic, N.Y., 1972, 1974 and 1976.

[52] T. Y. NA, *Computational Methods in Engineering Boundary Value Problems*, Academic, N.Y., 1979.

[53] T. R. F. NONWEILER, *Computational Mathematics: An Introduction to Numerical Approximation*, Horwood, Chichester, 1984.

[54] F. W. J. OLVER, *Asymptotics and Special Functions*, Academic, N.Y., 1974.

[55] J. M. ORTEGA AND W. C. RHEINBOLDT, *Iterative Solutions of Nonlinear Equations in Several Variables*, Academic, N.Y., 1970.

[56] A. M. OSTROWSKI, *Solution of Equations and Systems of Equations*, Academic, N.Y., 1966.

[57] M. J. D. POWELL, *Approximation Theory and Methods*, Cambridge University Press, Cambridge, 1981.

[58] R. D. RICHTMYER AND K. W. MORTON, *Difference Methods for Initial Value Problems*, Interscience, N.Y., 1964.

[59] T. R. RIVLIN, *An Introduction to the Approximation of Functions*, Dover, N.Y., 1969.

[60] S. M. ROBERTS AND J. G. SHIPMAN, *Two-Point Boundary Value Problems: Shooting Methods*, Elsevier, N.Y., 1972.

[61] P. ROZSA, ed., *Numerical Methods*, North-Holland, Amsterdam, 1980.

[62] A. SARD AND S. WEINTRAUB, *A Book of Splines*, Wiley, N.Y., 1971.

[63] L. F. SHAMPINE AND M. K. GORDON, *Computer Solutions of Ordinary Differential Equations: The Initial Value Problem*, Freeman, San Francisco, 1975.

[64] O. D. SMITH, *Numerical Solution of Partial Differential Equations: Finite Difference Methods*, Clarendon Press, Oxford, 1978.

[65] S. K. STEIN, *"The error of the trapezoidal method for a concave curve,"* AMM (1976) pp. 643–645.

[66] F. STUMMEL, *"Rounding error analysis of elementary numerical algorithms,"* Computing, Suppl. (1980) pp. 169–195.

[67] J. L. SYNGE, *The Hypercircle Method in Mathematical Physics: A Method for the Approximate Solution of Boundary Value Problems*, Cambridge University Press, Cambridge, 1957.

[68] F. SZIDAROVSZKY AND S. YAKOWITZ, *Principles and Processes of Numerical Analysis*, Plenum Press, N.Y., 1976.

[69] J. TODD, ed., *A Survey of Numerical Analysis*, McGraw-Hill, N.Y., 1962

[70] H. S. WALL, *"A modification of Newton's method,"* AMM (1948) pp. 90–94.

[71] D. M. YOUNG AND R. D. GREGORY, *A Survey of Numerical Mathematics I, II*, Addison-Wesley, Reading, 1973.

9. Mathematical Economics

THE BEST TIME TO DIE*

MARGARET W. MAXFIELD† AND NAOMI J. McCARTY‡

Abstract. When a new tax law is subjected to a simple mathematical analysis, some strange effects are discovered.

This elementary mathematical analysis will not qualify anyone as a tax expert, even in the area of Material Participation. However, the analysis does reveal some strange effects of the law that may not have been anticipated when the law was enacted.

Recently the Congress [1] undertook to improve the situation of a surviving spouse who, despite "material participation" in a family farm or business, might see most of the property included in the gross taxable estate of the first spouse to die. The Material Participation (MP) exclusion is an optional election, in case various eligibility requirements are met. It calls for the exclusion from the gross estate of the first spouse to die of the sum of two quantities:

(i) the excess of the joint interest over the adjusted (at 6% simple interest, it turns out) consideration for acquisition of the property (usually what the property cost), multiplied by 2% for each year of material participation, but not more than 50%;

(ii) the adjusted consideration furnished by the surviving spouse.

The excluded amount is subject to two limitations, one relative, the other absolute: the amount excluded may not exceed half the joint interest and it may not exceed $500,000.

Since even if no MP election is made the surviving spouse is entitled to share in the joint interest in proportion to the consideration that spouse furnished initially, we deduct this amount and ask whether the net (remaining) excluded amount is positive, that is, whether MP election will result in any (additional) exclusion.

It is useful to state quantities in units of C, the initial "consideration" for acquiring the property. For instance, instead of V, the "joint interest" or value of the property as appraised at time of death of the first spouse to die (or alternate valuation date), we use the "appreciation" $A = V/C$.

With quantities expressed in this scale, and assuming that neither limitation is reached, we then have the inequality

$$Y = (x - s)(A - t) \geq 0,$$

where:

$s =$ proportion of initial consideration provided by the surviving spouse. This proportion is often zero or very small, and is always less than 0.5 for MP to be elected. $0 \leq s \leq 0.5$.

$t = 1 + 0.06T$, where T is the number of years of the joint tenancy, from initial acquisition of the property to the death of the first spouse to die. In parts of this paper the variable corresponding to time will be t, and in other parts T, depending on notational convenience.

---

* Received by the editors October 22, 1979.

† Department of Management, CBA, Kansas State University, Manhattan, Kansas 66506.

‡ Department of Accounting, University of Tulsa, Tulsa, Oklahoma 74104.

$x = \min\,(0.02U, 0.5)$, where U is the number of years of MP. In the important special case treated here, U is equal to T, so that x and t are related by $x = \min\,((t-1)/3, 0.5)$.

A = value of joint interest, in C units, where C was the initial consideration.

Y = net exclusion due to MP election expressed in C units, provided the limitations do not apply. The net exclusion is made up of (i) $x(A-t)$, plus (ii) st, less the survivor's own share, sA.

Consider the Y-surface in (t, s, A, Y)-space given by the function

$$Y = \begin{cases} Y_1 = \left(\dfrac{t-1}{3}-s\right)(A-t) & \text{for } t \leq 2.5, \\[2mm] Y_2 = (0.5-s)(A-t) & \text{for } t > 2.5. \end{cases}$$

The hyperplanes $s =$ a constant and $A =$ a constant, for choices of these parameters, intersect the Y-surface in a trace in the (t, Y)-plane.

If $A \leq 2.5$, then the trace is the parabolic arc

$$Y_1 = -\frac{t^2}{3} + \left(s + \frac{1}{3} + \frac{A}{3}\right)t - sA,$$

from $(t, Y_1) = (3s+1, 0)$ to $(A, 0)$. See Fig. 1.

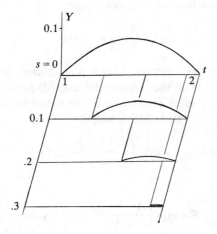

FIG. 1. *Constant A less than 2.5. A = 2. For fixed s, Y = Y_1 is parabolic.*

If $A > 2.5$, the trace is made up of a parabolic segment

$$Y = -\frac{t^2}{3} + \left(s + \frac{1}{3} + \frac{A}{3}\right)t - sA,$$

from $(t, Y) = (3s+1, 0)$ to $(2.5, (0.5-s)(A-2.5))$, and a linear segment

$$Y_2 = -(0.5-s)t + (0.5-s)A,$$

from $(2.5, (0.5-s)(A-2.5))$ to $(A, 0)$. See Fig. 2.

We note that the limits in Y-value, as t approaches 2.5 from the left and from the right, respectively, are equal:

$$\lim_{t \to 2.5} Y = (0.5-s)(A-2.5).$$

Then the Y function of t for fixed A and s is continuous between $t = 3s + 1$ and $t = A$, its domain of positivity.

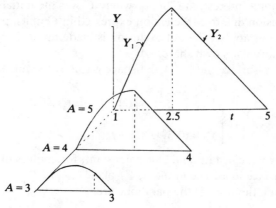

FIG. 2. *Constant A greater than 2.5. $A = 3, 4, 5$, with $s = 0$. $Y = Y_1$ is parabolic for $t \leqq 2.5$; $Y = Y_2$ is linear for $t > 2.5$.*

In case $A > 2.5$, however, Y fails to be differentiable with respect to t at $t = 2.5$, since the left- and right-derivatives are not equal:

$$s + \frac{A}{3} - \frac{4}{3} \neq s - 0.5.$$

From the fact that the function Y has a maximum with respect to t, where $t = 1 + 0.06T$, we see that from the (admittedly limited) point of view of MP exclusion there is an optimal time T after acquisition of the property for the first spouse to die! This optimal T value is $\hat{T} = (\hat{t} - 1)/0.06$, where

$$\hat{t} = \min\left(\frac{A + 1 + 3s}{2}, \, 2.5\right);$$

that is,

$$\hat{T} = \min\left(\frac{A - 1 + 3s}{0.12}, \, 25\right) \quad \text{years.}$$

The maximum excluded value for this time is

$$\hat{Y} = \begin{cases} \dfrac{(A - 1 - 3s)^2}{12} & \text{if } A + 3s \leqq 4, \\ (0.5 - s)(A - 2.5) & \text{if } A + 3s > 4. \end{cases}$$

This state of affairs—an optimal time T to death, with net MP exclusion Y decreasing for longer times—is unusual in tax. We would expect the exclusion to increase with T up to a plateau and then remain constant for larger T values. In fact, we note that the law calls for the percentage x to do just that—to increase linearly, by 2% for each year of MP, up to a value of 50%. If that percentage, x, were applied to a constant, we would have the usual tax picture. The decrease in exclusion for $T > \hat{T}$ is caused by the fact that the percentage x is applied to the excess, in C units, $(A - t)$, where the deducted base value t depends on T.

In our first analysis the appreciation A was held fixed, but what if A increases with T? Suppose that the joint interest measured in C units increases at a simple interest rate $(0.06 + a)$, greater than the 6% adjustment provided for in the law:

$$A = 1 + (0.06 + a)T, \qquad a > 0.$$

Then we have

$$Y = \begin{cases} Y_1 = (0.02T - s)aT & \text{for } T \leq 25, \\ Y_2 = (0.5 - s)aT & \text{for } T > 25. \end{cases}$$

The parabolic segment Y_1 of Y as a function of T is negative between $T = 0$ and $T = 50s$, with a minimum at $T = 25s$. It is positive and increasing for T between $50s$ and 25 years. The linear segment Y_2 for $T > 25$ has positive slope $(0.5 - s)a$. Again, Y is continuous for $T > 50s$, but is not differentiable at $T = 25$. See Fig. 3.

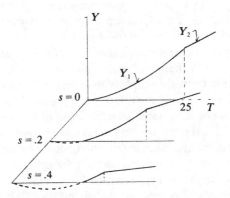

FIG. 3. *Increasing A, simple interest model.* $A = 1 + (0.06 + a)T$, *with* $a = 0.1$.

By assuming a simple interest growth for A, we have obtained a more familiar tax picture. The excluded amount Y does not decrease with T after becoming positive. However, there is no horizontal plateau, since Y continues to increase after $T = 25$.

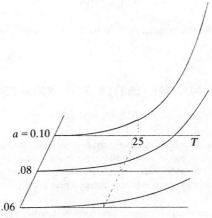

FIG. 4. *Increasing A, compound interest model.* $A = (1 + a)^T$, *for* $a = 0.06, 0.08, 0.10$.

If A grows at a compound interest rate, a reasonable hypothesis for a successful farm or business, then we encounter exponential functions. Let $A = (1+a)^T$. Then

$$Y = \begin{cases} Y_1 = (0.02T - s)[(1+a)^T - (1+0.06T)] & \text{for } T \leq 25, \\ Y_2 = (0.5 - s)[(1+a)^T - (1+0.06T)] & \text{for } T > 25. \end{cases}$$

In Fig. 4 the Y functions are graphed for $a = 0.06$, 0.08 and 0.10, with $s = 0$. Again, Y is continuous within the domain of positivity, but it is not differentiable at $T = 25$.

It is tempting to try second-guessing Congress. What *did* they have in mind? It is instructive to go back over this analysis and question how far we could have advanced without mathematical symbolism. The law was written in English, or what passes for English in law jargon. The semiotics of that language, even allowing for a few numerical percentages, are inadequate for such an analysis.

One question that arises whenever a tax concession is introduced is the effect on revenues. How much revenue must the government forego because of MP election?

In case the MP election is made in a particular instance, the amount of the net exclusion depends on s, T, A and x, and these variables are interrelated. For instance, there is probably a negative correlation between s and T because women, who have longer life expectancies than their husbands, have recently had greater potential for contributing larger shares s toward property than formerly. Since wealth and longevity tend to support each other, A and T probably are positively correlated.

The MP election is not automatic, so estimation of lost revenue would require a probability measure, and that measure would have to be conditioned on the probability that no formal arrangement had been established before death.

Since it looks impossible to obtain even a ballpark estimate of lost revenue a priori, we might question whether an estimate can be found empirically now that the law is in effect. This, too, appears hopeless, as there is no way to isolate effects of the MP provision from the effects of other changes in the tax picture, to say nothing of changes in the ambient economy.

REFERENCE

[1] *Internal Revenue Code of* 1954, §§ 2040c, added to by the Revenue Act of 1978, § 511.

POSTSCRIPT

The Material Participation provision was dropped from the tax law shortly after it was instituted.

INFLATION MATHEMATICS FOR PROFESSIONALS*

EDWARD W. HEROLD†

Abstract. The personal finances of professional scientists, engineers and mathematicians are greatly affected by inflation. A simple mathematical model is of considerable value in planning ahead. To prevent financial collapse before the end of life, the model should be used many years before retirement and re-applied annually. A numerical example shows the need for a financial plan and the utility of the model.

* Received by the editors June 3, 1980, and in final form November 18, 1980.

† Consultant, 332 Riverside Drive East, Princeton, NJ 08540.

The writer's 1979 paper [1] on the effect of inflation was directed to the engineer facing retirement, and created considerable interest. Other professionals confront financial problems similar to those of the engineer and are equally able to interpret a mathematical approach. We here supplement the 1979 paper by expanding the mathematical solution to improve its utility during pre-retirement years. The approach taken uses a differential equation, rather than the more awkward conventional mathematics of finance which uses discrete-interval relationships. Use of continuous functions permits greater sophistication and an analysis which is better suited to the inflation considerations covered herein.

The appropriate differential equation states that the change in capital with time equals the after-tax income minus the expense. With inflation, if living expense rises exponentially, it will often be found that both capital and income will eventually be used. The objective of planning is to prevent the financial endpoint from occurring before the end of life. The solution to the differential equation provides the planning tool needed. Because inflation rates, sources of income and life styles change, each family or individual should make appropriate assumptions and review their finances annually.

The notation to be used is as follows. Passage of time will be measured in years, designated as n.

S_n = capital accumulation at year n.

S_0 = capital at start of calculation.

i_v = after-tax return on capital.

$I(n)$ = after-tax income from all other sources, as a function of n.

$E(n)$ = living expense as a function of n.

The differential equation is

(1)
$$\frac{dS_n}{dn} = i_v S_n + I(n) - E(n),$$

which has a solution

(2)
$$S_n = e^{ni_v}\left\{ S_0 + \int_0^n [I(n) - E(n)] e^{-ni_v}\, dn \right\}.$$

During earning years, $I(n)$ ordinarily exceeds $E(n)$ and capital accumulates, with the exponential outside of the largest brackets the dominant term. After retirement, because of inflation, $E(n)$ may rise faster with time than does $I(n)$; capital is used up and, in time, the integral becomes negative and S_n goes to zero. The reader may make his or her own prognosis as to the future form for the income and expense functions, taking inflation into account. Before retirement, (2) is applied to find the accumulated capital up to the point of retirement. A second application, with new initial S_0, income function and expense function, will determine conditions for financial survival thereafter.

For illustrative purposes, it is useful to assume simple, easy-to-integrate functions, which we will now undertake. If inflation causes living expense to rise at the continuously compounded rate i_f, then the expense function is

(3)
$$E(n) = E_0 e^{i_f n},$$

where E_0 is the initial living expense at $n = 0$. It should be noted that the continuously-compounded inflation rate herein used, i_f, differs slightly from the "effective rate" commonly found in discussions of inflation. The effective rate is defined as the increase in a full year, which is $(e^{i_f}-1)$. For an $i_f = 0.10$, the effective rate is 0.1052, which should be kept in mind in the numerical example given below. The uncertainty of specific numbers involved in future planning makes the difference inconsequential.

The income function is often composed of a fixed part (such as a fixed salary, pension or annuity) and a part which is partially or wholly indexed to inflation (as with an increasing salary or an indexed Social Security benefit). Thus the income function may be expressed as

(4) $$I(n) = F + A_0 e^{i_a n},$$

where F is the fixed after-tax income, A_0 is the initial after-tax variable income, and i_a is the assumed after-tax rate of increase of the variable income.

Inserting (3) and (4) in (2) and integrating gives the useful solution

(5) $$S_n = e^{n i_v}\left[S_0 + F\frac{1-e^{-n i_v}}{i_v} + A_0\frac{e^{(i_a-i_v)n}-1}{i_a-i_v} - E_0\frac{e^{(i_f-i_v)n}-1}{i_f-i_v}\right].$$

Inspection shows that if the inflation rate i_f exceeds both the after-tax return on investment and the after-tax increase in variable income, capital will *always be used up*, i.e., S_n will go to zero. This is the financial endpoint with which the planner is concerned. Equation (5) is readily evaluated on a scientific calculator; if the calculator is programmable, the equation can be set up with a program loop to give S_n vs. n by a single key stroke for each point. Sample calculations show the insidious nature of inflation which, in its effect on finances, is analogous to terminal cancer.

Sensitivity to the inflation rate. Both the rate of return on investment and the rate of increase in variable income tend to track the inflation rate. Thus, capital accumulation as well as expense are sensitive to the inflation rate, varying exponentially therewith. However, and for the same reason, i.e., the tracking of the rates, inspection of the bracketed term of (5) suggests that the point in time at which capital goes to zero is relatively *insensitive* to the inflation rate. This feature is important because it is exactly this period of time with which advance planning is concerned.

As an illustration of the insensitivity, assume that all the rates are equal, $i_v = i_a = i_f = i$. Inserting equal rates in (5), gives

(6) $$S_n = e^{in}\left[S_0 + F\frac{1-e^{-in}}{i} + n(A_0 - E_0)\right].$$

Before retirement, earnings will usually exceed expense, $A_0 > E_0$, so that capital steadily accumulates. After retirement, A_0 may consist of Social Security (indexed to inflation) possibly plus some part-time earnings, but $A_0 < E_0$. It is then that capital goes to zero as n increases. As (6) shows, the $S_n = 0$ point has only one term involving the rate factor, the multiplier of the fixed income. Unless F is exceptionally high, survival endpoints are little affected by the rate factor.

A numerical example of planning. Assume a salaried professional family of 1980, 15 years before expected retirement in 1995, who wish to plan for financial survival at least as long as their probable lifetimes. The earlier paper [1] includes a table of life survival probabilities which apply to most technical professions. Retirement between ages 65 and 70, with a 30% survival probability, suggests that a financial plan should encompass about 20 years more.

Equation (5) will be used twice, first for the 15 years between 1980 and 1995, and then for the 20 years of retirement thereafter. The two equations must then be made consistent with each other. We will assume an inflation rate of $i_f = 0.10$, an after-tax return on investment $i_v = 0.07$ and that after-tax salary increases are fully indexed to inflation, $i_a = 0.10$. After retirement, Social Security benefits will also be assumed to be fully indexed, so that $i_a = 0.10$ applies to this income also.

Inserting the above rates in (5) with $n = 15$, and assuming no fixed income $(F = 0)$, gives the 1995 accumulation of capital as

(7) $$S_{15} = 2.85S_0 + 54.1(A_0 - E_0).$$

After retirement, the survival condition is found by using $n = 20$ and equating (5) to zero. This gives

(8) $$S_0' + 10.8F' + 27.4(A_0' - E_0') = 0$$

where the primed symbols are the 1995 initial conditions. The two equations are combined by recognizing that $S_{15} = S_0'$.

To further simplify, let the 1995 fixed income, F', be an after-tax pension equal to half of the 1995 after-tax salary. Then, from (4),

$$F' = \tfrac{1}{2}A_0 e^{1.5} = 2.24A_0.$$

Social Security income in 1980 is tax free, fully indexed, and covers perhaps $\frac{1}{3}$ of the living expense of a retired middle-class professional. Assuming the same conditions for 1995 gives A_0' (Social Security) $= E_0'/3$. Combining (7) and (8) with these conditions gives the final survival condition

(9) $$E_0 = 0.053S_0 + 1.45A_0 - 0.34E_0'.$$

If the planner wishes to continue the 1980 life style into retirement, then (3) shows that

$$E_0' = E_0 e^{1.5} = 4.48E_0$$

and (9) becomes

(9a) $$E_0 = 0.021S_0 + 0.58A_0.$$

A hypothetical professional family with a 1980 salary of $45,000 would, after taxes retain about $A_0 = \$30,000$. If the capital, S_0, is $100,000, the family may adopt a life style equivalent to $E_0 = \$19,500$ in 1980 dollars. Above this, their capital accumulation will be insufficient to permit continuation of the life style into retirement. An alternative is to plan to reduce retirement life style to 80% of that during earning years. Equation (9) shows that they may then increase 1980 living expense to $E_0 = \$22,000$. Without planning, most professionals will be tempted to live more expensively than the life style suggested by the model. If used annually as assumptions change, the model greatly reduces the likelihood of financial failure after retirement.

POSTSCRIPT

Readers familiar with financial terminology will note that equations 2, 5, and 6 give the "future value" and, in each equation, the large-bracket expression is the "present value" of the future changes in capital. In the paragraph on income tax, the specific tax rates given are out-of-date, but the general caution on tax effects remains valid.

THE OPTIMAL SIZE OF A SUBSTITUTE
TEACHER POOL*

MICHAEL A. GOLBERG† AND JOHN MOORE‡

1. Introduction. In most large school districts, substitute teachers are pro-vided when regular teachers are absent. This requires that the district have access to a large pool of available part-time teachers. In many systems, these teachers are on a purely stand-by basis; that is, they get paid only when they are called. Under this procedure, the optimal policy is to hire as many teachers as meet minimum certification standards. However, this frequently leads to a very small number of paychecks per teacher, thus forcing them to leave the pool. As a consequence, morale tends to be low and turnover high, leaving many poorly trained and unmotivated people as teachers [1].

To overcome this, Bruno [1] proposed the adoption of an alternate system of utilizing substitute teachers. He indicated, through a Monte Carlo simulation of the needs of the Los Angeles City School District, that considerable savings could be achieved if one were to use a fixed-size pool of teachers who would be paid whether or not they actually taught. The problem then posed was to determine the optimal size of the pool. He was able to solve the problem through Monte Carlo methods.

It is the purpose of this paper to give a simple mathematical model for the theory Bruno presents. We are, therefore, able to obtain analytically the results Bruno gets via computer simulation. Consequently a district can utilize our results with zero cost of implementation (and no computers). Our method closely resembles simple inventory methods which are common in operations research [2] and requires only a knowledge of elementary calculus and probability, and as such should serve as an interesting classroom model in courses on either of these subjects.

2. The model. It was observed in [1] that the daily demand for teachers is stochastic and follows a different distribution for each day of the week. As a first approximation, it was assumed that there was no carry over of demand from one day to the next. As a consequence, the total cost of hiring substitutes will be minimized if the daily cost is minimized. This will generally lead to a different pool size for each day.

Suppose that on day "i" the demand for teachers is a random variable X with density $\varphi(x)$. That is, $P\{X \leq x\} = \int_0^x \varphi(x)\, dx$. Let S be the size of the pool. Let the cost of a teacher be r_1 dollars per day. If a substitute teacher is not available, we assume that the demand can be made up by using regular teachers and paying them overtime. This, according to Bruno, is the practice of the Los Angeles City School System and other large urban school districts [1, p. 417]. This is done at a rate of r_2 dollars per day and $r_2 > r_1$.

Let $C(x/S)$ be the cost when the demand is x. $C(x/S)$ then satisfies

---

* Received by the editors March 30, 1975.

† University of Nevada, Las Vegas, Las Vegas, Nevada 89154.

‡ Environmental Protection Agency, Las Vegas, Nevada 89154.

(1) $$C(x/S) = \begin{cases} r_1 S, & 0 \leq x \leq S, \\ r_1 S + r_2(x - S), & x \geq S. \end{cases}$$

The expected cost $C(S) = E(C(x/S))$ is then

(2) $$C(S) = r_1 S + r_2 \int_S^\infty (x - S)\varphi(x)\, dx.$$

The optimal pool size is the one that minimizes $C(S)$. Differentiating (2) gives S as the solution to

(3) $$r_1 - r_2 \int_S^\infty \varphi(x)\, dx = 0.$$

If we let $F(S) = P\{X \leq S\}$ be the cumulative distribution function of X, then the integral in (3) is $1 - F(S)$, so that S satisfies the optimality condition

(4) $$F(S) = 1 - r_1/r_2.$$

To implement (4) a school administrator would only need to know the distribution of demand, a quantity which should be readily available. No complicated simulation need be done as in [1].

3. Comparison with Bruno's results. To see how our model compares with Bruno's, we will evaluate (4) using the statistics in his paper. The only detailed figures in [1] are for Monday, so we use those. Table 4 in [1], which we reproduce below as our Table A, gives the substitute demand distribution for Monday.

TABLE A

No. of Teachers	%	Cumulative %
200–275	.027	.027
276–350	.027	.054
351–425	.027	.081
426–500	.027	.108
501–575	.162	.270
576–650	.108	.378
651–725	.486	.864
726–800	.081	.945
801–875	.027	.972
876–950	.027	1.000

r_1 is taken as \$30 and r_2 as \$54. According to (4), S satisfies $F(S) = 24/54 = .444$. That is, S is the 44th percentage point of the distribution. Note that this means that on approximately 44% of the school days, no regular teachers would have substitute duties.

Since we do not have an analytic form for the distribution in Table A, we interpolate in essentially the same way as Bruno does. For simulation purposes, he assumes that the distribution is uniform in each of the intervals. Noticing that the

38th percentile occurs at 650, S lies between 651 and 725. Using the above assumptions, $S = 650 + x$, where x is determined from the equation $.444 = .378 + .486x/75$. This gives $x = 10.2$, so that $S = 660$, which agrees with the figure obtained by simulation.

Although (4) gives the pool size of minimum cost, it is not the only parameter of interest in the problem. Since the model assumes that excess demand can be made up by using regular teachers, it is of interest to determine what fraction of the load must be borne by them. A quantity called the "level of service" is introduced in Bruno's paper to measure this. His definition is

$$
(5) \qquad L(S) = 1 - \frac{\text{number of man-hours short}}{\text{total number of substitute man-hours demanded}},
$$

so that 1-L(S) is essentially the fraction of the substitute load that must be supplied by regular teachers. No analytic formulas are given in [1], and again $L(S)$ is obtained via Monte Carlo. Using the quantities defined in § 2, we give $L(S)$ as

$$
(6) \qquad L(S) = 1 - \left\{ \int_S^\infty (x - S)\varphi(x)\, dx \Big/ \int_0^\infty \varphi(x) x\, dx \right\},
$$

where $\int_0^\infty \varphi(x) x\, dx$ is the expected demand. This can be simplified to

$$
(7) \qquad L(S) = \int_0^S x\varphi(x)\, dx + S \int_S^\infty \varphi(x)\, dx \Big/ \int_0^\infty x\varphi(x)\, dx.
$$

From this we see that $L(S) \geq 0$, $L(0) = 0$ and $\lim_{S \to \infty} L(S) = 1$ and also that $L(S)$ is increasing and thus has the shape given by Fig. 2 in [1].

To illustrate the use of (6), suppose that S is the optimal pool size, so that $\int_S^\infty \varphi(x)\, dx = r_1/r_2$. Then

$$
L(S) = \frac{\int_0^S x\varphi(x)\, dx + S r_1/r_2}{\int_0^\infty x\varphi(x)\, dx}.
$$

Evaluating the integrals using Table A, we get that $L(S) = .95$, which again agrees with the results in [1]. A fuller discussion of the use of $L(S)$ in actual problems is given by Bruno and the reader is referred to his paper for these ideas.

4. Conclusion. We have shown how a simple mathematical model can be effective in studying the utilization of substitute teachers. The model is cheaper and easier to use than the Monte Carlo methods in [1]. That such models have practical value is shown by the results of Bruno. He indicates that in 1970 these ideas could have produced $600,000 yearly savings for the Los Angeles City School District.

REFERENCES

[1] J. E. BRUNO, *The use of Monte Carlo techniques for determining the optimal size of substitute teacher pools in large urban school districts*, Socio-Econ. Plan. Sci., 4 (1970), pp. 415–428.
[2] F. S. HILLIER AND G. J. LIEBERMAN, *Introduction to Operations Research*, Holden-Day, San Francisco, 1972.
[3] R. SISSON, *School administration—An area for mathematical analysis*, SIAM News, 6 (1973).

A SIMPLE PROOF OF THE RAMSEY SAVINGS EQUATION*

MOHAMED EL-HODIRI†

In [2] Ramsey considers the problem of minimizing the accumulated difference between bliss B and net utility $U(x) - V(a)$, where $U(x(t))$ is the instantaneous utility of consuming $x(t)$ and where $V(a)$ is the instantaneous disutility of working at the rate $a(t)$.[1] Capital $c(t)$ and labor $a(t)$ are used to produce the output flow according to the production function $f(c(t), a(t))$. Output in turn is divided between capital accumulation, $c(t)$ and consumption $x(t)$. Thus the problem is to minimize the integral[2] $\int_0^T (B - U(x) + V(a))\, dt$ subject to $\dot{c} + x = f(c, a)$. In other words, we are to minimize the integral:

$$\int_0^T (B - U(f(c, a) - \dot{c}) + V(a))\, dt.$$

Setting $a = \dot{y}$, the problem becomes:

$$\text{minimize} \int_0^T L(Z, \dot{Z})\, dt$$

where $Z = (Z_1, Z_2) = (c, y)$.

Since L does not depend explicitly on t, the Euler equations have the form (see Akhiezer [1]):

(1)
$$\frac{d}{dt} L_{\dot{Z}} = L_Z,$$

(2)
$$L - \dot{Z}_1 L_{\dot{Z}_1} - \dot{Z}_2 L_{\dot{Z}_2} = 0.$$

Writing these in terms of our problem we have:

(3.1)
$$\frac{d}{dt} U' = U' f_c,$$

(3.2)
$$\frac{d}{dt} (U' f_a - V') = 0,$$

(3.3)
$$B - U + V - \dot{c} U' - a(U' f_a - V') = 0.$$

Clearly,

(3.4)
$$U' f_a - V' = 0.$$

---

* Received by the editors, April 15, 1976.

† Department of Economics, University of Kansas, Lawrence, Kansas 66045.

[1] The purpose of this note is purely pedantic since it reveals no new "truths." It merely makes it easier to understand what is already a widely held belief.

[2] Ramsey considers the case where $T = \infty$. By taking the limit, and arguing around a bit, we can show that the Euler equations in our simple problem will still hold. That, however, will make the proof far from simple. It should also be noted that we don't discuss the existence problem, thus embracing the risk of an empty intersection between the set of arcs for which a solution exists and the set of arcs for which Euler's equations are necessary conditions for an extremum.

Hence, (3.3) becomes

(3.5) $$B - U + V - \dot{c}U' = 0.$$

Consequently,

(4) $$\dot{c} = \frac{B - U + V}{U'}.$$

Equation (4) is Ramsey's savings equation and has the usual interpretation that the "optional" policy is to save more if the net satisfaction is below bliss level and to dissave if that level is surpassed. Equations (3.1) and (3.4) express the usual market conditions for capital and labor.

REFERENCES

[1] N. I. AKHEIZER, *The Calculus of Variations*, Blaisdell, New York, 1962.
[2] F. RAMSEY, *A mathematical theory of savings*, Economic J., 38 (1928), pp. 543–559.

ADJUSTMENT TIME IN THE NEOCLASSICAL GROWTH MODEL: AN OPTIMAL CONTROL SOLUTION*

STEPHEN D. LEWIS†

Abstract. The long adjustment times implied by the neoclassical growth model are reviewed. The model is transformed into state-space notation, and the results of the minimum-time problem from optimal control theory are used as a basis for decreasing these adjustment times. This note illustrates the dynamic behaviour of a popular economic model and provides an example of applying optimal control theory to theoretical economics.

1. Introduction. Models of economic growth attempt to explain how economies grow over time. It is common to represent the complexities of actual behaviour by a number of basic assumptions that reduce the problem to manageable proportions while still maintaining the essential characteristics of economic growth. We begin by considering a hypothetical economy that produces one good which represents output as a whole. This good is produced by capital and labour, the only two factors of production, both of which are homogeneous. Output is malleable and either is consumed or else becomes a part of the capital stock. While very restrictive, these assumptions, nonetheless, provide the basis for examining many questions concerning economic growth. More specifically these basic assumptions when combined with further details presented below specify a model that describes the conditions under which a growing economy may be in steady-state equilibrium. When the key variables of the model are expressed in per capita terms, it is easy to examine the type of dynamic behaviour expected when steady-state conditions are not satisfied. This dynamic behaviour and the modifications that can be produced through the application of optimal control theory are the major focus of this paper. A very readable treatment of economic growth theory and the motivation for the assumptions summarized above is available in Jones [3].

The next section sets forth the basic assumptions of the growth model to be analyzed.

*Received by the editors December 14, 1981, and in revised form July 10, 1982.
†Department of Economics, University of Alberta, Edmonton, Alberta, Canada T6G 2H4.

In §3, steady-state and dynamic properties of the model are summarized. The adjust-ment-time problem is discussed next. The last section considers the use of optimal control theory to shorten excessive adjustment times.

2. Model specification. One of the major and most popular explanations of a growing economy that builds on these basic assumptions is the neoclassical growth model based upon the work of Solow [8]. In his model, the *labour force, N*, grows at a constant proportional *rate, n*. Full employment prevails at all times. N and the *capital stock, K*, combine to produce *output, Y*. The production process is summarized by an aggregate *production function, F(K, N)* which is continuous and linear homogeneous. As discussed previously, Y is identically equal to expenditure for consumption plus *investment, I*. By definition I represents changes to K and, using a "˙" over a variable to represent differentiation with respect to time, then $I \equiv \dot{K}$. Following the Keynesian tradition, *saving behaviour, S*, is specified as a constant proportion, s, of Y where s *is the marginal propensity to save* and is constrained by $0 < s < 1$. A final condition imposed on the model is the equilibrium condition that I is always equal to S. All of these assumptions are summarized by equations (1)–(5):

(1) $$N = e^{nt}, \qquad n > 0,$$

(2) $$Y = F(K, N),$$

(3) $$I \equiv \dot{K},$$

(4) $$S = sY, \qquad 0 < s < 1,$$

(5) $$S = I.$$

It is common in models of economic growth to represent key variables in ratio form. The per capita variables output-labour ratio, $y = Y/N$, and capital-labour ratio, $k = K/N$, are useful in the discussion that follows. With these new variables the aggregate production function can be represented by

$$\frac{Y}{N} = F\left[\frac{K}{N}, 1\right],$$

or more simply by

(6) $$y = f(k).$$

To ensure the existence, uniqueness and stability of a steady-state equilibrium, it is further assumed that $f(k)$ is "well-behaved."[1]

In terms of growth rates and per capita variables, equations (1)–(6) can now be condensed into

(7) $$\frac{\dot{N}}{N} = n,$$

(8) $$\frac{\dot{k}}{k} = \frac{\dot{K}}{K} - \frac{\dot{N}}{N} \quad \text{or} \quad \dot{k} = \frac{\dot{K}}{N} - \frac{\dot{N}}{N}k,$$

---

[1]The additional assumptions, all of which have plausible economic interpretations, are $f'(k) > 0$, $f''(k) < 0$, $f'(0) = \infty$, $f'(\infty) = 0$, $f(0) = 0$ and $f(\infty) = \infty$.

$$(9) \qquad \frac{\dot{K}}{N} = sf(k).$$

3. Steady-state and dynamic properties. The basic dynamic equation of the neoclassical growth model based upon Solow [8] is derived by combining (7)–(9) to obtain

$$(10) \qquad \dot{k} = sk^{\alpha} - nk$$

with

$$k \geq 0, \qquad n > 0,$$
$$0 < s < 1, \qquad 0 < \alpha < 1.$$

The expression k^{α} is the per capita form of a Cobb–Douglas production function which is commonly used in models of economic growth in place of $f(k)$ when explicit solutions to dynamic behaviour are required. It is easily verified that this function is "well-behaved." The parameter α represents the elasticity of output with respect to capital (i.e., $\alpha = dy/dk \cdot k/y$). Thus the first expression in (10) represents how much is saved and invested out of per capita output. The second expression determines the amount of investment required to keep the capital-labour ratio constant.

In a steady-state equilibrium (i.e., $\dot{k} = 0$), the economy grows at rate n and the equilibrium capital-labour ratio, k_e, is given by

$$(11) \qquad k_e = \left(\frac{s}{n}\right)^{1/(1-\alpha)}.$$

One of the implicit influences on s is the tax rate as determined by fiscal policy. The equilibrium change in k_e for a change in s is obtained from

$$(12) \qquad \frac{dk_e}{ds} = \frac{k^{\alpha}}{n - s\alpha k^{\alpha-1}}.$$

The "well-behaved" properties of the function k^{α} ensure that at an equilibrium point $dk_e/ds > 0$. These properties also guarantee that in the neighborhood of k_e dynamic behaviour is stable (i.e., $\partial \dot{k}/\partial k = sf'(k) - n < 0$).

4. The adjustment problem. In the previous section, the steady-state equilibrium was described, and it was determined that dynamic behaviour was stable. The interesting question to be considered now is how the model behaves when the steady-state condition summarized in (11) is not satisfied. By solving the differential equation (10) the adjustment problem can be characterized in terms of the time required to move from an arbitrary initial position to the steady-state implied by a given set of parameter values.

We start by considering the movement from an initial equilibrium k_0 based upon s_0 to a new equilibrium k_e determined by s_e. From (11),

$$k_0 = \left(\frac{s_0}{n}\right)^{1/(1-\alpha)}, \qquad k_e = \left(\frac{s_e}{n}\right)^{1/(1-\alpha)}$$

where $s_0 \gtrless s_e \Rightarrow k_0 \gtrless k_e$. Dynamic behaviour is examined by first defining a new variable $z = k^{1-\alpha}$. Substitution into (10) yields

$$(13) \qquad \dot{z} + n(1-\alpha)z = (1-\alpha)s.$$

Solving (13) and transforming back to k's yields

(14)
$$k(t) = \left[\frac{s}{n} + \left(k_0^{1-\alpha} - \frac{s}{n}\right) e^{-n(1-\alpha)t}\right]^{1/(1-\alpha)}.$$

From (14), it is evident for $s = s_e$ and starting from $k_0 \neq k_e$ that $k(t)$ only approaches k_e asymptotically. To find the time t_λ required to adjust $k(t)$ a fraction λ of the distance between k_0 and k_e, let

$$\lambda = \frac{k_\lambda - k_0}{k_e - k_0},$$

where $k_\lambda = k(t_\lambda)$.
   Then from (14)

(15)
$$t_\lambda = \frac{-1}{n(1 - \alpha)} \ln\left\{\frac{s_e/n - k_\lambda^{1-\alpha}}{s_e/n - k_0^{1-\alpha}}\right\}.$$

TABLE 1
Adjustment times for $k(t)$*

	Percent adjustment			
	30%	50%	70%	90%
t(years):	12.9	25.0	43.4	82.7

*Parameter values: $k_0 = 5.0$, $k_e = 5.5$, $\alpha = 0.3$, $n = 0.04$, $s_0 = 0.15$, $s_e = 0.1603$.

Table 1 gives sample values for λ and t_λ and illustrates for parameter values typical of actual economics that adjustment times in the neoclassical growth model are exceedingly slow. This adjustment problem has been amply noted by others ([2], [6], [7]). The usual attack on excessive adjustment times is to introduce modifications to the basic model. Often adjustment times have been reduced by a factor of two or three, but they still remain uncomfortably long. In the next section the structure of the model is retained but optimal control theory is used as a method of altering adjustment times.

   5. An optimal control solution. Long adjustment times can be shortened and $k(t)$ can be made equal to k_e in a finite period of time by considering results from the minimum-time problem [1] of optimal control theory. These results are most easily obtained by first converting the model into standard state-space notation. For z as defined above, let

$$x(t) = z(t) - z_e, \quad \dot{x}(t) = \dot{z}(t), \qquad s(t) = s_e + cu(t)$$

with $|u(t)| \leq 1, c > 0$. The saving rate has been split up into an equilibrium component, s_e, and a policy component, $cu(t)$. This latter component represents the extent to which fiscal policy can influence the saving rate. Empirical evidence suggests that $0 < s(t) < 1$ so that values of c should not violate these restrictions.
   The optimal control problem is to minimize the performance index

$$\int_0^{t_f} dt,$$

subject to

$$\dot{x}(t) = -n(1 - \alpha)(x(t) + z_e) + (1 - \alpha)(s_e + cu(t)),$$

$$x(0) = k_0^{1-\alpha} - k_e^{1-\alpha}, \qquad x(t_f) = 0,$$

$$c > 0,$$

$$0 < (s_e + cu(t)) < 1.$$

To solve this problem, consider the Hamiltonian

(16) $H = 1 + p(t)[-n(1 - \alpha)(x(t) + z_e)] + p(t)[(1 - \alpha)(s_e + cu(t))],$

where $p(t)$ is a costate variable. Pontryagin's minimum principle is expressed as

$$p^*(t)[(1 - \alpha)c]u^*(t) \leq p^*(t)[(1 - \alpha)c]u(t).$$

Since by assumption $(1 - \alpha)c > 0$, the control which minimizes (16) is given by

$$u(t) = -\text{sgn}\{p(t)\}$$

where sgn denotes the signum function. It is necessary for $p(t)$ to satisfy the differential equation

(17) $$\dot{p}(t) = -\frac{\partial H}{\partial x(t)}.$$

Equation (17) implies that $p(0) \neq 0$ and $\text{sgn}\{p(t)\} = \text{sgn}\{p(0)\}$. Consequently no switching of controls can occur and therefore the policy that minimizes H is given by $u(t) = +1$ or $u(t) = -1$ for all $t > 0$. It is further clear from (17) that $p(t) \neq 0$ for finite t. Therefore there is no time interval for which Pontryagin's minimum principle fails to determine the minimum-time relationship between $x(t)$ and $u(t)$. Since the minimum-time problem is based upon dynamics of a first-order linear equation and since $\partial \dot{x}/\partial x < 0$, it is well known that an optimal control solution exists. The control law which moves any $x(t) \neq 0$ to zero in minimum time is given by

(18) $$u^* = -\text{sgn}\{x(t)\}.$$

The optimal control, u^*, is seen to be of the "bang-bang" type and with linear first-order dynamics, $x(t)$ does not change sign and no switching in the control will occur [1].

The solution to the minimum-time problem is also easily presented in a diagram using (10). Consider an initial situation in which $k_0 < k_e$ (i.e., $x(0) < 0$). From (18), maximum capital formation at each moment of time and the quickest movement from k to k_e are obtained when $u^* = +1$. By setting $s(t) = s_e + c$, the different adjustment paths are illustrated by a comparison of the two phase lines

(19) $$\dot{k}_e = s_e k^\alpha - nk,$$

(20) $$\dot{k}_+ = s(t)k^\alpha - nk,$$

in Fig. 1. Asymptotic adjustment according to (19) is represented by segment AD whereas the minimum-time adjustment path is calculated from (20) and is illustrated by BC. Fiscal policy as determined by $s = s(t)$ causes greater capital formation than is the case when $s = s_e$ with the result that k_e can be reached in a finite period of time.

Finally for given initial conditions and parameter values, the solution of this problem determines the minimum time t^* to bring about complete adjustment of $k(t)$ to k_e. In terms of k's this expression is obtained by substituting $s(t)$ for s and k_e for k_λ into (15)

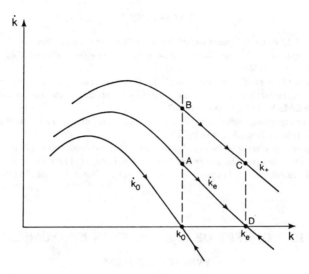

FIG. 1. *Phase diagram.*

with the result

$$t^* = \frac{-1}{n(1-\alpha)} \ln \left| \frac{(s_e + cu^*)/n - k_e^{1-\alpha}}{(s_e + cu^*)/n - k_0^{1-\alpha}} \right|.$$

This substitution allows the explicit determination of the time required for complete adjustment. t^* is also seen to be a function of c which reflects the impact that fiscal policy can have on $s(t)$. For the parameter values used previously, this relationship is summarized in Table 2. The conclusion is that the more powerful is fiscal policy the quicker the movement from k_0 to k_e. This of course is what the minimum-time problem is designed to determine.

TABLE 2
Minimum time for complete adjustment*

	Fiscal policy parameter c			
	0.01	0.05	0.1	0.15
t^*(years):	25.4	6.7	3.5	2.4

*Initial values as in Table 1.

While optimal control theory has been applied to the most basic version of the neoclassical growth model, the techniques employed in this paper can be applied to other versions when dynamic behaviour can be described by a single first-order differential equation. If more than a first-order dynamic system results, then the adjustment problem becomes more complicated and a solution to the minimum-time problem may not exist. In Lewis [4] [5], some second-order systems are considered. Depending on the particular model examined analytical solutions for t^* may not be possible, but computer simulation can always be used.

Acknowledgments. Computing support from the University of Alberta is appreciated. Referee comments were also very helpful.

REFERENCES

[1] M. ATHANS AND P. L. FALB, *Optimal Control*, McGraw-Hill, New York, 1966.
[2] J. CONLISK, *Unemployment in a neoclassical growth model: the effect on speed of adjustment*, Economic J., 76 (1966), pp. 550–566.
[3] H. JONES, *Modern Theories of Economic Growth*, Thomas Nelson and Sons, London, 1975.
[4] S. D. LEWIS, *Adjustment time and optimal control of neoclassical monetary growth models*, Optim. Control Appl. Meth., 2 (1981), pp. 251–267.
[5] S. D. LEWIS, *Price expectations and adjustment time in neoclassical monetary growth models*, Internat. J. Policy Anal. Inform. Syst., 5 (1981), pp. 125–137.
[6] K. SATO, *On adjustment time in neoclassical growth models*, Rev. Econ. Stud., 33 (1966), pp. 263–268.
[7] R. SATO, *Fiscal policy in a neoclassical growth model*, Rev. Econ. Stud., 30 (1963), pp. 16–23.
[8] R. M. SOLOW, *A contribution to the theory of economic growth*, Quart. J. Econ., 70 (1956), pp. 65–94.

THE CONCEPT OF ELASTICITY IN ECONOMICS*

YVES NIEVERGELT†

Abstract. The following pages introduce the notion of elasticity and present its main mathematical and graphical properties, illustrating them on real economic cases, such as the consumption of marijuana and the demand for hospital beds.

1. Introduction. Economists often use the notion of *elasticity* of a function, which they define to be (see [4, pp. 116–117])

$$(1) \qquad \eta_F(x) := \lim_{\Delta x \to 0} \frac{(\Delta F)/F}{(\Delta x)/x} = \lim_{\Delta x \to 0} \frac{x}{F(x)} \cdot \frac{F(x + \Delta x) - F(x)}{\Delta x},$$

where F is usually a positive function defined on all or part of the positive real axis $\mathbb{R}_+^*$. In this situation, $\eta_F(x)$ exists if and only if the derivative $F'(x)$ does, and then

$$(2) \qquad \eta_F(x) = \frac{x}{F(x)} \cdot F'(x).$$

Thus, $\eta_F(x)$ measures the ratio of the *relative* change in $F(x)$ to the corresponding *relative* change in x, whereas $F'(x)$ measures the ratio of their absolute changes.

This article explains how the concept of elasticity relates to real life, shows how to visualize η on logarithmic graph paper, and presents a few calculus rules governing elasticities. This may provide useful suggestions to those whose interests lie in fields other than economics, but who teach calculus to business students. First, we illustrate the above definition by the two kinds of functions most frequently encountered in the classroom and in practice.

Example 1.1. Power functions of the type $F : \mathbb{R}_+^* \to \mathbb{R}_+^*, x \mapsto A \cdot x^\alpha$ have a constant elasticity, equal to α (and conversely, by (9) below):

$$(3) \qquad \eta_F(x) = \frac{x}{F(x)} \cdot F'(x) = \frac{x}{A \cdot x^\alpha} \cdot A \cdot \alpha \cdot x^{\alpha-1} = \alpha.$$

Example 1.2. Consider an affine function $D :]0, b/m[\to \mathbb{R}_+^*, p \mapsto -m \cdot p + b$ with positive coefficients m and b. Then (see Fig. 1):

---

*Received by the editors March 26, 1982, and in revised form June 17, 1982. The work of this author was partially supported by a grant from the Swiss National Research Fund.
†Department of Mathematics, University of Washington, Seattle, Washington 98195.

$$\text{(4a)} \qquad \eta_D(p) = \frac{p}{D(p)} \cdot D'(p) = \frac{p}{-mp + b} \cdot (-m) = \frac{p}{p - b/m} = -\frac{OP}{AP}.$$

Since $-mp = D(p) - b$, one can also write:

$$\text{(4b)} \qquad \eta_D(p) = \frac{p}{D(p)} \cdot D'(p) = \frac{-mp}{D(p)} = \frac{D(p) - b}{D(p)} = -\frac{BQ}{OQ}.$$

2. Application. Suppose that p represents the unit price of some commodity, say in dollars per item, and $q := D(p)$ the quantity sold at price p in one time period, say in items per week. (Economists call D a *demand curve*.) Denote P the inverse function of D, so that $p = P(q)$; then η_D determines the maximum of the total revenue function TR as follows:

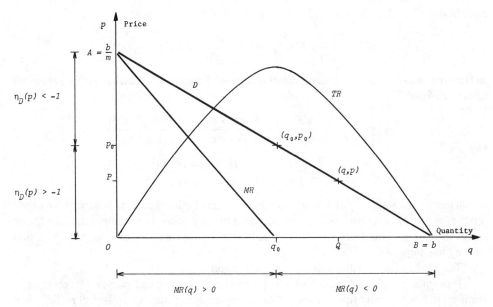

FIG. 1. *The domain of D lies on the vertical axis, whereas* TR *and* MR *have theirs on the horizontal axis.*

since $\text{TR}(q) := p \cdot q = P(q) \cdot q$ (price times quantity), differentiating with respect to q yields:

$$\text{TR}' = P + q \cdot P' = P \cdot \left(1 + \frac{q}{P} \cdot \frac{dP}{dq}\right) = P \cdot \left(1 + \left[\frac{P}{q} \cdot D' \circ P\right]^{-1}\right) = P \cdot \left(1 + \frac{1}{\eta_D \circ P}\right).$$

Consequently, the marginal revenue $\text{MR}(q) := \text{TR}'(q)$ is positive where $\eta_D(p) < -1$ and negative where $\eta_D(p) > -1$ (with $p = P(q)$). For instance, if D is an affine function as in Example 1.2, then total revenues reach their maximum at the midpoint (q_0, p_0), where $\eta_D(p_0) = -1$ by (4a), and therefore $\text{TR}'(q_0) = 0$ (see Fig. 1). Students will enjoy working out the following two real cases:

Example 2.1. Belinfante and Davis [1] found that the demand for record albums was $q = D(p) = -88.3 \cdot p + 1821$, so that $p = -0.01133 \cdot q + 20.62$. Hence, total revenues will be largest with a price tag $p_0 = 20.62/2$, i.e. \$10.31 per album, and then $q_0 = 1821/2 \cong 910$ records will be sold.

Example 2.2. Hogarty and Elzinga [3] estimated the price elasticity of demand for beer to be a constant $\eta_D \equiv -1$, with $q = D(p) = 123/p$ (where p is in cents per can and q in cans per adult per day). In this case, total revenues do not depend on price: $\text{TR}(q) = \text{TR} \circ D(p) = p \cdot D(p) = 123\cent$ (per adult per day) regardless of price and quantity.

3. Visualizing η. Plotting a function F against logarithmic scales produces a curve whose slope equals the elasticity η_F of F: performing the changes of coordinates

$$(5) \qquad \phi : u \mapsto e^u \quad \text{and} \quad \psi : v \mapsto \ln(v)$$

and forming the composite (which amounts to changing scales)

$$(6) \qquad f := \psi \circ F \circ \phi,$$

one obtains $f' = (\psi' \circ (F \circ \phi)) \cdot (F' \circ \phi) \cdot \phi'$. Hence,

$$(7) \qquad f'(u) = \frac{1}{F(e^u)} \cdot F'(e^u) \cdot e^u = \frac{e^u}{F(e^u)} \cdot F'(e^u) = \eta_F(e^u),$$

so that the slope of f at u is indeed the elasticity of F at $x = e^u$. An alternative argument runs as follows:

$$(8) \qquad \eta_F(x) = \frac{x}{F(x)} \cdot \frac{dF}{dx} = \frac{(1/F(x)) \cdot dF}{(1/x) \cdot dx} = \frac{d[\text{Log}(F(x))]}{d[\text{Log}(x)]}$$

$$= \lim_{\Delta x \to 0} \frac{\text{Log}[F(x + \Delta x)] - \text{Log}[F(x)]}{\text{Log}(x + \Delta x) - \text{Log}(x)}$$

which equals the slope of f at $u = \ln(x)$. Observe that both arguments hold with any base for the logarithm, e.g. 10 or e. Of course, the slope of f must be measured against linear scales to yield η_F. This gives an easy way to locate the values of x where $\eta_F(x) = -1$: find where f has slope -1.

The next example will catch students' attention.

Example 3.1. Nisbet and Vakil [6] proposed two demand curves for marijuana among UCLA students (quantities in "lids" per month, prices in dollars per "lid"; one "lid" equals one dried ounce):

(A) $q = D_1(p) := 15.4 \cdot p^{-1.013}$, i.e. $p = 14.8 \cdot q^{-0.987}$,

(B) $q = D_2(p) := -0.225 \cdot p + 3.74$, i.e. $p = -4.44 \cdot q + 16.6$.

D_1 has constant elasticity -1.013 and therefore would simply appear as a straight line with slope -1.013 on logarithmic scales. At the going price of $10 per ounce, D_2 has elasticity -1.51, the slope of the tangent to $\psi \circ D_2 \circ \phi$ at $(q^*, p^*) := (1.49, 10)$. Sliding a straight edge with slope -1 until it is tangent to the curve shows that D_2 has elasticity -1 at $(q_0, p_0) := (1.87, 8.30)$, where revenues would be maximal (see Fig. 2).

4. Calculus with elasticities. The elasticities η_F and η_G of two given positive functions $F, G : \mathbb{R}_+^* \to \mathbb{R}_+^*$ follow rules similar to those of elementary calculus:

(a) $\eta_{FG} = \eta_F + \eta_G$,

(b) $\eta_{F/G} = \eta_F - \eta_G$,

(c) $\eta_{\lambda F} = \eta_F$,

(d) $\eta_{F^\alpha} = \alpha \cdot \eta_F$,

(e) $\eta_{F \circ G} = (\eta_F \circ G) \cdot \eta_G$ (chain rule),

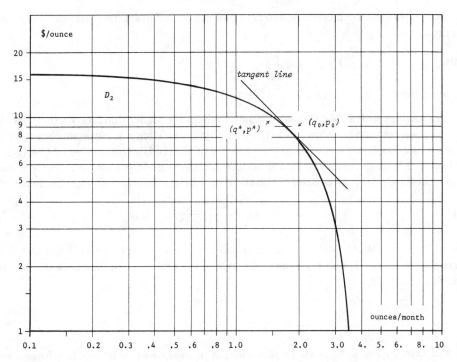

FIG. 2. $D_2(p) = -0.225 \cdot p + 3.74$, *on logarithmic scales.* $(q^*, p^*) = (1.49, 10)$ *shows the prevailing market situation, while* $(q_0, p_0) = (1.87, 8.30)$ *indicates the optimal point, where the tangent line to* D_2 *(drawn in) has slope* -1.

where α and λ are constants. Let us prove (a) and (e) as examples:

(a) $\quad \eta_{FG} = \dfrac{x}{F \cdot G} (F \cdot G)' = \dfrac{x}{F \cdot G} \cdot (F'G + F \cdot G') = \eta_F + \eta_G,$

(e) $\quad \eta_{F \circ G} = \dfrac{x}{F \circ G} \cdot (F \circ G)' = \dfrac{x}{F \circ G} \cdot (F' \circ G) \cdot G'$

$$= \dfrac{G}{F \circ G} \cdot (F' \circ G) \cdot \dfrac{x}{G} \cdot G' = (\eta_F \circ G) \cdot \eta_G.$$

The proof of (b) is similar to that of (a), while (c) and (d) constitute particular cases of (e).

A function F can also be recovered from its elasticity as follows: since $\eta_F = (x/F) \cdot dF/dx$ then $dF/F = (\eta_F/x) \cdot dx$, and an integration yields

$$\ln |F(x)| = \int \frac{\eta_F(x)}{x} \cdot dx + \ln |K|,$$

(9)

$$F(x) = K \cdot \exp \left[\int \frac{\eta_F(x)}{x} \cdot dx \right]$$

for some constant K. Here is an application of this formula.

Example 4.1. Cullis, Forster and Frost [2] studied the demand for inpatient treatment, whose elasticity can be written $\eta_D(S) = -2.19 \cdot S$, where S denotes the number of available hospital beds, and D that of deaths and discharges per year. Equation (9) then yields

$$(10) \qquad D(S) = K \cdot \exp\left[\int \frac{-2.19 \cdot S}{S} \cdot dS\right] = K \cdot e^{-2.19 \cdot S}.$$

(K was estimated to be about 5.17.)

5. Analogy with an elastic rubber band. Perhaps the reader wonders whether there is a connection between economic elasticity and physical elasticity. It turns out that there is a formal similarity between the price elasticity of demand, written $\eta_D = (d \, \mathrm{Log}\,(D))/(d \, \mathrm{Log}\,(p))$, and the tensor σ that describes the stress undergone by an isotropic and homogeneous elastic rubber band of volume V subject to a deformation gradient tensor λ:

$$(11) \qquad \sigma = -\frac{k \cdot T}{V} \cdot \frac{d \, \mathrm{Log}\,(Z)}{d \, \mathrm{Log}\,(\lambda)},$$

where Z represents the partition function and $\Psi = -kT \cdot \mathrm{Log}\,(Z)$ the Helmholtz free energy. (See [5, p. 62] and [7, p. 64].) Of course, the present paper does not tell how to stretch a dollar bill.

REFERENCES

[1] A. BELINFANTE AND R. R. DAVIS, JR., *Estimating the demand for record albums*, Rev. Business and Economic Research, XIV/2 (winter 1978–1979), pp. 47–53.

[2] J. G. CULLIS, D. P. FORSTER AND C. E. B. FROST, *The demand for inpatient treatment: some recent evidence.* Appl. Economics, 12 (1980), pp. 43–60.

[3] T. F. HOGARTY AND K. G. ELZINGA, *The demand for beer*, Rev. Economics and Statistics, LIV/2 (1972), pp. 195–198.

[4] E. MANSFIELD, *Microeconomics, Theory and Applications*, 3rd. ed., W. W. Norton, New York, 1979.

[5] F. D. MURNAGHAN, *Finite Deformation of an Elastic Solid*, John Wiley, New York, 1951.

[6] C. T. NISBET AND F. VAKIL, *Some estimates of price and expenditure elasticities of demand for marijuana among U.C.L.A. students*, Rev. Economics and Statistics, LIV/4 (1972), pp. 473–475.

[7] C. J. THOMPSON, *Mathematical Statistical Mechanics*, Macmillan, New York, 1972.

Supplementary References
Mathematical Economics

[1] W. J. BAUMOL, *Economic Theory and Operations Analysis*, Prentice-Hall, N.J., 1972.

[2] J. V. BAXLEY AND J. C. MOORHOUSE, "*Lagrange multiplier problems in economics*," AMM (1984) pp. 404–412.

[3] R. A. BREALEY, *An Introduction to Risk and Return from Common Stocks*," M.I.T. Press, Cambridge, 1969.

[4] R. E. BEARD, "*Non-life insurance*," BIMA (1972) pp. 21–28.

[5] G. BEER, "*The Cobb-Douglas production function*," MM (1980) pp. 44–48.

[6] A. R. BERGSTROM, *Selected Economic Models and their Analysis*, Elsevier, N.Y., 1967.

[7] J. BLINN, *Patterns and Configurations in Economic Science*, Reidel, Dordrecht, 1973.

[8] M. BRENNER, ed., *Option Pricing: Theory and Applications*, Lexington Books, Boston, 1983.

[9] F. R. BUIANOUCKAS, "*A survey: Non-cooperative games and a model of the business cycle*," AMM (1978) pp. 146–155.

[10] E. BURMEISTER AND A. R. DOBELL, *Mathematical Theories of Economic Growth*, Macmillan, N.Y., 1970.

[11] J. W. S. CASSELS, *Economics for Mathematicians*, Cambridge University Press, Cambridge, 1982.

[12] A. C. CHIANG, *Fundamental Methods of Mathematical Economics*, McGraw-Hill, N.Y., 1974.

[13] C. W. CLARK, *Mathematical Bioeconomics: The Optimal Management of Renewable Resources*, Wiley, N.Y., 1976.

[14] ———, "*Mathematical models in the economics of renewable resources*," SIAM Rev. (1979) pp. 81–99.

[15] D. L. CLIMENTS, *An Introduction to Mathematical Models in Economic Dynamics*, Polygonal, Passaic, N.J., 1984.

[16] E. H. DAVIS, "*Cauchy and economic cobwebs*," MM (1980) pp. 171–173.

[17] R. H. DAY AND T. GROVES, eds., *Adaptive Economic Models*, Academic, N.Y., 1975.

[18] R. H. DAY AND S. M. ROBINSON, eds., *Mathematical Topics in Economic Theory and Computation*, SIAM, 1972.

[19] R. DOBBINS AND S. F. WITT, *Portfolio Theory and Investment Management*, M. Robertson, Oxford, 1983.

[20] P. J. DHRYMES, *Econometrics*, Harper and Row, N.Y., 1970

[21] E. J. ELTON AND M. J. GRUBER, *Modern Portfolio Theory and Investment Analysis*, Wiley, N.Y., 1981.

[22] P. C. FISHBURN AND H. O. POLLAK, "Fixed-route cost allocation," AMM (1983) pp. 366–377.

[23] J. FRANKLIN, "Mathematical methods of economics," AMM (1983) pp. 229–243.

[24] ———, *Methods of Mathematical Economics*, Springer-Verlag, N.Y., 1980.

[25] D. GALE, *The Theory of Linear Economic Models*, McGraw-Hill, N.Y., 1970.

[26] G. GANDOLFO, *Mathematical Methods and Models in Economic Dynamics*, North-Holland, Amsterdam, 1971.

[27] J. HADAR, *Mathematical Theory of Economic Behavior*, Addison-Wesley, Reading, 1971.

[28] G. HEAL, G. HUGHES AND R. TARLING, *Linear Algebra and Linear Economics*, Macmillan, London, 1974.

[29] J. M. HENDERSON AND R. E. QUANDT, *Microeconomic Theory: A Mathematical Approach*, McGraw-Hill, N.Y., 1971.

[30] R. HENN AND O. MOESCHLIN, eds., *Mathematical Economics and Game Theory*, Springer-Verlag, N.Y., 1977.

[31] J. HIRSCHLEIFER, *Price Theory and Applications*, Prentice-Hall, N.J., 1980.

[32] P. HUDSON, "Simple mathematics in simple economic planning," Math. Gaz. (1977) pp. 105–119.

[33] R. A. JARROW AND A. RUDD, *Option Pricing*, R. D. Irwin Inc., Homewood, Il., 1983.

[34] E. KALAI, "Proportional solutions to bargaining situations: Interpersonal utility comparisons," Econometrica (1977) pp. 1623–1630.

[35] J. N. KAPUR, "Optimal portfolio selection," Int. J. Math. Educ. Sci. Tech. (1983) pp. 313–332.

[36] M. G. KENDALL, ed., *Mathematical Model Building in Economics and Industry*, Hafner, N.Y., 1968.

[37] J. M. KEYNES, *General Theory of Employment, Interest and Money*, Macmillan, London, 1973.

[38] E. KLEIN, *Mathematical Models in Theoretical Economics: Topological and Vector Space Foundations of Equilibrium Analysis*, Academic, N.Y., 1973.

[39] D. O. KOEHLER, "The cost accounting problem," MM (1980) pp. 3–12.

[40] T. C. KOOPMANS AND A. F. BAUSCH, "Selecting topics in economics involving mathematical reasoning," SIAM Rev. (1959) pp. 79–148.

[41] A. KOUTSOYIANNIS, *Theory of Econometrics*, Macmillan, London, 1973.

[42] W. LEONTIEF, *Essays in Economics: Theories, Facts and Policies*, Sharpe, N.Y., 1977.

[43] S. C. LITTLECHILD, "Common costs, fixed charges, club and games," Review Econ. Studies (1975) pp. 117–124.

[44] R. LUCE AND H. RAIFFA, *Games and Decisions*, Wiley, N.Y., 1957.

[45] V. L. MAKAROV AND A. M. RUBINOV, *Mathematical Theory of Economic Dynamics and Equilibria*, Springer-Verlag, N.Y., 1977.

[46] A. G. MALLIARIS, "Martingale methods in financial decision-making," SIAM Rev. (1981) pp. 434–443.

[47] L. J. MIRMAN AND Y. TAUMAN, "Demand compatible equitable cost sharing prices," Math. Oper. Res. (1982) pp. 40–56.

[48] Y. MURATA, *Mathematics for Stability and Optimization of Economic Systems*, Academic, N.Y., 1977.

[49] D. G. NEWMAN, *Engineering Economic Analysis*, Engineering Press, San Jose, 1983.

[50] W. NICHOLSON, *Microeconomic Theory*, Dryden, Illinois, 1978.

[51] H. NIKAIDO, *Convex Structures and Economic Theory*, Academic, N.Y., 1968.

[52] E. R. PIKE, "Mathematics and investment," BIMA (1984) pp. 162–164.

[53] J. A. RICKAND AND A. M. RUSSELL, "Interest rates in perspective; and a new inequality," Math. Sci. (1981) pp. 111–121.

[54] P. A. SAMUELSON, *Economics*, McGraw-Hill, N.Y., 1973.

[55] ———, "Mathematics of speculative price," SIAM Rev. (1973) pp. 1–42.

[56] H. SCARF, *The Computation of Economic Equilibria*, Yale University Press, New Haven, 1974.

[57] H. E. SCARF, "Fixed-point theorems and economic analysis," Amer. Sci. (1983) pp. 289–296.

[58] P. SHEPHEARD, "A test of the random walk theory of stock market prices," BIMA (1977) pp. 2–7.

[59] R. W. SHEPARD, *The Theory of Cost and Production Functions*, Princeton University Press, Princeton, 1970.

[60] R. SOLOW, "The economics of resources or the resources of economics," Amer. Econ. Rev. (1974) pp. 1–14.

[61] J. E. SPENCER AND R. C. GEARY, *Exercises in Mathematical Economics and Econometrics*, Griffin, Great Britain, 1974.

[62] H. E. THOMPSON, *Applications of Calculus in Business and Economics*, Benjamin, Reading, 1972.

[63] R. C. THOMPSON, "The true growth rate and the inflation balancing principle," AMM (1983) pp. 207–210.

[64] A. G. WILSON, "The evolution of urban spatial structure: A review of progress and research problems using S.I.A. models," BIMA (1982) pp. 90–100.

10. Optimization (*Including linear, nonlinear, dynamic, and geometric programming, control theory, games and other miscellaneous topics*)

COMPLEMENTARY SLACKNESS AND DYNAMIC PROGRAMMING*

MORTON KLEIN†

Abstract. It is shown that a policy improvement algorithm for a discrete discounted cost dynamic programming problem can be obtained from the complementary slackness theorem of linear programming.

Introduction. Consider a finite state discounted cost deterministic (for simplicity) infinite horizon dynamic programming problem. It is well known that the policy improvement procedure for solving such a problem coincides with the simplex method solution to an equivalent linear program, except that with policy improvement more than one variable change can occur at each iteration [1], [2]. What is perhaps not as well known is that a slight variation of the policy improvement algorithm can be obtained from the complementary slackness theorem of linear programming. This observation, although extremely simple, does not seem to have appeared in the literature. It provides, in a linear programming course, another use of complementary slackness.

Discrete dynamic programming. The dynamic programming problem of interest here is that of solving the functional equation

$$(1) \qquad v_i = \min_{j \in K(i)} \{c_{ij} + \alpha v_j\}, \qquad i = 0, \cdots, n,$$

where v_i is the total discounted cost over an unbounded discrete time horizon associated with an optimal policy when a system starts in state i, α ($0 \leq \alpha < 1$) is a given one period discount factor, c_{ij} is the cost associated with the decision to make a one period transition from state i to j, and $K(i)$ is the set of states reachable in one step from state i. An optimal policy for this problem is known [1] to consist of a set of (deterministic) transitions, one for each state, which is independent of time.

Linear programming. As shown in [2, pp. 448–449] an optimal solution to the dynamic programming problem can also be found by solving the linear program

I: Maximize $\sum_{i=0}^{n} v_i$, subject to the constraints:

$$(2) \qquad v_i - \alpha v_j \leq c_{ij}, \qquad i = 0, \cdots, n, \quad j \in K(i).$$

The idea behind the transformation is that (1) implies that the v's must satisfy (2) and in order for (2) to satisfy (1), for each state i, (2) must hold with equality for at least one (ij). Any positive numbers can be used as the coefficients of the v_i in the objective function [2, p. 449]; we have used ones for computational convenience.

Given problem I, an optimal policy can be also obtained by solving its dual problem

II: Find nonnegative x_{ij}'s which minimize $\sum_i \sum_j x_{ij} c_{ij}$ subject to the constraints

$$(3) \qquad \sum_{j \in K(i)} x_{ij} - \alpha \sum_{j \in K(i)} x_{ji} = 1, \qquad i = 0, \cdots, n.$$

In [1] it is shown that if the simplex method is used to solve problem II, each basic feasible solution will have exactly one positive variable, say x_{ij}, for each state and hence is never degenerate. Each positive variable can be interpreted as an instruction to make a

---

*Received by the editors September 20, 1981, and in revised form June 10, 1982.

†Department of Industrial Engineering and Operations Research, Columbia University, New York, New York 10027.

transition to state $\bar{j}$ whenever the system is in state i. Thus, a basic feasible solution corresponds to a feasible policy for the dynamic programming problem; although we don't use the simplex method, we will take advantage of this one-to-one correspondence.

The dual variables also carry another interpretation. This interpretation depends on the coefficients attached to the v's in the primal objective function (and consequently on the values of the right side constants in the dual's constraints). In our formulation using *ones* as the coefficients of the v's, the x_{ij}'s represent the discounted number of times transition (ij) will occur if the process is started in each stage *once*. In a stochastic setting, the positive coefficients associated with each v would represent the probabilities of starting the process in that state. In such a case, each x_{ij} could be thought of as the expected discounted number of times each transition from state i to state j is made.

As indicated earlier, our approach uses the complementary slackness theorem, which states that feasible solutions to I and II are optimal if and only if

(4) $$x_{ij}(c_{ij} - v_i + \alpha v_j) = 0 \quad \text{for all } i = 0, \cdots, n, \quad j \in K(i).$$

A policy improvement method can now be constructed:

Step 0. Choose any feasible policy for the dynamic programming problem. That is, select a set of $n + 1$ x_{ij}'s, one for each state i, each of which will be positive. (Note: Actual values of the x_{ij}'s are not needed. We only have to decide which ones will be positive because of the 1–1 correspondence between feasible policies and the basic feasible solutions of problem II.)

Step 1. Construct a solution to primal problem I, using (4), by setting $c_{ij} - v_i + \alpha v_j = 0$ for each instance in which x_{ij} is positive. This involves solving a system of $n + 1$ equations in the same number of variables for which a unique solution can always be found [1].

Step 2. Calculate the quantities

(5) $$c_{ij} - v_i + \alpha v_j \quad \text{for each } x_{ij} = 0.$$

Two cases can occur:

A. $c_{ij} - v_i + \alpha v_j \geq 0$ for all $x_{ij} = 0$. If this happens the primal solution is feasible and the policy being evaluated is optimal since the x_{ij}'s are feasible and complementary slackness holds.

B. One or more of the quantities $c_{ij} - v_i + \alpha v_j < 0$. If this happens, the primal solution is not feasible and the current policy can be improved.

Suppose that

(6) $$c_{ik} - v_i + \alpha v_k < 0,$$

and the current policy calls for transition (ij), that is, $x_{ij} > 0$. Then because of (6), an improved policy calls for transition (ik). Such improvements can be made for each instance in which (6) occurs. It may be noted that several of the calculated quantities can be negative for the same state i, that is, for a given state i, we can have $c_{ij} - v_i + \alpha v_j < 0$, for $j = k^{(1)}$, $j = k^{(2)}$, etc. If this occurs any one of the associated policy changes can be made.

To see that such a policy change leads to a strict improvement, consider the dual objective function. Using (3), since the dual solution is feasible, we write it's value in the form

$$\sum_{i=0}^{n} \sum_{j \in K(i)} x_{ij} c_{ij} + \sum_{i=0}^{n} v_i \left(1 - \sum_{j \in K(i)} x_{ij} + \alpha \sum_{j \in K(i)} x_{ij} \right).$$

After some rearrangement, this reduces to

$$\sum_{i=0}^{n} \sum_{j \in K(i)} x_{ij}(c_{ij} - v_i + \alpha v_j) + \sum_{i=0}^{h} v_i.$$

Hence, since the v_i's are constants for a given policy, if e.g. $c_{ik} - v_i + \alpha v_k$ is negative, the value of the dual can be reduced by changing the associated zero-valued dual variable x_{ik} to a positive value and changing the value of the current positive-valued variable, say $x_{i.}$, to zero.

A numerical example. Find: v_0, v_1, and v_2 to satisfy $v_i = \min_j \{c_{ij} + .8v_j\}$, with transition costs

$[c_{ij}] =$	$i \backslash j$	0	1	2
	0	5	8	11
	1	0	6	9
	2	M	1	7

(Here M is a very large number and the discount factor $\alpha = .8$.)

The equivalent (primal) linear program is to find v_0, v_1, v_2 which are not sign-constrained and which

maximize $v_0 + v_1 + v_2$,

subject to $.2v_0 \qquad\qquad\qquad \le 5$,

$\qquad\qquad\qquad v_0 - .8v_1 \qquad\qquad \le 8$,

$\qquad\qquad\qquad v_0 \qquad - .8v_2 \le 11$,

$\qquad\qquad\qquad -.8v_0 + v_1 \qquad\qquad \le 0$,

$\qquad\qquad\qquad\qquad .2v_1 \qquad\qquad \le 6$,

$\qquad\qquad\qquad\qquad v_1 - .8v_2 \le 9$,

$\qquad\qquad\qquad\quad - .8v_1 + v_2 \quad \le 1$,

$\qquad\qquad\qquad\qquad\qquad .2v_2 \le 7$.

Its dual is to find nonnegative x_{ij}'s, $i, j = 0, 1, 2$, which

minimize $5x_{00} + 8x_{01} + 11x_{02} + 6x_{11} + 9x_{12} + x_{21} + 7x_{22}$,

subject to $.2x_{00} + x_{01} + x_{02} - .8x_{10} = 1$,

$\qquad\qquad -.8x_{01} + x_{10} + .2x_{11} + x_{12} - .8x_{21} = 1$,

$\qquad\qquad -.8x_{02} - .8x_{12} + x_{21} + .2x_{22} = 1$.

Calculations. *Step* 0. Let (i, j) represent a decision to make a transition from state i to state j. Then $\{(0, 0), (1, 0), (2, 1)\}$ represents an initial policy. This policy implies that the dual variables x_{00}, x_{10} and x_2 will be positive and all others will be equal to zero.

Step 1. For the positive x_{ij}'s, set

$$5 - .2v_0 = 0 \qquad \text{(since } x_{00} > 0\text{),}$$
$$0 - v_1 + .8v_0 = 0 \qquad \text{(since } x_{10} > 0\text{),}$$
$$1 - v_2 + .8v_1 = 0 \qquad \text{(since } x_{21} > 0\text{).}$$

Solving yields $v_0 = 25$, $v_1 = 20$, $v_2 = 17$ with associated value $= 62$.

Step 2. For each dual $x_{ij} = 0$, evaluate $c_{ij} - v_i + \alpha v_j$:

$$x_{01}: \ 8 - v_0 + .8v_1 = 8 - 25 + 16 = -1,$$
$$x_{12}: \ 9 - v_1 + .8v_2 = 9 - 20 + 13.6 = 2.6,$$
$$x_{02}: 11 - v_0 + .8v_2 = 11 - 25 + 13.6 = -.4,$$
$$x_{11}: \ 6 - .2v_1 = 6 - 4 = 2,$$
$$x_{22}: \ 7 - .2v_2 = 7 - 3.4 = 3.6$$

Since the values associated with x_{01} and x_{02} are negative the policy being evaluated is not optimal; either of the associated policy changes would lead to an improvement. In this case, since both changes involve transitions from state 0, they cannot be made at the same time. A simple choice is the most negative valued transition $(0, 1)$; thus an improved policy is $\{(0, 1), (1, 0), (2, 1)\}$.

We now return to

Step 1. For the new policy, set

$$8 - v_0 + .8v_1 = 0 \qquad \text{(since } x_{01} > 0\text{),}$$
$$0 - v_1 + .8v_0 = 0 \qquad \text{(since } x_{10} > 0\text{),}$$
$$1 - v_2 + .8v_1 = 0 \qquad \text{(since } x_{21} > 0\text{).}$$

The solution is $v_0 = 22.22$. $v_1 = 17.78$, $v_2 = 15.22$ with associated value $= 55.22$.

Step 2. For each dual $x_{ij} = 0$, evaluate $c_{ij} - v_i + \alpha v_j$:

$$x_{00}: \ 5 - .2v_0 = 5 - .2 (22.22) = .56$$
$$x_{11}: \ 6 - .2v_1 = 6 - .2 (17.78) = 2.44$$
$$x_{22}: \ 7 - .2v_2 = 7 - .2 (15.22) = 3.96$$
$$x_{20}: 11 - v_0 + .8v_2 = 11 - 22.22 + .8 (22.22) = 6.56$$
$$x_{12}: \ 9 - v_1 + .8v_2 = 9 - 17.78 + .8 (22.22) = 9.$$

Since all evaluations are nonnegative, the policy tested is optimal.

The values of the dual variables may also be found by solving the system of dual constraints. Since we know that x_{00}, x_{11}, x_{22}, x_{20} and x_{12} are equal to zero, the system reduces to

$$x_{01} - .8x_{10} \qquad\qquad = 1,$$
$$-.8x_{01} + \quad x_{10} - .8x_{21} = 1,$$
$$x_{21} = 1.$$

Its solution is $x_{21} = 1$, $x_{10} = 7 \tfrac{2}{9}$, and $x_{01} = 6\tfrac{7}{9}$.

Remarks. 1. Step 1 is the same as the "value determination" step of the standard policy improvement procedure described in [2].

2. Step 2 is slightly different from the policy improvement step described in [2]. However, if for a given state i there is more than one possible improvement and we choose the new transition from i to be to that state j^* satisfying $\min_{j \in K}(i) \{c_{ij} - v_i + \alpha v_j\}$, then the standard policy improvement method and the one described here call for the same changes.

3. Our method is also very similar to the simplex technique for the transportation problem (see [2, pp. 186–188]). A feasible policy in our problem corresponds to a set of routes in the transportation problem associated with a basic feasible solution. Step 1 is similar to the simplex transportation problem method of finding a set of dual variables which satisfy complementary slackness conditions. Step 2 corresponds to evaluating the "reduced cost" transportation cells to see if the introduction of a nonbasic variable will give rise to an improved solution; at the same time a negative value of $c_{ij} - v_i + \alpha v_j$ implies that the v_i's are not feasible. An optimal solution for the transportation problem (and an optimal policy for our problem) is indicated when a feasible dual (primal) solution is obtained.

REFERENCES

[1] C. DERMAN, *Finite State Markovian Decision Processes* Academic Press, New York, 1970.
[2] H. WAGNER, *Principles of Operations Research,* 2nd ed. Prentice-Hall, Englewood Cliffs, NJ 1975.

The Game of Slash

Problem 76-1, by D. N. BERMAN (University of Waterloo).

The board used here consists of a single row of positions $(1, 2, \cdots, n)$, ordered from left to right, in which a given number of pieces are placed in some fashion among the positions. Only one piece may ever occupy a given position. Alternating play between two players is made by moving any one of the pieces as far to the left as desired but still remaining to the right of the piece immediately on its left. The winner is the player who leaves his opponent no possible move.

Another variation of the game allows the players to move as far to the left as desired to an unoccupied position.

Determine a winning strategy for the game.

Solution by D. J. WILSON (University of Melbourne).

Considering each unoccupied position in turn, count the number of *pieces* lying to its *right*. If this number is odd, place a match in the position as a marker; if it is even, leave the position empty. Thus, between any pair of successive pieces either every position will be occupied by a match or they will all be empty. Regard every collection of matches lying between a pair of successive pieces as a single pile, and consider the position which these piles form in the game of *Nim* [1]. The player whose turn it is to move can make sure of winning if and only if this Nim position is *unsafe*.

A winning move is to take matches away from one of the piles so as to leave a *safe* position in Nim and then make the number of spaces between the corresponding pair of pieces equal to the number of matches in the reduced pile. This can always be done by moving the right-hand piece of the pair.

After the next move the number of matches in exactly one pile will have to be changed. Whether this number is increased or decreased, it is easy to show the position left by the move must again correspond to an unsafe position of Nim. Since the final position of the game corresponds to a safe position of Nim it follows that a player can be sure of winning if he always leaves his opponent with such a position.

REFERENCE

[1] J. C. HOLLADAY, *Cartesian products of termination games*, Contributions to the Theory of Games, vol. III, Annals of Mathematics Studies, no. 39, Princeton University Press, Princeton, N.J., 1957, pp. 189–200.

W. C. Davidon indicated that a program for playing Nim or Slash with an HP-65 pocket calculator is being submitted to the Hewlett-Packard User's Library. It analyses Nim with up to five heaps or Slash positions with up to ten pieces; the number of each Nim heap, or in alternate Slash intervals, can be from zero through 99.

Partial solution of the second variation of Slash by the proposer.

A winning strategy for the second variation of Slash in the case of four pieces can be given in terms of Nim. This includes the strategy for two (three) pieces simply by placing two (one) pieces initially in positions 1, 2(1).

Suppose the pieces are in positions P_1, P_2, P_3, P_4, $P_i \neq P_j$, $i \neq j$. Call this a Slash position. Construct four Nim piles of sizes $P_1 - 1$, $P_2 - 1$, $P_3 - 1$, $P_4 - 1$ and call the Slash position safe if the corresponding Nim position is balanced. The player faced with an unsafe position will win by moving to a safe position at each turn. The proof is a consequence of the following three assertions.

(a) Any move from a safe position always results in an unsafe position.

(b) For any unsafe position, there is a move that makes it safe.

(c) The final Slash position is safe.

Assertion (a) follows directly from Nim. Each move in Slash corresponds to reducing a Nim pile thereby making an unbalanced Nim position. If we start with an unbalanced Nim position we can always balance the piles by reducing one pile. But Nim allows two piles to be the same size and this is not allowed in Slash. However the four piles corresponding to any unsafe position must be of different sizes, and it is easily verified that the resulting balanced position cannot have two equal piles. This proves (b). The Nim piles 0, 1, 2, 3 corresponding to the final Slash position 1, 2, 3, 4 is balanced as asserted in (c).

The above strategy is not a winning strategy for more than four pieces since (b) is not always true. forming two equal Nim piles may be required in moving from an unsafe position to a safe position, even though the original Nim piles were of different sizes.

Comment by R. Silber (North Carolina State University.)

This problem appears in J. Conway's new book, *On Numbers and Games*. It also was discussed by M. Gardner in the September, 1976 issue of Scientific American, where it is called the "silver dollar game without the dollar." The generalized game is deep and is also solved (see C. P. Welter, *The theory of a class of games on a sequence of squares in terms of the advancing operation on a special group*, Proc. Roy. Acad. Amsterdam, Ser. A, 57 (1954), pp. 194–200). Welter's game is discussed briefly in T. H. O'Beirne, *Puzzles and Paradoxes*, Chap. 9, as well as in the aforementioned book of Conway, where Welter's results are presented and simplified.

Optimum Sorting Procedure

Problem 59-3, by Paul Brock.

A fundamental procedure in all business operations is that of filing information. Whenever the information in the file is to be updated, the updating items
are first sorted in accordance with the key of the file. This is the standard alternative to a direct use of the file which is a random access procedure. In many
cases, particularly in multiple file problems, the sorting procedure must be done
by a computer. This is an expensive and time consuming operation. Many
different procedures have been suggested, and each takes a certain amount of
time. It would be useful to determine a minimum computer time procedure.

A general investigation of this problem is under way. In this investigation the
following general problem arises: Given a sequence of positive integers, what is
the expected length of its maximal monotonic nondecreasing subsequence.

The solution depends upon the length of the sequence and the number of allowable integers. The special case for two integers has been solved. The solution
depends upon the following lemma which is proposed as a problem: Among
sequences of fixed length M consisting the p 1's and q 2's, the number of sequences whose maximal monotonic nondecreasing subsequences are of length n
is the same for all p, q such that $0 < p, q < n \leq M$.

Solution by the proposer.

Let $F(p, q, n)$ be the totality of F sequences of length n for a given p, q choice
where an F sequence of length n is defined as a sequence containing a maximum
monotone nondecreasing subsequence of length n. The proof is by induction on
M. The smallest value of M for which

(1) $0 < p, q < n \leq M$

holds is $M = 2$. For this case the theorem is obvious. Now consider the theorem
for sequences of length $\bar{M} < M$: choose a p, q satisfying $p + q = M$ and (1),
and fix $n \leq M$.

Case 1: $n = M$: For any values of $p, q > 0$, there is only one F sequence, namely,
        $11 \cdots 122 \cdots 2$.

Case 2: $n < M$: Assume $q < n - 1$. Since $n < M$, it is clear that $p, q > 2$ for
        (1) to hold. Consider those sequences starting with a 2. Since $q < n$,
        the 2 cannot be counted towards any F sequence. Thus, the total
        number of F sequences starting with a 2 is $F(p, q - 1, n)$. Now con-
        sider those sequences starting with a 1. The 1 must be contained in
        every F sequence. Hence, the totality with this condition satisfied is
        $F(p - 1, q, n - 1)$. Consequently, $F(p, q, n) = F(p, q - 1, n) +$
        $F(p - 1, q, n - 1)$.

Since

$$p + q - 1 < M,$$

$$p, q - 1 < n < M,$$

$$p - 1, q < n - 1 < M,$$

the conditions on the right satisfy the inductive hypothesis. To remove the condition $q < n - 1$, we consider the case of sequence ending in 1 or 2. This yields:

$$F(p, q, n) = F(p - 1, q, n) + F(p, q - 1, n - 1).$$

Here, the condition $p < n - 1$ must be imposed to satisfy the inductive hypothesis. Finally if $p = q = n - 1$, this would be the only combination of p 1's and q 2's satisfying (1) and hence the theorem is true trivially.

J. Gilmore refers to the paper *Sorting, trees and measures of order*, by W. H. Burge, Information and Control, vol. 1 (1958), pp. 181–197, which is concerned with finding the sorting method which takes the least time to keep a large quantity of data in an orderly array for easy reference. Burge finds that the optimal strategy for sorting a set of data depends upon the amount of order (in the information-theoretic sense) already existing in the data; and he shows how to find an optimal sorting strategy given a measure of this order.

Another Sorting Problem

Problem 60-8, by J. H. VAN LINT (Technische Hogeschool, Te Eindhoven, Nederland).

Consider all sequences of length M consisting of p_1 1's, p_2 2's, $\cdots p_k$ k's ($k \geq 2$) whose maximal monotonic non-decreasing subsequences of contiguous numbers are of length n where n satisfies the inequalities

(1) $$n > M - p_i \qquad\qquad i = 1, 2, \cdots, k.$$

Determine the number $N(p_1, p_2, \cdots, p_k; n)$ of sequences with this property.

Editorial note: This problem is similar to that of Brock's Optimum Sorting Procedure, Problem 59-3. However, in the latter problem, contiguity is not required.

Solution by the proposer. As a consequence of (1) the subsequence must contain every number $1, 2, \cdots, k$ at least once and hence it starts with the number 1 and ends with the number k.

First consider $n \leq M - 2$.

If the monotonic subsequence of length n is at the beginning (at the end) of the sequence there are $M - n$ elements left, of which $M - n - 1$ can have arbitrary values, because $M - n < p_i$ for all i, while the first (last) of these elements can have all values except k (except 1). If we choose these elements the monotonic subsequence is fully determined. Hence there are $2 \cdot k^{M-n-1} \cdot (k - 1)$ sequences of this type.

Now consider the case where the monotonic subsequence is preceded by elements and followed by $M - n - l$ elements. Here l can have the values 1, 2 $\cdots, M - n - 1$. The last of the preceding elements can have all values except 1. The first of the elements following the subsequence can have all values except k. The other $M - n - 2$ preceding and following elements can have arbitrary values. (We again use (1)). After choosing these elements the subsequence is fully determined. We can still choose l.

So there are $(M - n - 1) \cdot (k - 1)^2 \cdot k^{M-n-2}$ sequences of this type. Adding the two results we find:

(2)
$$N = 2 \cdot k^{M-n-1}(k - 1) + (M - n - 1) \cdot (k - 1)^2 \cdot k^{M-n-2}$$
$$= (k - 1) \cdot k^{M-n-2} \cdot \{2k + (k - 1)(M - n - 1)\}$$

We see that N is the same for all p_1, p_2, $\cdots$, p_k as long as (1) is satisfied and of course $p_1 + p_2 + \cdots + p_k = M$.

For $k = 2$ we find for (2): $N = 2^{M-n-2}(M - n + 3)$.

We still have 2 cases to discuss: $n = M$ and $n = M - 1$. Both are trivial. For $n = M$ we have $N = 1$ and for $n = M - 1$ the above holds but we only have the first term i.e., $N = 2(k - 1)$.

The Ballot Problem

Problem 59-1, by MARY JOHNSON (American Institute of Physics)
AND M. S. KLAMKIN (University of Alberta).

A society is preparing 1560 ballots for an election for three offices for which there are 3, 4, and 5 candidates, respectively. In order to eliminate the effect of the ordering of the candidates on the ballot, there is a rule that each candidate must occur an equal number of times in each position as any other candidate for the same office. What is the least number of different ballots necessary?

It is immediately obvious that 60 different ballots would suffice. However, the following table gives a solution for 9 different ballots:

No. of Ballots	312	78	130	234	182	104	208	286	26
Office									
1........	A	A	A	B	B	B	C	C	C
2........	D	D	E	E	F	G	F	G	E
3........	H	I	K	I	K	J	J	L	L

Another solution (by C. Berndtson) is given by

No. of Ballots	260	182	78	234	52	130	104	312	208
Office									
1........	A	A	A	B	B	B	B	C	C
2........	D	F	E	G	G	D	G	E	F
3........	H	I	J	J	H	I	K	L	K

The above tables just give the distribution for the first position on the ballot for each office. The distributions for the other positions are obtained by cyclic permutations.

We now show that 9 is the least possible number of ballots. Let us consider the distribution for office 3 using only 8 different ballots. We must have the following (for simplicity we consider a total of 60 ballots):

No. of Ballots	x	$12 - x$	y	$12 - y$	z	$12 - z$	12	12
Office								
3........	H	H	I	I	J	J	K	L

Now to get a total of 15 representations for each position for office 2, we must have $x = y = 3$, $z = 6$. But this does not satisfy the requirements for office 1. Similarly no number of ballots fewer than 8 will suffice.

It would be of interest to solve this problem in general. The problem is to determine a distribution of the candidates such that the system of linear equations for the number of each type of ballot, which contains more equations than unknowns, is solvable in positive integers.

A trick solution to the problem can be obtained using 5 different ballots: add two fictitious names to the group of 3 and one to the group of 4. We then have 3 offices for which there are 5 "candidates" for each. This would also provide a survey on the effect of ordering of the candidates on the ballot.

POSTSCRIPT

For a solution when there are only two offices, see "The Ballot Problem," Math. Modelling (1984) pp. 1–6 and a subsequent simplification to appear (in the same journal).

An Extremal Problem

Problem 67-17, by J. Neuringer (AVCO Corporation) and D. J. Newman (Temple University).

Consider the differential equation

$$\{D^2 - F(x)\}y = 0, \qquad y(0) = 1, \qquad y'(0) = 0.$$

Choose $F(x)$ subject to the conditions

$$F(x) \geq 0, \qquad \int_0^1 F(x)\,dx = M,$$

so as to maximize $y(1)$.

This problem has arisen in connection with the construction of an optimal refracting medium.

Solution by J. Ernest Wilkins, Jr. (Gulf General Atomic Incorporated).

Let $\mathfrak{M}$ be the class of integrable, almost everywhere nonnegative, functions $F(x)$ defined on the interval $(0, 1)$ whose integral over $(0, 1)$ is M. For each such $F(x)$ there exists a unique "associated" function $y(x)$ with an absolutely continuous first derivative $y'(x)$ which satisfies the differential equation $y'' = Fy$ almost everywhere on $(0, 1)$ and for which $y(0) = 1$, $y'(0) = 0$. Let A be the least upper bound of the values $y(1)$ as $F(x)$ ranges over $\mathfrak{M}$. I shall prove that $A = M + 1$ if $0 \leq M \leq 1$, $A = 2\exp(M^{1/2} - 1)$ if $M \geq 1$.

The proof of this assertion will depend on the following lemma, which is suggested by the usual calculus of variations devices.

LEMMA 1. *If a constant v and two absolutely continuous functions $u(x)$ and $\lambda(x)$ can be found such that* (i) $\lambda \leq 0$, (ii) $\lambda(u' + u^2) \geq 0$, (iii) $\lambda' = 2u(\lambda + v) - 1$, (iv) $\lambda(1) + v = 0$, (v) $v\left\{\int_0^1 (u' + u^2)\,dx - M + u(0)\right\} + \lambda(0)u(0) = 0$, *then*

$$A \leq \exp\int_0^1 u(x)\,dx.$$

Suppose that $F(x)$ is an element of $\mathfrak{M}$ and that $y(x)$ is associated with $F(x)$. It is easy to show that $y(x) \geq 1$ and hence the function $V(x) = y'(x)/y(x)$ is absolutely continuous. Moreover, $V(0) = 0$ and $F = V' + V^2$. Therefore, by virtue of the hypotheses (i) through (iv) and the definition of $\mathfrak{M}$,

$$\log y(1) = \int_0^1 V(x)\,dx$$

$$\leq \int_0^1 \{V - \lambda(V' + V^2) - v(V' + V^2 - M)\}\,dx$$

$$= \int_0^1 \{u - \lambda(u' + u^2) - v(u' + u^2 - M)\}\,dx$$

$$\quad - \int_0^1 \{(V - u)[2u(\lambda + v) - 1] + (V' - u')(\lambda + v)\}\,dx$$

$$\quad - \int_0^1 (V - u)^2(\lambda + v)\,dx$$

$$= \int_0^1 u\,dx - v\int_0^1 (u' + u^2 - M)\,dx - (V - u)(\lambda + v)\Big|_0^1$$

$$\quad - \int_0^1 (V - u)^2(\lambda + v)\,dx$$

$$= \int_0^1 u\,dx - \int_0^1 (V - u)^2(\lambda + v)\,dx.$$

The conclusion of the lemma now follows from the inference from (iii) and (iv) that

$$\lambda + v = \int_x^1 \exp\left\{-2\int_x^t u(s)\,ds\right\}dt \geq 0.$$

When $0 \leq M \leq 1$, the quantities

$$v = \frac{1}{M + 1}, \qquad u(x) = \frac{M}{Mx + 1}, \qquad \lambda(x) = -\frac{x(Mx + 1 - M)}{M + 1}$$

are easily seen to satisfy the hypotheses of the lemma, and hence

$$A \leq \exp\int_0^1 u(x)\,dx = M + 1.$$

When $M \geq 1$, the quantities $v = 1/(2M^{1/2})$,

$$u(x) = M^{1/2}, \qquad \lambda(x) = 0 \quad \text{if} \quad 0 \leq x \leq 1 - M^{-1/2},$$

$$u(x) = \frac{M^{1/2}}{2 - M^{1/2}(1 - x)},$$

$$\lambda(x) = -\frac{\{M^{1/2}(1 - x) - 1\}^2}{2M^{1/2}} \quad \text{if} \quad 1 - M^{-1/2} \leq x \leq 1,$$

also satisfy the hypotheses of the lemma, and hence

$$A \leq \exp \int_0^1 u(x)\, dx = 2 \exp(M^{1/2} - 1).$$

To see that the upper bounds just established are in fact least upper bounds, it is sufficient to exhibit a sequence of absolutely continuous functions $V_n(x)$ defined for sufficiently large n when $0 \leq x \leq 1$, such that $V_n(0) = 0$, $F_n = V'_n + V_n^2$ is in $\mathfrak{M}$ and

$$\exp\left\{ \lim_{n \to \infty} \int_0^1 V_n(x)\, dx \right\} = \exp \int_0^1 u(x)\, dx.$$

When $M \leq 1$, define

$$V_n(x) = \begin{cases} n \tanh nx & \text{if } \ 0 \leq x \leq M/n^2, \\ (x + \alpha_n)^{-1} & \text{if } \ M/n^2 \leq x \leq 1, \end{cases}$$

in which

$$\alpha_n = -\frac{M}{n^2} + \frac{1}{n} \coth \frac{M}{n},$$

so that V_n is absolutely continuous on $(0, 1)$. Since $F_n = n^2$ on $(0, M/n^2)$ and $F_n = 0$ on $(M/n^2, 1)$, it is clear that F_n is in $\mathfrak{M}$. Moreover $V_n(0) = 0$ and

$$\exp \int_0^1 V_n(x)\, dx = \left\{ n\left(1 - \frac{M}{n^2}\right) \sinh \frac{M}{n} + \cosh \frac{M}{n} \right\} \to M + 1.$$

When $M > 1$, we define

$$V_n(x) = \begin{cases} M^{1/2} nx & \text{if } \ 0 \leq x \leq 1/n, \\ M^{1/2} & \text{if } \ 1/n \leq x \leq 1 - M^{-1/2} + 2/(3n), \\ (x + \alpha_n)^{-1} & \text{if } \ 1 - M^{-1/2} + 2/(3n) \leq x \leq 1, \end{cases}$$

for values of n such that $n \geq 2M^{1/2}/3$, $n \geq M^{1/2}/(3(M^{1/2} - 1))$. The function $V_n(x)$ is absolutely continuous if

$$\alpha_n = 2M^{-1/2} - 1 - 2/(3n)$$

and $V_n(0) = 0$. Moreover, $F_n = M^{1/2} n + Mn^2 x^2$, M, and 0 on the indicated subintervals of $(0, 1)$, and hence F_n is in $\mathfrak{M}$. Finally,

$$\exp \int_0^1 V_n(x)\, dx = 2\left(1 - \frac{M^{1/2}}{3n}\right) \exp\left\{ M^{1/2}\left(1 + \frac{1}{6n}\right) - 1 \right\}$$

$$\to 2 \exp(M^{1/2} - 1).$$

Minimum-Loss Two Conductor Transmission Lines

Problem 59-4, by GORDON RAISBECK (Bell Telephone Laboratories).

Let C_1 and C_2 be two closed curves in a plane, one totally surrounding the other, bounding an annular region. Let ψ be a harmonic function within the annular region, having a constant value on C_1 and a constant value on C_2. If this configuration is regarded as the cross-section of a transmission line carrying a TEM wave, with cylindrical conductors of section C_1 and C_2, with a lossless dielectric between them, then the attenuation[1] is proportional to

$$\alpha = \frac{\int |\nabla\psi|^2 \, ds}{\iint |\nabla\psi|^2 \, da},$$

where the single integral is taken over the boundary $C_1 + C_2$ and the double integral over the area between them.

It has been shown[1] that under the constraint

$$\left\{\int_{C_1} ds\right\}^{-1} + \left\{\int_{C_2} ds\right\}^{-1} = \text{constant},$$

i.e., that the harmonic mean of the perimeters of the conductors is fixed, the minimum of the attenuation α is attained when the boundaries are concentric circles. Prove (or disprove) that the same conclusion holds under the alternative constraint that the area bounded between the curves is fixed.

[1] Gordon Raisbeck, *Minimum-Loss Two-Conductor Transmission Lines*, Trans. IRE PGCT, Sept. 1958.

Supplementary References
Optimization

I. Linear, Nonlinear, Dynamic, and Geometric Programming

[1] J. ABADIE, ed., *Nonlinear Programming*, Interscience, N.Y., 1967.

[2] R. ARIS, *Discrete Dynamic Programming*, Blaisdell, Waltham, 1964.

[3] E. M. L. BEALE, ed., *Applications of Mathematical Programming Techniques*, English Universities Press, London, 1970.

[4] C. S. BEIGHTLER AND D. T. PHILLIPS, *Applied Geometric Programming*, Wiley, N.Y., 1976.

[5] R. BELLMAN, *Dynamic Programming*, Princeton University Press, Princeton, 1957.

[6] R. E. BELLMAN AND S. E. DREYFUS, *Applied Dynamic Programming*, Princeton University Press, Princeton, 1962.

[7] R. G. BLAND, "The allocation of resources by linear programming," Sci. Amer. (19181) pp. 126–144.

[8] J. BRACKEN AND G. P. McCORMICK, *Selected Applications of Nonlinear Programming*, Wiley, N.Y., 1968.

[9] A. CHARNES AND W. W. COOPER, *Management Models and Industrial Applications of Linear Programming I, II*, Wiley, N.Y., 1961.

[10] L. COOPER AND M. W. COOPER, *Introduction to Dynamic Programming*, Pergamon, Oxford, 1981.

[11] G. B. DANTZIG, *Linear Programming and Extensions*, Princeton University Press, Princeton, 1963.

[12] G. B. DANTZIG AND B. C. EAVES, eds., *Studies in Optimization*, MAA, 1974.

[13] J. B. DENNIS, *Mathematical Programming and Electrical Networks*, Technology Press, N.Y., 1959.

[14] S. E. DREYFUS, *Dynamic Programming and the Calculus of Variations*, Academic Press, N.Y., 1965.

[15] S. DREYFUS AND A. M. LAW, *The Art and Theory of Dynamic Programming*, Academic Press, N.Y., 1977.

[16] R. J. DUFFIN, E. L. PETERSON AND C. M. ZENER, *Geometric Programming*, Wiley, N.Y., 1967.

[17] J. C. ECKER, "Geometric programming: Methods, computers and applications," SIAM Rev. (1980) pp. 338–362.

[18] A. V. FIACCO AND G. P. McCORMICK, *Nonlinear Programming, Sequential Unconstrained Minimization Techniques*, Wiley, N.Y., 1968.

[19] G. HADLEY, *Nonlinear and Dynamic Programming*, Addison-Wesley, Reading, 1964.

[20] D. H. JACOBSON AND D. Q. MAYNE, *Differential Dynamic Programming*, Elsevier, N.Y., 1970.

[21] S. KARLIN, *Mathematical Methods and Theory in Games, Programming and Economics*, Addison-Wesley, Reading, 1959.

[22] A. KAUFMANN, *Graphs, Dynamic Programming, and Finite Games*, Academic, N.Y., 1967.

[23] E. L. PETERSON, *"Geometric programming—a survey,"* SIAM Rev. (1976) pp. 1–51.

[24] S. M. ROBERTS, *Dynamic Programming in Chemical Engineering and Process Control,* Academic, N.Y., 1964.

[25] H. THEIL AND C. VAN DE PANNE, *"Quadratic programming as an extension of classical quadratic maximization,"* Management Sci. (1960) pp. 1–20.

[26] L. E. WARD, JR., *"Linear programming and approximation problems,"* AMM (1961) pp. 46–52.

[27] G. ZOUTENDIJK, *"Nonlinear programming: A numerical survey,"* SIAM J. Control (1966) pp. 194–210.

II. Control Theory

[28] D. J. BELL, ed., *Recent Mathematical Developments in Control,* Academic, London, 1973.

[29] R. BELLMAN, *Adaptive Control Processes: A Guided Tour,* Princeton University Press, Princeton, 1961.

[30] ———, *Introduction to the Mathematical Theory of Control Processes I, II,* Academic Press, N.Y., 1967 and 1971.

[31] L. D. BERKOVITZ, *"Optimal control theory,"* AMM (1976) pp. 225–239.

[32] D. N. BURGHES AND A. GRAHAM, *Introduction to Control Theory, Including Optimal Control,* Horwood, Chichester, 1980.

[33] F. CSAKI, *Modern Control Theories,* Akademiai Kiado, Budapest, 1972.

[34] I. FLUGGE-LOTZ, *Discontinuous and Optimal Control,* McGraw-Hill, N.Y., 1968.

[35] C. M. HAALAND AND E. P. WIGNER, *"Defense of cities by antiballistic missiles,"* SIAM Rev. (1977) pp. 279–296.

[36] C. S. JONES AND A. STRAUSS, *"An example of optimal control,"* SIAM Rev. (1968) pp. 25–55.

[37] M. S. KLAMKIN, *"A student optimum control problem and extensions,"* Math. Modelling (1985) pp. 49–64.

[38] G. KNOWLES, *An Introduction to Applied Optimal Control Theory,* Academic, N.Y., 1981.

[39] P. KOKOTOVIC, *"Application of singular perturbation techniques to control problems,"* SIAM Rev. (1984) pp. 501–550.

[40] J. MACKI AND J. STRAUSS, *An Introduction to Optimal Control Theory,* Springer-Verlag, N.Y., 1982.

[41] P. C. PARKS, *"Mathematicians in control,"* BIMA (1977) pp. 180–188.

[42] L. S. PONTRYAGIN, V. G. BOL'TANSKII, R. S. GAMKRELIDZE AND E. F. MISCHENKO, *The Mathematical Theory of Optimal Processes,* Pergamon, Oxford, 1964.

[43] L. C. YOUNG, *Lectures on the Calculus of Variations and Optimal Control Theory,* Saunders, Philadelphia, 1969.

III. Games

[44] R. BARTON, *A Primer on Simulation and Gaming,* Prentice-Hall, N.J., 1970.

[45] E. R. BERLEKAMP, J. H. CONWAY AND R. K. GUY, *Winning Ways I, II,* Freeman, San Francisco, 1977.

[46] I. BUCHLER AND H. NOTINI, *Game Theory in the Behavioral Sciences,* University of Pittsburgh Press, Pittsburgh, 1969.

[47] J. H. CONWAY, *"All games bright and beautiful,"* AMM (1977) pp. 417–434.

[48] ———, *On Numbers and Games,* Academic, London, 1976.

[49] H. S. M. COXETER, *"The golden section, phyllotaxis, and Wythoff's game,"* Scripta Math. (1953) pp. 135–143.

[50] W. H. CUTLER, *"An optimal strategy for pot-limit poker,"* AMM (1975) pp. 368–376.

[51] M. DRESHER, A. W. TUCKER AND P. WOLFE, eds., *Contributions to the Theory of Games III,* Princeton University Press, Princeton, 1957.

[52] N. V. FINDLER, *"Computer poker,"* Sci. Amer. (1978) pp. 112–119.

[53] O. HAJEK, *Pursuit Games,* Academic, N.Y., 1975.

[54] J. C. HOLLADAY, *"Some generalizations of Wythoff's game and other related games,"* MM (1968) pp. 7–13.

[55] H. W. KUHN AND A. W. TUCKER, eds., *Contributions to the Theory of Games I, II,* Princeton University Press, Princeton, 1950, 1953.

[56] R. D. LUCE AND H. RAIFFA, *Games and Decisions,* Wiley, N.Y., 1958.

[57] R. D. LUCE AND A. W. TUCKER, eds., *Contributions to the Theory of Games IV,* Princeton University Press, Princeton, 1959.

[58] J. C. C. McKINSEY, *Introduction to the Theory of Games,* McGraw-Hill, N.Y., 1952.

[59] W. H. RUCKLE, *Geometric Games and their Applications,* Pitman, London, 1983.

[60] ———, *"Geometric games of search and ambush,"* MM (1979) pp. 195–206.

[61] C. A. B. SMITH, *"Graphs and composite games,"* J. Comb. Th. (1966) pp. 51–81.

[62] J. VON NEUMANN AND O. MORGENSTERN, *Theory of Games and Economic Behaviour,* Princeton University Press, Princeton, 1947.

[63] J. D. WILLIAMS, *The Compleat Strategyst,* McGraw-Hill, N.Y., 1954.

IV. Miscellaneous

[64] M. AURIEL AND D. J. WILDE, *"Optimality proof for the symmetric Fibonacci search technique,"* Fibonacci Quart. (1966) pp. 265–269.

[65] B. AVI-ITZHAK, ed., *Developments in Operations Research I, II,* Gordon and Breach, N.Y., 1971.

[66] A. V. BALAKRISHMAN, *Kalman Filtering Theory,* Optimization Software Inc., N.Y., 1984.

[67] E. BALAS, *"An additive algorithm for solving linear programs with zero-one variables,"* Oper. Res. (1965) pp. 517–546.

[68] M. L. BALINSKY, *"Integer programming: Methods, uses, computation,"* Management Sci. (1965) pp. 253–313.

[69] C. BANDLE, *"A geometric isoperimetric inequality and applications to problems of mathematical physics,"* Comment. Math. Helv. (1974) pp. 253–261.

[70] R. BELLMAN, A. O. ESOGBUE AND I. NABESHIMA, *Mathematical Aspects of Scheduling & Applications,* Pergamon, Oxford, 1982.

[71] M. BELLMORE AND G. NEMHAUSER, *"The travelling salesman problem,"* Oper. Res. (1968) pp. 538–558.

[72] H. J. BRASCAMP AND E. H. LIEB, *"On extentions of the Brunn-Minkowski and Prekopa-Leindler theorems, including inequalities for log concave functions, and with an application to the diffusion equation,"* J. Funct. Anal. (1976) pp. 366–389.

[73] S. T. BROOKS, *"A comparison of maximum-seeking methods,"* Oper. Res. (1959) pp. 430–457.

[74] S. S. BROWN, *"Optimal search for a moving target in discrete space and time,"* Oper. Res. (1980) pp. 1275–1289.

[75] E. G. COFFMAN, ed., *Computer and Job-Shop Scheduling Theory,* Wiley, N.Y., 1976.

[76] L. COLLATZ AND W. WETTERLING, *Optimization Problems,* Springer-Verlag, N.Y., 1975.

[77] L. COOPER AND D. I. STEINBERG, *Introduction to Methods of Optimization,* Saunders, Philadelphia, 1970.

[78] L. COOPER AND M. W. COOPER, *"Nonlinear integer programming,"* Comp. Math. Appl. (1975) pp. 215–222.

[79] COURANT, R., *"Soap film experiments with minimal surfaces,"* AMM (1940) pp. 167–174.

[80] H. G. DAELLENBACH AND J. A. GEORGE, *Introduction to Operations Research Techniques,* Allyn and Bacon, Boston, 1978.

[81] N. J. FINE, *"The jeep problem,"* AMM (1946) pp. 24–31.

[82] L. R. FOULDS, *Combinatorial Optimization for Undergraduates,* Springer-Verlag, N.Y., 1984.

[83] ———, *"Graph theory: A survey of its use in operations research,"* New Zeal. Oper. Res. (1982) pp. 35–65.

[84] J. N. FRANKLIN, *"The range of a fleet of aircraft,"* SIAM J. Appl. Math. (1960) pp. 541–548.

[85] D. R. FULKERSON, *"Flow networks and combinatorial operations research,"* AMM (1966) pp. 115–137.

[86] D. GALE, *"The jeep once more or jeeper by the dozen,"* AMM (1970) pp. 493–500.

[87] S. I. GASS, *"Evaluation of complex models,"* Comp. Oper. Res. (1977) pp. 27–35.

[88] D. P. GAVER AND G. L. THOMPSON, *Programming and Probability Models in Operations Research,* Brooks/Cole, Monterey, 1973.

[89] R. L. GRAHAM, *"The combinatorial mathematics of scheduling,"* Sci. Amer. (1978) pp. 124–132, 154.

[90] G. Y. HANDLER AND P. B. MIRCHANDANI, *Location on Networks: Theory and Algorithms,* M.I.T. Press, Cambridge, 1979.

[91] M. R. HESTENES, *Optimization Theory,* Interscience, N.Y., 1975.

[92] F. HILLIER AND G. J. LIEBERMAN, *Introduction to Operations Research,* Holden-Day, San Francisco, 1967.

[93] J. M. HOLTZMAN AND H. HALKIN, *"Directional convexity and the maximum principle for discrete systems,"* SIAM J. Control (1966) pp. 263–275.

[94] R. JACKSON, *"Some algebraic properties of optimization problems in complex chemical plants,"* Chem. Eng. Sci. (1964) pp. 19–31.

[95] P. B. JOHNSON, *"The washing of socks,"* MM (1966) pp. 77–83.

[96] A. KAUFMANN AND R. FAURE, *Introduction to Operations Research,* Academic, N.Y., 1968.

[97] H. J. KELLEY, *"Gradient theory of optimal flight paths,"* Amer. Rocket Soc. J. (1960) pp. 947–954.

[98] Z. J. KIEFER, *"Sequential minimax search for a max,"* Proc. AMS (1953) pp. 502–506.

[99] S. KIRKPATRICK, et al., *"Optimization by simulated annealing,"* Science (1983) pp. 671–680.

[100] M. S. KLAMKIN, *"A physical application of a rearrangement inequality,"* AMM (1970) pp. 68–69.

[101] ———, *"On Chaplygin's problem,"* SIAM J. Math. Anal. (1977) pp. 288–289.

[102] ———, *"Optimization problems in gravitational attraction,"* J. Math. Anal. Appl. (1972) pp. 239–254.

[103] M. S. KLAMKIN AND A. RHEMTULLA, *"The ballot problem,"* Math. Modelling (1984) pp. 1–6.

[104] R. C. LARSON AND A. R. ODON, *Urban Operations Research,* Prentice-Hall, N.J., 1981.

[105] E. LAWLER, *Combinatorial Optimization: Networks and Matroids,* Holt, Rinehart & Winston, N.Y., 1976.

[106] G. LEITMANN, ed., *Topics in Optimization,* Academic Press, N.Y., 1967.

[107] A. G. J. MACFARLANE, *"An eigenvector solution of an optimal linear regulator problem,"* J. Elec. Control (1963) pp. 643–653.

[108] A. MIELE, ed., *Theory of Optimum Aerodynamic Shapes,* Academic Press, N.Y., 1965.

[109] D. J. NEWMAN, *"Location of the maximum on a unimodal surface,"* J. Assn. Comp. Mach. (1965) pp. 395–398.

[110] J. C. C. NITSCHE, *"Plateau's problem and their modern ramifications,"* AMM (1974) pp. 945–968.

[111] I. NIVEN, *Maxima and Minima Without Calculus,* MAA, 1981.

[112] L. T. OLIVER AND D. J. WILDE, *"Symmetric sequence minimax search for a maximum,"* Fibonacci Quart. (1964) pp. 169–175.

[113] R. OSSERMAN, "The isoperimetric inequality," Bull. AMS (1978) pp. 1182–1238.

[114] C. H. PAPADIMITRIOU AND K. STEIGLITZ, Combinatorial Optimization: Algorithms and Complexity, Prentice-Hall, N.J., 1982.

[115] L. E. PAYNE, "Isoperimetric inequalities and their applications," SIAM Rev. (1967) pp. 453–488.

[116] C. G. PHIPPS, "The jeep problem: A more general solution," AMM (1946) pp. 458–462.

[117] G. POLYA AND G. SZEGO, Isoperimetric Inequalities in Mathematical Physics, Princeton University Press, Princeton, 1951.

[118] G. RAISBECK, "An optimum shape for fairing the edge of an electrode," AMM (1961) pp. 217–225.

[119] V. J. RAYWARD-SMITH AND M. SHING, "Bin packing," BIMA (1983) pp. 142–148.

[120] T. L. SAATY, Mathematical Methods of Operational Research, McGraw-Hill, N.Y., 1959.

[121] ———, Optimization in Integers and Related Extremal Problems, McGraw-Hill, N.Y., 1970.

[122] B. V. SHAH, R. J. BUEHLER AND O. KEMPTHORNE, "Some algorithms for minimizing a function of several variables," SIAM J. Appl. Math. (1964) pp. 74–92.

[123] D. K. SMITH, Network Optimization Practice: A Computational Guide, Horwood, Chichester, 1982.

[124] D. R. SMITH, Variational Methods in Optimization, Prentice-Hall, N.J., 1974.

[125] L. J. STOCKMEYER AND A. K. CHANDRA, "Intrinsically difficult problems," Sci. Amer. (1979) pp. 140–159.

[126] L. D. STONE, Theory of Optimal Search, Academic, N.Y., 1975.

[127] H. A. TAHA, Integer Programming, Macmillan, N.Y., 1978.

[128] ———, Operations Research, An Introduction, Macmillan, N.Y., 1976.

[129] H. WACKER, ed., Applied Optimization Techniques in Energy Problems," Teubner, Stuttgart, 1985.

[130] H. M. WAGNER, Principles of Operations Research with Applications to Managerial Decisions, Prentice-Hall, N.J., 1969.

[131] D. J. WILDE, Optimum-Seeking Methods, Prentice-Hall, N.J., 1964.

[132] D. J. WILDE AND C. S. BEIGHTLER, Foundations of Optimization, Prentice-Hall, N.J., 1967.

[133] D. WILDFOGEL, "The maximum brightness of Venus," MM (1984) pp. 158–164.

10.1 Music

OPTIMAL TEMPERAMENT*

A. A. GOLDSTEIN†

We show that musical scales can be constructed by finding solutions of inconsistent systems of linear equations with various criteria of optimality. As examples, several typical historical scales are reconstructed.

There is a dual purpose to this note. One is to amuse the reader. For this reason our style will be semi-expository and the paper will be reasonably self-contained. The other purpose is to provide a method for generating scales for possible musical systems.

Information on historical scales may be found in Barbour [1]. The work in [1] has been made practical by Jorgenson [2], I am grateful to him and the referee for helpful suggestions.

A (musical) scale is an increasing sequence of positive numbers $f_1, f_2, \cdots$ called *notes* or *frequencies*. A scale is *n-tempered* if for some positive n,

$$(\text{I}) \qquad\qquad f_{i+n} = 2f_i, \qquad\qquad i = 1, 2, 3, \cdots.$$

The ratio $f_{i+n}|f_i$ is called the *octave*. For keyboard instruments $n = 12$. Our interest lies mainly for 12-tempered scales.

A scale is called *regular*, if for each j, $f_{i+j}|f_i$ is independent of i. Ratios $f_i|f_j$, $f_k|f_i$, $j < i < k$ are said to be *consecutive*.

Remark. If a scale is n-tempered, then for every integer $k > 0$, there exist positive integers l and m such that $\prod_{i=1}^{m} (f_{1+ki}|f_{1+k(i-1)}) = f_{1+nl} = 2^l f_1$. Thus, for this m and l, the product of m consecutive ratios of the form $f_{1+ki}|f_{1+k(i-1)}$ spans l octaves.

Proof. $\prod_{i=1}^{m} (f_{1+ki}|f_{1+k(i-1)}) = f_{1+mk}|f_1$. Take l so that if $m = ln|k$, m is an integer. By (I) $f_{nl+1} = f_{1+n(l-1)+n} = 2f_{1+n(l-1)} = 2^l f_1$. □

When $n = 12$, a regular tempered scale is called *equal-tempered*. There exists a unique such scale. The numbers m and l showing Table 1 apply for this case.

TABLE 1

k	m	l
1	12	1
2	6	1
3	4	1
4	3	1
5	12	5
6	2	1
7	12	7
8	3	2
9	4	3
10	6	5
11	12	11

---

* Supported by N.S.F. Grant MPS 72-04787-A02.

† Department of Mathematics, University of Washington, Seattle, Washington 98195.

Since $(f_{k+i}/f_i)^m = 2^l$, it follows then that $f_{k+i}/f_i = 2^{k/12}$. This scale was discovered in 1533 by Lanfranco. For reasons discussed below it had limited use until the 1800's. Since then it has displaced all its competitors, and is in almost universal use today.

In what follows we also discuss nonregular and partially regular scales. A *partially regular* scale satisfies $f_{i+j}/f_i = $ const. for all $i \in I$, where I is a proper subset of the positive integers.

The early Greeks constructed scales by compounding the ratios 2/1 and 3/2. Later Ptolemy, the founder of trigonometry and refiner of Greek astronomy, investigated the superparticular series $2/1,\ 3/2,\ 4/3,\ 5/4, \cdots$. The ratios $2/1, \cdots, 6/5$ and compoundings of these; e.g. $8/5 = (2/1)|(5/4)$ and $5/3 = (4/3)(5/4)$, are especially consonant. By this we mean that when frequencies of these ratios are simultaneously sounded, the result is pleasing.

There are two musical aspect to scales—the melodic and the harmonic. The melodic is the impression received when notes are sounded in succession, while the harmonic, the impression when the sounding of notes is simultaneous.

There is experimental justification for the theses that the melodic aspects of scales are conventional and learned, while there are convincing theories initiated by Helmholtz and based on experiment and deduction that there is a real basis for the harmonic aspect, that is independent of individual experience. The theory predicts that the superparticular ratios with small integers yield the most consonant simultaneous combination of sounds.

The ratios based on superparticulars and shown in Table 2 were proposed by Ptolemy. They are called *just ratios*.

TABLE 2

Ratio	Musical name	Ratio	Compound ratio
f_{i+1}/f_i	minor second	16/15	—
f_{i+2}/f_i	major second	10/9	—
f_{i+3}/f_i	minor third	6/5	—
f_{i+4}/f_i	major third	5/4	—
f_{i+5}/f_i	perfect fourth	4/3	—
f_{i+6}/f_i	diminished fifth	45/32	(9/8)(5/4)
f_{i+7}/f_i	perfect fifth	3/2	—
f_{i+8}/f_i	minor sixth	8/5	(2/1)|(5/4)
f_{i+9}/f_i	major sixth	5/3	(4/3)(5/4)
f_{i+10}/f_i	minor seventh	16/9	(2/1)|(9/8)
f_{i+11}/f_i	major seventh	15/8	(5/4)(3/2)

If the just ratio is not superparticular, the compound ratio listed shows its generation from superparticulars. If we use the values from Table 1 and the above Remark, we find Table 3. From Table 3 we see that the deviation of just ratios from equal-tempered ratios is surprisingly small. These differences however are not negligible. Because of the "beat phenomena," (discussed in books on acoustics) the trained ear can detect pitch errors in the ratio of 2 simultaneously sounding notes to 1 cent or less, where $\mu = 1$ cent $= 2^{1/1200} \doteq 1.0005778$ has been adapted as the unit for measuring musical intervals.

TABLE 3

k	$(f_{k+i}\|f_i)^m$	2^l	Equal tempered ratios	Just ratios
1	$(16/15)^{12} \doteq 2.17$	2	$2^{1/12} \doteq 1.0595$	$16/15 \doteq 1.0667$
2	$(10/9)^{6} \doteq 1.88$	2	$2^{1/6} \doteq 1.1225$	$10/9 \doteq 1.1111$
3	$(6/5)^{4} \doteq 2.07$	2	$2^{1/4} \doteq 1.1892$	$6/5 \doteq 1.2000$
4	$(5/4)^{3} \doteq 1.95$	2	$2^{1/3} \doteq 1.2599$	$5/4 \doteq 1.2500$
5	$(4/3)^{12} \doteq 31.57$	32	$2^{5/12} \doteq 1.3348$	$4/3 \doteq 1.3333$
6	$(45/32)^{2} \doteq 1.98$	2	$2^{1/2} \doteq 1.4142$	$45/32 \doteq 1.4063$
7	$(3/2)^{12} \doteq 129.75$	128	$2^{7/12} \doteq 1.4983$	$3/2 \doteq 1.5000$
8	$(8/15)^{3} \doteq 4.10$	4	$2^{2/3} \doteq 1.5874$	$8/5 \doteq 1.6000$
9	$(5/3)^{4} \doteq 7.72$	8	$2^{3/4} \doteq 1.6818$	$5/3 \doteq 1.6666$
10	$(16/9)^{6} \doteq 31.57$	32	$2^{5/6} \doteq 1.7818$	$16/9 \doteq 1.7777$
11	$(15/8)^{12} \doteq 1888$	2048	$2^{11/12} \doteq 1.8877$	$15/8 \doteq 1.8750$

The equal tempered pure 4ths (4/3) and 5ths (3/2) differ from the corresponding just ratios by approximately 2 cents, which is a small error. But the major 3rds (5/4) and minor 6ths (8/5) have errors of approximately 14 cents, while the minor 3rds (6/5) and major sixths (5/3) have errors of approximately 16 cents. The resulting dissonance, (very distinct on the organ) was intolerable to many in the 16th, 17th, and 18th centuries, and delayed the universal adoption of equal temperament. Literally hundreds of alternative temperaments were explored [1]. We shall discuss a few of these alternatives after a brief digression.

DEFINITION. An *optimal temperament* is a temperament in which the ratios f_i/f_k approximate just ratios according to some criterion of optimality.

Consider then the system

(1) $f_{i+k}/f_i = r_{i+k,i}, \qquad 1 \leq i \leq 12, \qquad 1 \leq k \leq 11, \qquad f_{i+12} = 2f_i$

where $r_{i+k,i}$ is the kth just ratio.

This is a nonlinear system of 132 equations in 11 unknowns (we fix $f_1 = 1$): musically speaking, there are 12 distinct ratios of each kind (not counting repetition in octaves) and there are 11 kinds of ratios.

The above system is nonlinear. The musician's approach is to examine a companion system to (1) linearized by use of logarithms. Recall that $\mu = 2^{1/1200}$ and define F_j by the equations, $F_1 = 0$ and $\mu^{F_j - F_i} = r_{ji} = f_j/f_i$. Then

$$F_{i+k} - F_i = \log r_{i+k,i}| \log \mu,$$

(2) $$1 \leq i \leq 12, \qquad i \leq k \leq 11,$$

$$F_{i+12} = 1200 + F_i, \qquad F_1 = 0.$$

The numbers $r_{i+k,i}$ are the just ratios of Table 2. We shall call the differences $F_j - F_i$, *intervals*. Table 4 compares equal-tempered and just intervals in cents.

The fundamental ratios 3/2 and 4/3 (perfect fifth and perfect fourth) are well approximated. The most irritating of these errors is the major 3rd and minor 6th, followed by the minor 3rds and major 6th. The remaining errors are relatively innocuous.

TABLE 4

Cents (rounded)

Just interval	Equal tempered	Error
F_2–F_1 (minor 2nd) 112	100	−12
F_3–F_1 (major 2nd) 182	200	+18
F_4–F_1 (minor 3rd) 316	300	−16
F_5–F_1 (major 3rd) 386	400	+14
F_6–F_1 (perfect 4th) 498	500	+2
F_7–F_1 (diminished 5th) 590	600	+10
F_8–F_1 (perfect 5th) 702	700	−2
F_9–F_1 (minor 6th) 814	800	−14
F_{10}–F_1 (major 6th) 884	900	+16
F_{11}–F_1 (minor 7th) 996	1000	+4
F_{12}–F_1 (major 7th) 1088	1100	+12
F_{13}–F_1 (octave) 1200	1200	0

Since $F_{i+k} - F_i + F_{i+12} - F_{i+k} = F_{i+12} - F_i = 1200$, the pairs $F_{i+k} - F_i$ and $F_{i+12} - F_{i+k}$ are called complements. They are:

minor second: major seventh,

major second: minor seventh,

minor 3rd: major sixth,

major 3rd: minor sixth,

perfect 4th: perfect 5th,

diminished fifth: diminished fifth.

For example, for each perfect 4th there is a complementary perfect 5th, and their errors are dependent. Moreover the errors in the seconds, sevenths and diminished 5th are of negligible importance. We shall consider therefore optimization over the intervals of the major and minor 3rds and pure fifths. With respect to perfect fifths, minor and major thirds are complementary, since $F_{i+3} - F_i + F_{i+7} - F_{i+3} = F_{i+7} - F_i$. We see that errors in these intervals are also dependent. Observe that, however, if any 2 of the intervals $F_{i+3} - F_i$, $F_{i+7} - F_{i+3}$, and $F_{i+7} - F_i$, are just, the remaining one is just. For optimization, therefore, we consider only the intervals of the pure fifth and major 3rd.

Observe that 3 consecutive major thirds must sum to 1200 cents, and just major 3rds are 386 cents. Equal temperament approximates with each major 3rd at 400 cents. If all these consecutive 3rds are in equal use in a musical composition, these 14 cent errors are optimal. If some major 3rds are rarely used, we can make them larger than 400 cents and render the errors in the consecutive neighbors smaller. Also, we may wish the errors to be unequal to change "color"—have some major thirds "smooth" (small errors) and others rough (large errors).

Since the renaissance almost all of Western music has been constructed of chords based on triads. A *chord* is the simultaneous sounding of 3 or more notes. A *major triad* on F_i is the chord $\{F_i, F_{i+4}, F_{i+7}\}$, here $F_{i+4} - F_i$ is a major 3rd, $F_{i+7} - F_{i+4}$ a minor third and $F_{i+7} - F_i$ a perfect 5th. A *minor triad* on F_i is the chord $\{F_i, F_{i+3}, F_{i+7}\}$. Here $F_{i+3} - F_i$ is a minor third and $F_{i+7} - F_{i+3}$ a major third. The *minor triad relative to* F_i is the minor triad $\{F_{i+9}, F_{i+12}, F_{i+15}\}$.

A *well-tempering* is a tempered scale in which all major and minor triads on F_i, $1 \leq i \leq 12$, are acceptable. This means that errors in the thirds from just intervals are less in absolute value than approximately 25 cents, and errors in perfect 5ths less than approximately 12 cents in absolute value.

Examples of historical temperaments. We shall reproduce the major historical temperaments. To do this we shall specify all the major thirds and optimize over the perfect 5ths.

Let

$$(3) \qquad a_i = F_{4+1} - F_i, \qquad\qquad i = 1, 2, \cdots, 12 \quad \text{(major thirds)}.$$

Since $F_{12+i} = F_i + 1200$,

$$a_i + a_{4+i} + a_{8+i} = 1200, \qquad\qquad i = 1, 2, 3, 4.$$

Given a_i, $1 \leq i \leq 12$, specifying 4 of the F_i determines the remaining 8. Since $F_1 = 0$, there are 3 free variables: F_2, F_3, and F_4. We use these to optimize over the fifths:

$$(4) \qquad F_{7+i} - F_i = 702, \qquad F_{12+i} = 1200 + F_i, \qquad\qquad 1 \leq i \leq 12.$$

Using (3) we can rewrite (4) as:

$$\begin{aligned}
&\text{(I)} & F_4 + a_4 &= 702, \\
&\text{(II)} & F_1 - F_2 + a_1 + a_5 &= 702, \\
&\text{(III)} & F_2 - F_3 + a_2 + a_6 &= 702, \\
&\text{(IV)} & F_3 - F_4 + a_3 + a_7 &= 702, \\
&\text{(V)} & F_4 - F_1 + a_4 + a_8 - a_1 &= 702, \\
\text{(5)} \quad &\text{(VI)} & F_1 - F_2 - a_2 &= -498, \\
&\text{(VII)} & F_2 - F_3 - a_3 &= -498, \\
&\text{(VIII)} & F_3 - F_4 - a_4 &= -498, \\
&\text{(IX)} & F_4 - F_1 - a_1 - a_5 &= -498, \\
&\text{(X)} & F_1 - F_2 + a_1 - a_6 - a_2 &= -498, \\
&\text{(XI)} & F_2 - F_3 + a_2 - a_7 - a_3 &= -498, \\
&\text{(XII)} & F_3 - F_4 + a_3 - a_8 - a_4 &= -498.
\end{aligned}$$

The above is a system of 12 equations in 3 unknowns. We digress to discuss the "optimal" solution of such a system. Further information may be found in [5].

Let A be an $m \times n$ matrix, x and b column vectors of dimension n and m, respectively. Assume $m > n$. The system

$$Ax = b$$

is generally inconsistent. (It is consistent if and only if b lies on the range of A.)

Let $\|\cdot\|$ denote a distance function (norm) on m vectors. We shall use 2 such functions:

$$\|y\|_2 = \left[\sum_{i=1}^{m} y_i^2 \right]^{1/2}$$

and

$$\|y\|_\infty = \max_{1 \le i \le m} |y_i|.$$

An optimal solution of $Ax = b$ with respect to the norm $\|\cdot\|$ is an n-vector $\hat{x}$ such that

$$\|A\hat{x} - b\| \le \|Ax - b\| \quad \text{for all } n\text{-vectors } x.$$

The vector $\hat{x}$ always exists and it effects the best "compromise" solution in the sense that the m-vector $A\hat{x}$ is as close as possible to the m-vector b. (Closeness here is measured by our distance function $\|\cdot\|$.)

The problem when $\|\cdot\|_2$ is used is called "least squares" and $\hat{x}_2$ satisfies the $n \times n$ system of linear equations:

$$A^*Ax = A^*b,$$

where A^* is the transpose of A. These equations follow from the fact that b is closest to $A\hat{x}_2$ in the range of A. Hence $(b - A\hat{x}_2) \perp$ range of A. Since range of A is spanned by columns of A, $A^*(A\hat{x}_2 - b) = 0$.

When $\|\cdot\|_\infty$ is used it is possible to find $\hat{x}_\infty$ as a solution of a linear programming problem. This distance function is a natural one for musical scales, because the worst departure from just intonation is made as small as possible. The following condition ensures $\hat{x}_\infty$ is optimal [4, p. 35]. Let $(Ax - b)_i$ denote the ith component of $Ax - b$, and A_i the ith row of A. If $\|A\hat{x}_\infty - b\|_\infty = u_{i_k}(A\hat{x}_\infty - b)_{i_k}$, $u_i = \pm 1$, $1 \le k \le r \le m$, and 0 belongs to the convex hull $\{u_{i_k}A_{i_k} : 1 \le k \le r\}$, then $\|A\hat{x}_\infty - b\|_\infty \le \|Ax - b\|_\infty$ for all n-vectors x. We shall see below that because of special structure in the examples below $\hat{x}_2 = \hat{x}_\infty$, so that the least squares solution may be employed.

Again, because of the simple structure of the scale equations a criterion of optimality can be employed which would not be useful in general. We call x_0 an optimal interpolatory solution of the system

$$Ax = b$$

if the number of components of the vector $Ax_0 - b$ which vanish is maximal over all n-vectors, x.

In general we could not expect more than n components of $Ax - b$ to vanish for any x, but because of the special structure in the scale problem this can happen.

We first consider equal temperament from the point of view of optimizing over the perfect fifths.

Example 1. Equal temperament. For equal temperament we set $a_i \equiv 400$ and get from (5)

$$\text{(I)} \quad F_4 = 302,$$

$$\text{(II)} \quad F_2 = 98,$$

(6)

$$\text{(III)} \quad F_3 - F_2 = 98,$$

$$\text{(IV)} \quad F_4 - F_3 = 98.$$

(The set (6) is repeated 3 times.)

For system (6),

$$A = \begin{pmatrix} 0 & 0 & 1 \\ 1 & 0 & 0 \\ -1 & 1 & 0 \\ 0 & -1 & 1 \end{pmatrix}, \qquad x = \begin{pmatrix} F_2 \\ F_3 \\ F_4 \end{pmatrix}, \qquad b = \begin{pmatrix} 302 \\ 98 \\ 98 \\ 98 \end{pmatrix}.$$

The condition $A^*(A\hat{x} - b) = 0$ yields $F_2 = 100$, $F_3 = 200$, $F_4 = 300$, and

$$A\hat{x} - b = \begin{pmatrix} -2 \\ 2 \\ 2 \\ 2 \end{pmatrix}.$$

Let A^i denote the rows of A. Since $\|A\hat{x} - b\|_\infty = 2$ and 0 belongs to the convex hull of $\{-A^1, A^2, A^3, A^4\}$, $A\hat{x}$ is a best uniform approximation of b also. Using (3) we get $F_i = 100$.

Example 2. Meantone tuning. We next give an example of a non well-tempering, that is partially regular. This temperament called the common model meantone temperament, or simply meantone temperament, was the most used single temperament for approximately 200 years. It was invented by Pietro Aron in 1523. It is the acme of Baroque harmoniousness. With this temperament, 8 major triads are very consonant with just major 3rds. The 8 relative minor triads are very consonant with minor 3rds in absolute error of 5.5 cents. The perfect 5ths in these triads have an absolute error of 5.5 cents. Two out of 3 consecutive major 3rds are just. A typical assignment of major 3rds would be:

$$a_1 = 386, \qquad a_5 = 386, \qquad a_9 = 428,$$

$$a_2 = 428, \qquad a_6 = 386, \qquad a_{10} = 386,$$

$$a_3 = 386, \qquad a_7 = 428, \qquad a_{11} = 386,$$

$$a_4 = 386, \qquad a_8 = 386, \qquad a_{12} = 428.$$

We now have the equations

$$F_2 = 70,$$

$$F_3 - F_2 = 112,$$

(7)

$$F_4 - F_3 = 112,$$

$$F_4 = 316.$$

(The set (7) is repeated 3 times.)
   As before we set

$$A = \begin{pmatrix} 1 & 0 & 0 \\ -1 & 1 & 0 \\ 0 & -1 & 1 \\ 0 & 0 & 1 \end{pmatrix}, \quad x = \begin{pmatrix} F_2 \\ F_3 \\ F_4 \end{pmatrix}, \quad b = \begin{pmatrix} 70 \\ 112 \\ 112 \\ 316 \end{pmatrix}.$$

The condition $A^*(A\hat{x} - b) = 0$ yields $F_2 = 75.5$, $F_3 = 193$ and $F_4 = 310.5$ and

$$A\hat{x} - b = \begin{pmatrix} 5.5 \\ 5.5 \\ 5.5 \\ -5.5 \end{pmatrix}.$$

Since 0 belongs to the convex hull of $\{A^1, A^2, A^3, -A^4\}$ $A\hat{x}$ is an optimal uniform approximation to b also. The remaining F_i can be found from (3).

   Example 3. Werkmeister temperament #1. The urge for the general use of more than 8 major and minor triads led to the development of well-tempering. The most famous well-tempering, other than equal temperament is called Werkmeister's temperament #1, invented by Andreas Werkmeister in 1697. According to Barbour [1], J. S. Bach was reported to have said he was not in favor of equal temperament, so that his "well-tempered Clavier" must have had a temperament such as the Werkmeister #1 in mind. On the other hand it appears that C. P. E. Bach, the son of J. S. Bach, was in favor of equal temperament. (This statement is made in his book on the art of playing the keyboard.) Temperaments of the Werkmeister type divide the triads in 3 categories—smooth, moderate and rough. In comparison, in equal temperament all the triads are moderate. The calculation of this temperament is somewhat less trivial. The assignment of major thirds for this temperament is

$$a_1 = 390, \quad a_5 = 402, \quad a_9 = 408,$$
$$a_2 = 408, \quad a_6 = 390, \quad a_{10} = 402,$$
$$a_3 = 396, \quad a_7 = 408, \quad a_{11} = 396,$$
$$a_4 = 402, \quad a_8 = 396, \quad a_{12} = 402.$$

The equations become

(S)

1) $F_4 = 300,$ 7) $F_3 - F_2 = 102,$
2) $F_2 = 90,$ 8) $F_4 - F_3 = 96,$
3) $F_3 - F_2 = 96,$ 9) $F_4 = 294,$
4) $F_4 - F_3 = 102,$ 10) $F_2 = 90,$
5) $F_4 = 294,$ 11) $F_3 - F_2 = 102,$
6) $F_2 = 90,$ 12) $F_4 - F_3 = 96.$

CLAIM. *The optimal interpolatory solution of the above system is unique and* 8 *equations are satisfied.*

Proof. There are only 4 distinct linear functions in (S) namely $F_2, F_4, F_4 - F_3, F_3 - F_2$. Given numbers n_1, n_2, n_3 and n_4 the system

(T)

$$F_2 = n_1,$$

$$F_4 = n_2,$$

$$F_4 - F_3 = n_3,$$

$$F_3 - F_2 = n_4,$$

is, at best, consistent. Clearly 9 is the maximum number of equations that could be satisfied in (S). These 9 are the remaining equations after 1), 3), and 4) are cast out.

For these 9 equations to be satisfied (T) must be consistent with $n_1 = 90$, $n_2 = 294$, $n_3 = 96$ and $n_4 = 102$. But (T) is inconsistent.

Consider next assignments which could possibly yield 8 pure fifths. They are

n_1	n_2	n_3	n_4
90	300	96	102
90	294	96	96
90	294	102	102

There are no others. Only the last is consistent. This is the Werkmeister temperament. The scale is 0, 90, 192, 294, 390, 498, 588, 696, 729, 888, 996, 1092, 1200.

A simpler temperament is due to Marpurg. Here $a_i = 400$ $1 \leq i \leq 12$. Similar reasoning yields 9 pure fifths. The scale is 0, 98, 204, 302, 400, 498, 604, 702, 800, 898, 1004, 1102, and 1200.

BIBLIOGRAPHY

[1] J. MURRAY BARBOUR, *Tuning and Temperament*, Michigan State University Press, East Lansing, MI, 1961.
[2] OWEN JORGENSON, *Tuning historical temperaments by ear*, North Michigan University Press, to appear.
[3] SIR JAMES JEANS, *Science and Music*, Cambridge University Press, Cambridge, England, 1961.
[4] E. WARD CHENEY, *An Introduction to Approximation Theory*, McGraw-Hill, New York, 1967.

Supplementary References
Music

[1] T. BACHMAN AND P. J. BACHMAN, *"An analysis of Bela Bartok's music through Fibonaccian numbers and the golden mean,"* Music Quarterly (1979) pp. 72–82.
[2] J. BACKUS, *The Acoustical Foundation of Music*, Norton, N.Y., 1969.
[3] J. M. BARBOUR, *A geometrical approximation to the roots of numbers,"* AMM (1957) pp. 1–9.
[4] ———, *"Music and ternary continued fractions,"* AMM (1948) pp. 545–555.
[5] ———, *"Synthetic musical scales,"* AMM (1929) pp. 155–160.
[6] A. H. BENADE, *Fundamentals of Musical Acoustics*, Oxford University Press, Oxford, 1976.
[7] ———, *Horns, Strings and Harmony*, Anchor-Doubleday, N.Y., 1960.
[8] A. D. BERUSKIN, *"Tuning the ill-tempered clavier,"* Amer. J. Phys. (1978) pp. 792–795.

[9] R. S. BRINDLE, *Contemporary Percussion*, Oxford University Press, Oxford, 1970.

[10] J. D. BROWN, "*Music and mathematicians since the seventeenth century*," Math. Teach. (1968) pp. 783–787.

[11] F. J. BUDDEN, "*Modern mathematics and music*," Math. Gaz. (1967) pp. 204–215.

[12] H. M. S. COXETER, "*Music and mathematics*," Math. Teach. (1968) pp. 312–320.

[13] D. J. DICKINSON, "*On Fletcher's paper on 'Campanological groups*,'" AMM (1957) pp. 331–332.

[14] T. J. FLETCHER, "*Campanological groups*," AMM (1956) pp. 610–626.

[15] A. FORTE, *The Structure of Atonal Music*, Yale University Press, New Haven, 1973.

[16] D. GALE, "*Tone perception and decomposition of periodic functions*," AMM (1979) pp. 36–42.

[17] J. M. GRAY, "*Scaling the musical timbre*," J. Acoust. Soc. Amer. (1977) pp. 1270–1277.

[18] W. L. GULICK, *Hearing: Physiology and Psychophysics*, Oxford University Press, Oxford, 1971.

[19] D. E. HALL, *Musical Acoustics: An Introduction*, Wadsworth, Belmont, 1980.

[20] G. D. HALSEY AND E. HEWLITT, "*More on the superparticular ratios in music*," AMM (1972) pp. 1096–1099.

[21] H. L. F. HELMHOLTZ, *Sensation of Tone*, Dover, N.Y., 1954.

[22] J. J. JOSEPHS, *The Physics of Musical Sound*, Van Nostrand, N.J., 1967.

[23] F. LAWLIS, "*The basis of music—mathematics*," Math. Teach. (1967) pp. 593–596.

[24] A. LAWRENCE, *Architectural Acoustics*, Elsevier, N.Y., 1970.

[25] T. S. LITTLER, *Physics of the Ear*, Pergamon, N.Y., 1965.

[26] M. V. MATHEWS, *The Technology of Computer Music*, M.I.T. Press, Cambridge, 1969.

[27] M. V. MATTHEWS AND J. KOHUT, "*Electronic stimulation of violin resonances*," J. Acoust. Soc. Amer. (1973) pp. 1620–1626.

[28] V. O'KEEFE, "*Mathematical-musical relationship: A bibliography*," Math. Teach. (1972) pp. 315–324.

[29] H. F. OLSON, *Music, Physics, and Engineering*, Dover, New York, 1967.

[30] J. M. H. PETERS, "*The mathematics of barbershop quartet singing*," BIMA (1981) pp. 137–143.

[31] J. R. PIERCE, *The Science of Musical Sound*, Freeman, San Francisco, 1983.

[32] D. L. REINER, "*Enumeration in music theory*," AMM (1985) pp. 51–54.

[33] J. ROGERS AND B. MITCHELL, "*A problem in mathematics and music*," AMM (1968) pp. 871–872.

[34] T. D. ROSSING, *Acoustics of Bells*, Van Nostrand-Reinhold, N.J., 1981.

[35] ———, "*The acoustics of bells*," Amer. Sci. (1984) pp. 440–447.

[36] ———, "*The physics of kettledrums*," Sci. Amer. (1982) pp. 172–178.

[37] ———, *The Science of Sound*, Addison-Wesley, Reading, 1982.

[38] M. SCHECTER, "*Tempered scales and continued fractions*, AMM (1980) pp. 40–42.

[39] C. E. SEASHORE, *Psychology of Music*, Dover, N.Y., 1967.

[40] A. L. L. SILVER, "*Musimatics or the nun's fiddle*," AMM (1971) pp. 351–356.

[41] C. TAYLOR, "*Music, mathematics and physics*," BIMA (1978) pp. 114–120.

[42] H. VON FOERSTER AND J. W. BEAUCHAMP, eds., *Music by Computers*, Wiley, N.Y., 1969.

[43] D. A. WALLER, "*Some combinatorial aspects of the musical chords*," Math. Gaz. (1978) pp. 12–15.

[44] G. WEINREICH, "*The coupled motion of piano strings*," Sci. Amer. (1979) pp. 118–127.

[45] ———, "*Coupled piano strings*," J. Acoust. Soc. Amer. (1977) pp. 1474–1484.

[46] A. WOOD., *The Physics of Music*, Methuen, London, 1962.

[47] I. ZENAKIS, *Formalized Music*, Indiana University Press, Bloomington, 1971.

11. Stochastic Models

THE PIG HOUSE PROBLEM
OR
THE DETECTION AND MEASUREMENT OF
PAIRWISE ASSOCIATIONS*

R. C. GRIMSON†

Suppose that a box contains 2 A's, 2 B's, 2 C's, etc. If k letters are drawn at random and without replacement from the box, then the sample may contain some twin pairs. If many twin pairs result to the extent that the probability of that happening is small, then the randomness of the selection process is suspect and evidence for a pairwise association exists.

In this paper we introduce statistics for testing the null hypothesis that there is no pairwise association for this model and two companion models. Using only the assumption of randomness we find exact and asymptotic formulas for these statistics and we point to some applications.

In order to define the problem explicitly, it is instructive to exhibit the manner in which it arose. More important, however, is the fact that the formulas are applicable in several other settings where the model (sometimes subtly) fits.

For the past several years many swine herds throughout the United States have experienced a problem of tuberculosis lesions. In 1972 the Meat Inspection Regulations were amended so as to require the entire carcass condemned if lesions are found in more than one site. This change along with the increase in the price of hogs has brought this problem to its present significance; it particularly affects the farmers and meat packers.

The problem is to identify media of tuberculosis transmission.

As in most epidemiologic investigations, many issues surface including the necessity to account for about twenty-five host, parasite and environmental factors. But one of the specific questions concerns a suspected farrowing house (Fig. 1). Roughly, the empty house is filled with pregnant sows and when the piglets become a certain age, they are moved to a nursery. The house is then prepared for another group. This cycle repeats itself four times a year.

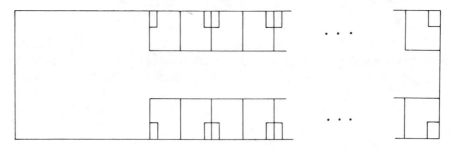

FIG. 1. *Farrowing house*

* Received by the editors August 3, 1976, and in revised form November 5, 1976.

† Department of Biostatistics, School of Public Health, University of North Carolina, Chapel Hill, North Carolina 27514. This work was partially supported by National Institutes of Health under Grant 5 F22 ES01633-02.

The small rectangles that straddle some of the pens are areas in which the piglets from both pens share heat lamps, spend most of their time and have some between pen contact with one another through slats in the sides. This basic arrangement is popular and has economic advantages.

In addition to several management–health questions, the following emerged: Is there any evidence based on the data that suggests that the health situation would significantly improve if (through design or renovation) there were no between-pen contact under the heat lamps? Implicit here is the idea that if there are lesions (one or more) from pigs farrowed in a given pen, then such are more likely to be found via its paired pen; i.e., some of the contagion is transmitted through this contact. Ideally, the pens contain nearly equal numbers of pigs.

We conceive of precisely three types of structures illustrated in Fig. 2 with lengths of 4, 5 and 4.

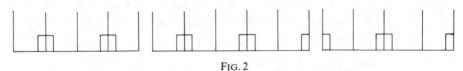

FIG. 2

The fundamental properties are portrayed by the following sequences of n odd-even symbols:

I. $\underset{\smile}{oe}oe \cdots e$ n even case

II. $\underset{\smile}{oe}oe \cdots o$ n odd case

III. $oe\underset{\smile}{oe} \cdots e$ n even shifted case.

First, we shall consider cases I and II together. In these cases the oe pairs represent the pairs for which we are testing for pairwise association. Let $\bar{X}$ be a random variable denoting the number of *marked* oe pairs (pairs sharing common facilities and for which infection is observed). Let $f(n, k)$ denote the number of distinct selections of k letters from the n letters of the sequence such that *no* oe pair is selected. If k is the total number of infected individuals in a group of n and if α is an integer, then the probability that $\bar{X}$ is larger than or equal to α is

$$(1) \qquad P(\bar{X} \geq \alpha) = \binom{n}{k}^{-1} \sum_{j \geq \alpha} \binom{r}{j} f(n - 2j, k - 2j)$$

where r is the number of oe pairs; $r = n/2$ if n is even and $r = (n - 1)/2$ if n is odd.

The hypothesis, H_0, of no pairwise association may be tested by setting α equal to the number of observed marked pairs so that $P(\bar{X} \geq \alpha)$ is the probability of getting an arrangement at least as extreme as the one we observe under the assumption of randomness. Thus, if $P(\bar{X} \geq \alpha)$ is less than a prescribed significance level, then we would not accept H_0.

For the n even case, all selections of k letters for which no oe part is present can be realized by first selecting k oe pairs and then choosing one letter from each pair. Hence,

$$(2) \qquad f(n, k) = \binom{n/2}{k} 2^k \qquad (n \text{ even}).$$

Also, we note that

$$f(n, k) = f(n-2, k-1) + f(n-1, k) \qquad (n \text{ even}).$$

Replacing n by $n+1$ and applying (2) we find

$$f(n, k) = f(n+1, k) - f(n-1, k-1) \qquad (n \text{ odd}).$$

Next, in considering case III we let $\bar{X}$ denote the number of marked eo pairs. If $g(n, k)$ denotes the numbers of distinct selections of k letters from the n letters of III such that no eo pair is selected then (1) with g replacing f remains valid. In this case $r = n/2 - 1$.

Selections of k elements from the shifted sequence may be put in 1–1 correspondence with selections from eo $\cdots$ eo, the n even sequence reversed. Associate the two endpoints of the shifted sequence with the last pair of the (reversed) n even sequence. Then including both e and o of this last pair there are $f(n-2, k-2)$ possible selections (this is legal since it corresponds to the end points of the shifted case, two letters not forming a marked pair). Otherwise there are $f(n, k)$ possibilities. Hence,

$$g(n, k) = f(n, k) + f(n-2, k-2) \qquad (n \text{ even shifted}).$$

The results are summarized as follows:
Result.

(3)
$$P(\bar{X} \geq \alpha) = \binom{n}{k}^{-1} \sum_{j \geq \alpha} \binom{r}{j} E(n-2j, k-2j)$$

where one of the following cases pertain:

$$E(n, k) = f(n, k) = \binom{n/2}{k} 2^k; \qquad r = n/2$$

$$(n \text{ even}, \bar{X} = \text{number marked oe pairs}),$$

$$E(n, k) = f(n+1, k) - f(n-1, k-1); \qquad r = (n-1)/2$$

$$(n \text{ odd}, \bar{X} = \text{number marked oe pairs}),$$

$$E(n, k) = f(n, k) + f(n-2, k-2); \qquad r = n/2 - 1$$

$$(n \text{ even shifted}, \bar{X} = \text{number marked eo pairs}).$$

For computation purposes we note the handy formula

$$f(n+2, k) = \frac{n/2+1}{n/2+1-k} f(n, k).$$

It is appropriate here to exhibit three new identities that follow from (3) and the definitions of E.

$$\sum_{j \geq 0} \binom{r}{j} E(n-2j, k-2j) = \binom{n}{k}.$$

For a related formula, see Gould [1, eqs. 3.22].

For the n even case we note that

$$\binom{n}{k} P(\bar{X} \geq \alpha) = \sum_{j \geq \alpha} \binom{n/2}{j} \binom{n/2-j}{k-2j} 2^{k-2j}$$

is a polynomial in n. The summand becomes

$$\frac{(n/2)_j(n/2-j)_{k-2j}2^{k-2j}}{j!(k-2j)!} = \frac{(n/2)^{k-j}2^{k-2j}+\text{lower order terms}}{j!(k-2j)!}.$$

The maximum exponent of n occurs where $j = \alpha$. This implies the asymptotic results that as n becomes large

(4) $$\binom{n}{k}P(\bar{X} \ge \alpha) \sim \frac{n^{k-\alpha}}{2^\alpha \alpha!(k-2\alpha)!}.$$

For the other two cases we could go through a similar argument but it is clear from the nature of the problem that as n becomes large, the distributions for the three cases become identical because what happens at endpoints of oe sequences becomes insignificant.

Formula (4) is to be used only for large values of n; on the other hand, the exact formula (3) is easy to apply by using the table for f mentioned in the last paragraph of the paper.

Fortunately, most applications are of the n even type, though the statistics were originally motivated by the n even shifted case. Because of the frequent occurrence of symmetry and binary processes in our environment, it is envisioned that in the proper setting the methods developed here are applicable to many infectious (or more generally, association) processes in which natural pairing pertains, e.g. eyes, breasts, lungs, two-to-a-room situations (dorm, hospital, cage), twins, married couples, matched pairs.

It becomes apparent that many generalizations can be defined. One interesting generalization which would have meaning in the farrowing house problem but in few others is realized when in addition to counting marked pairs, we want to count the other pairs with less weight, say half the weight. This and more straightforward generalizations are open and await further development. Also, improved approximations to the present case are sought. Even if pairwise contagion is known to exist, it is a simple matter to conceive of P (or $1-P$) being a *measure* of this association and perhaps this is the main merit of P. For example, we suggest that $1-P$ is an excellent measure of the extent of the tendency of, say, breast cancer or lung cancer to locate pairwise (on an individual).

Finally, P emerges as a natural statistic to apply for certain questions arising in some clinical trials and prospective studies. Here we have a curious overlap between the n even case of this paper and a special case ($r = 2$) of some formulas that materialized in a completely different context by different methods in [2]. When two treatments are being compared or when a study group is being compared with a control group, two series of trials, A and B, must be formed. Ideally, the assignment of a person to group A or B should be done by a strict randomization process unless a matched design is preferred. But due to limitations in time, numbers of patients, funds, etc., investigations often resort to systematic allocation or paired sampling or some other methods of "convenience matching". Such is often perceived to be independent of outcome; however, classical cases attest to the subtle fallacy of this in several studies. P may be applied in testing for this independence; in [2] further remarks are made along these lines.

As an aid to those who wish to use one of the exact formulas, this author would be pleased to supply a table of values for $f(n, k)$ for $n, k \le 30$.

Acknowledgments. I wish to express my appreciation to Dr. Caroline Becker, Epidemiologist, School of Public Health, University of North Carolina and to Dr. Billy Perryman, SE Regional Tuberculosis Epidemiologist in the Veterinarian Service, U.S. Department of Agriculture, for introducing me to the problem and for several enlightening discussions. Also, I would like to thank the referees for making several valuable suggestions.

REFERENCES

[1] H. W. GOULD, *Combinatorial Identities*, published by the author, Morgantown, WV., 1972.
[2] R. C. GRIMSON, *Column independence in* 0-1 *arrays: Implications in clinical trials and health surveys*, to appear.

RELIABILITY OF MULTICHANNEL SYSTEMS*

R. R. CLEMENTS†

Abstract. The results of simple probability theory are applied to the analysis of the reliability of multichannel redundant systems. The significant reductions in instantaneous probability of system failure achieved thereby are illustrated for the cases of nonageing and linearly ageing channel components. A model of component usage for systems with ageing components is derived.

1. Introduction. This material has been used by the author for a number of years as motivational and illustrative material with second year aeronautical engineering undergraduates. These students receive a 'core' mathematics course in common with students of other engineering disciplines and a number of additional lecture and tutorial sessions at which applications of the core mathematical content to aeronautical engineering topics are studied. In these sessions it is intended to introduce students to methods actually used in practice and an integral part of the illustrative material is the introduction of typical numerical values of physical properties to familiarize students with the range of values with which they will operate.

The basic probability theory the students have learnt is applied to the analysis of multichannel systems designed to achieve high system reliability (low instantaneous probability of failure) such as is required by aircraft control and stability augmentation systems, etc. The results obtained are seen to be qualitatively consistent with common sense (an important consideration in motivating engineering students to use mathematical methods to explore their own disciplines) and to enable quantitative design conclusions to be drawn.

2. Reliability. Consider N_0 identical components under test. After a time t let $N(t)$ be the number still operating. Then

$$R(t) = \lim_{N_0 \to \infty} \frac{N(N_0, t)}{N_0}$$

is called the reliability of the component. For a single component $R(t)$ is the probability that the component operates until a time greater than t.

---

* Received by the editors May 15, 1978 and in revised form March 16, 1979.
† Department of Engineering Mathematics, University of Bristol, BS8 1TR, England.

The mortality of a component, $m(t)$, is the probability of failure of a component at age t. Thus

$$m(t)\, \delta t = \text{P\{Component life} \in (t, t + \delta t)\}$$

$$= R(t) - R(t + \delta t)$$

$$= -\frac{dR}{dt}\delta t + O(\delta t^2).$$

Therefore

$$m(t) = -\frac{dR}{dt}.$$

Thus $m(t)$ is related to the probability density function of $R(t)$. (It is, in fact, the p.d.f. of $1 - R(t)$, the unreliability of the component.)

We also define the instantaneous failure rate $\lambda(t)$ as

$$\lambda(t)\, \delta t = \text{P \{failure} \in (t, t + \delta t) \text{ of a component operating at } t\},$$

i.e.

$$R(t)\lambda(t)\delta t = m(t)\delta t;$$

therefore

$$\lambda(t) = \frac{m(t)}{R(t)},$$

$$\lambda(t) = -\frac{1}{R(t)}\frac{dR}{dt},$$

$$\frac{dR}{R} = -\lambda\, dt,$$

and

$$R(t) = \exp\left(-\int_0^t \lambda(t)\, dt\right).$$

The essential difference that must be appreciated between $m(t)$ and $\lambda(t)$ is that $m(t)$ is the a priori probability of failure at a given age at the start of the operating period while $\lambda(t)$ is the conditional probability of failure given that the component has survived until age t.

The expected life of a component is

$$\text{E\{Life\}} = \int_0^\infty tm(t)\, dt$$

$$= -\int_0^\infty t\frac{dR}{dt}\, dt$$

$$= -[tR]_0^\infty + \int_0^\infty R(t)\, dt$$

$$= \int_0^\infty R(t)\, dt.$$

3. Parallel systems. Suppose, in order to improve, or to attain a minimum acceptable level of reliability in a system, several components are used in parallel so that a system failure occurs only if all the components fail simultaneously. This is common practice on systems requiring high reliability for safety reasons, e.g. command and/or stability augmentation systems on aircraft using 'fly by wire' philosophy, control surface actuator systems. Let $R_n(t)$ be the reliability of a system with n parallel channels. Then

$$R_n(t) = P\{\text{Life of system} > t\}$$
$$= 1 - P\{\text{Life of system} \leq t\}.$$

The system only fails if all the component channels fail, so, if the reliability of a single component is $R(t)$,

$$P\{\text{Life of system} \leq t\} = (P\{\text{Component life} \leq t\})^n$$
$$= (1 - R(t))^n;$$

therefore

$$R_n(t) = 1 - (1 - R(t))^n.$$

Hence we have, for $m_n(t)$ and $\lambda_n(t)$

$$m_n(t) = -\frac{dR_n}{dt}$$
$$= -nR'(t)(1 - R(t))^{n-1},$$
$$\lambda_n(t) = \frac{m_n(t)}{R_n(t)}$$
$$= \frac{-nR'(t)(1 - R(t))^{n-1}}{1 - (1 - R(t))^n}.$$

Now

$$\lambda(t) = \frac{m(t)}{R(t)}$$
$$= -\frac{R'(t)}{R(t)},$$

and

$$\frac{\lambda_n(t)}{\lambda(t)} = \frac{nR(t)(1 - R(t))^{n-1}}{1 - (1 - R(t))^n}.$$

4. The case when $\lambda(t)$ is constant. Certain components of systems can, to an adequate degree of accuracy, be assumed to have an instantaneous failure rate that is independent of the age of the component. Modern solid state electronic devices are commonly assumed to be of this type. Real time computer software is also judged by some experts to fall into this category. For systems composed of such components in parallel, the analysis can be carried further and predictions of system failure characteristics made. From these inspection schedules can be set which ensure an adequately small probability of system failure in service.

Let

$$\lambda(t) = \mu.$$

Then

$$R(t) = e^{-\mu t}, \qquad m(t) = \mu e^{-\mu t}.$$

Then

$$E\{\text{Life of single component}\} = \int_0^\infty R(t)\, dt$$

$$= \frac{1}{\mu},$$

i.e., $1/\mu$ is the expected life of the component (in hours say) and μ is the failure rate of the component per hour

$$R_n(t) = 1 - (1 - e^{-\mu t})^n.$$

Figure 1 illustrates the form of $R_n(t)$ for $n = 1, 2, 3, 4, 5$. For a given component failure rate μ, the reliability of the system increases with n, but not very sharply. The

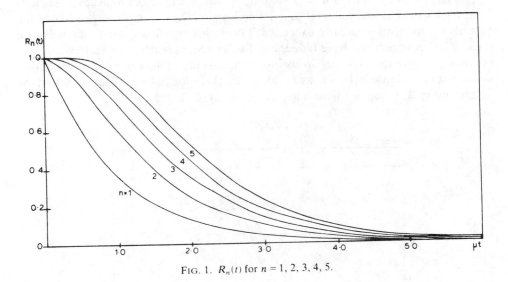

FIG. 1. $R_n(t)$ for $n = 1, 2, 3, 4, 5$.

significance of multichannel systems is seen fully when $\lambda_n(t)$, the instantaneous failure rate, is considered.

$$\frac{\lambda_n(t)}{\lambda(t)} = \frac{n e^{-\mu t}(1 - e^{-\mu t})^{n-1}}{1 - (1 - e^{-\mu t})^n}.$$

For $\mu t \ll 1$,

$$R(t) = e^{-\mu t} = 1 - \mu t + O(\mu t)^2.$$

Therefore

$$\frac{\lambda_n(t)}{\lambda(t)} = \frac{n(1-\mu t)(\mu t)^{n-1}}{1-(\mu t)^n} + O(\mu t)^{n+1}$$

$$= n(\mu t)^{n-1} + O(\mu t)^n$$

and

$$\log(\lambda_n(t)) \simeq (n-1)\log(\mu t) + \log(n) + \log\mu,$$

i.e., for small μt the graph of $\log(\lambda_n) - \log(\mu)$ against $\log(\mu t)$ is asymptotic to a straight line of slope $(n-1)$

For $\mu t \gg 1$,

$$R(t) \to 0,$$

$$\frac{\lambda_n(t)}{\lambda(t)} = \frac{nR(t)(1-(n-1)R(t))}{nR(t)} + O(R)$$

$$= 1 + O(R);$$

therefore

$$\log(\lambda_n(t)) \simeq \log\mu,$$

i.e., for large μt the graph of $\log(\lambda_n) - \log(\mu)$ asymptotes to zero. Physically this results from increasing channel failures. $\mu t \gg 1$ implies $t \gg 1/\mu$, i.e., the average life of the channel components is greatly exceeded. For such large times many channels in a multi-channel system will have failed. Finally the system reaches the state where it is operating on one channel only so its instantaneous probability of failure is that of a single channel. Hence $\lambda_n(t) \to \mu$ as $t \to \infty$, so $\log(\lambda_n) - \log(\mu) \to 0$. Figure 2 shows the graphs of $\log(\lambda_n) - \log(\mu)$ against $\log(\mu)$ for $n = 1, 2, 3, 4, 5$.

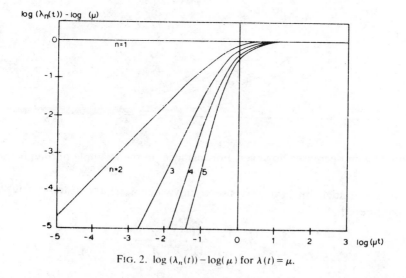

FIG. 2. $\log(\lambda_n(t)) - \log(\mu)$ for $\lambda(t) = \mu$.

Although the failure rate of the system components are independent of component age, the failure rate of the system as a whole is a function of age. For small μt, the failure rate of the system as a whole is very greatly reduced by the use of multiple channel systems. Small μt will be consistent with moderate ages, t, of the system if the single component failure rates are small. For a given acceptable system failure rate, using components of known constant failure rate, Fig. 2 enables the system life to be determined. For instance, using components with $\mu = 10^{-4}$ failures/hour a duplicated system would have a system failure rate less than 10^{-7} failures/hour for $\log (\mu t) < -3.3$, i.e., $t < 5.01$ hours whereas a triplicated system would have a failure rate less than 10^{-7} for $t < 2.01 \times 10^2$ hours. At the end of the system life the system can be renewed by independently checking each channel. If all the channels are found to be operative the system is effectively returned to a zero age state since the individual channel components do not age. Any number of channels less than the total number of channels may be found to have failed and replacement of such channels with similar operative components will again return the system to the zero age state.

An interesting contrast to the significant decreases in instantaneous failure rates achieved by multichannel design is the relatively small increase in expected life of the system. For a multichannel system

$$\text{E\{Life of } n\text{-channel system\}} = \int_0^\infty R_n(t) \, dt$$

$$= \int_0^\infty 1 - (1 - e^{-\mu t})^n \, dt$$

and putting $s = 1 - e^{-\mu t}$, we obtain

$$\text{E\{Life\}} = \int_0^1 \frac{1 - s^n}{1 - s} \frac{ds}{\mu}$$

$$= \frac{1}{\mu}\left(1 + \frac{1}{2} + \frac{1}{3} + \cdots + \frac{1}{n}\right).$$

Thus a duplicated system only has an expected life (i.e., mean time between failures) 1.5 times that of the basic system components. Of course, for sensitive systems, whose failure may lead to the loss of an aircraft, it is not the mean time between failures that is important but the instantaneous probability of failure, and multichannel systems offer a very satisfactory solution.

The choice between duplicated, triplicated or higher multichannel systems is affected by other criteria as well as achieving a satisfactorily low probability of failure. For a given channel failure rate and target system failure rate, the time between inspections can obviously be increased by using a larger number of channels. The proportionate gain decreases however. For instance, in the example quoted before, a quadruplicate system would only increase system life from 2.01×10^2 hours to 6.31×10^2 hours. Such an increase may not bring sufficient benefits to compensate for the additional weight and cost of the extra channel.

5. The case when components age linearly. Consider now systems composed of components whose instantaneous failure rate increases with time. Many mechanical and hydraulic components, over at least part of their working life, can be represented as ageing in this way. Such components are normally considered to have a "burn in" period when failure is more likely, followed by a period of relatively steady wear until old age when failure rate rises sharply again. Here we will consider multichannel systems composed of components whose instantaneous failure rate can be represented by $\lambda(t) = \mu^2 t$. Then $R(t) = e^{-(1/2)\mu^2 t^2}$, $m(t) = \mu^2 t\, e^{-(1/2)\mu^2 t^2}$,

$$E\{\text{Life of component}\} = \int_0^\infty R(t)\, dt$$

$$= \int_0^\infty e^{-(1/2)\mu^2 t^2}\, dt$$

$$= \left(\frac{\pi}{2}\right)^{1/2} \frac{1}{\mu}.$$

Thus μ is $(\pi/2)^{1/2}$ times the reciprocal of the mean time between failures of such components, or $(\pi/2)^{1/2}$ time the average failure rate.

For such systems we have

$$\frac{\lambda_n(t)}{\lambda(t)} = \frac{n\, e^{-(1/2)\mu^2 t^2}(1 - e^{-(1/2)\mu^2 t^2})^{n-1}}{1 - (1 - e^{-(1/2)\mu^2 t^2})^n}.$$

For $\mu t \ll 1$, $R(t) = 1 - \frac{1}{2}(\mu t)^2 + O(\mu t)^4$,

$$\frac{\lambda_n(t)}{\lambda(t)} = \frac{n(1 - \frac{1}{2}\mu^2 t^2)(\frac{1}{2}\mu^2 t^2)^{n-1}}{1 - (\frac{1}{2}\mu^2 t^2)^n} + O(\mu t)^{2n}$$

$$= n\frac{(\mu t)^{2(n-1)}}{2^{n-1}} + O(\mu t)^{2n},$$

$$\log\left(\frac{\lambda_n(t)}{\mu^2 t}\right) = 2(n-1)\log(\mu t) + \log\frac{n}{2^{n-1}} + \log(1 + O(\mu t)^2);$$

therefore

$$\log(\lambda_n(t)) = (2n-1)\log(\mu t) + \log\mu + \log\frac{n}{2^{n-1}} + O(\mu t)^2.$$

As for the $\lambda(t) = \mu$ case the graphs of $\log(\lambda_n)$ against $\log(\mu t)$ tend to a series of straight lines, but of slopes $(2n-1)$, when $\mu t \ll 1$.

For $\mu t \gg 1$, $R(t) \to 0$ and

$$\frac{\lambda_n(t)}{\lambda(t)} = 1 + O(R) \quad \text{as before;}$$

therefore

$$\log(\lambda_n(t)) \approx \log(\mu t) + \log\mu.$$

Figure 3 shows curves of $\log(\lambda_n) - \log\mu$ against $\log(\mu t)$ for this case. This figure, like Fig. 2, can be used for determining the life of multichannel systems composed of components with known average failure rate and linear ageing when the system must meet any instantaneous failure rate criterion. It is apparent that, if the acceptable failure rate of the system is of the same order as, or greater than, μ for the system components, no gain will result from multichannel design since, for $\log\lambda_n > \log\mu + 1$ the curves are indistinguishable and for $\log\lambda_n > \log\mu$ there is little difference. As before, the significant gains occur when the design calls for much higher reliability than a single channel can provide over significant times. For systems composed of components that age, independent inspection of the channels at the expiry of system life does not return the system to zero age as the channels have individually aged even if they have not yet

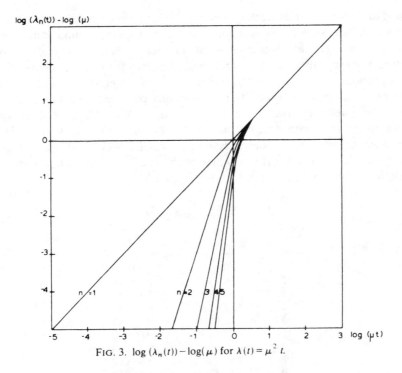

FIG. 3. $\log(\lambda_n(t)) - \log(\mu)$ for $\lambda(t) = \mu^2 t$.

failed. Normal practice with such systems is to remove them from service at the expiry of their reliability life and either scrap or renovate the components, replacing the system in service with a new zero age one. In this case a model of comparative component usage rates for different multichannel systems is of interest.

Let T_n be the life of an n-channel system with $\mu = \mu_0$ for each channel and a required system reliability of ϕ say, i.e. the system must satisfy $\lambda_n(t) < \phi \forall t < T_n$, so $\lambda_n(T_n) = \phi$ since $\lambda_n(t)$ is monotone increasing with T. Thus

$$\log(\lambda_n(T_n)) = \log \phi.$$

For $\mu_0 t \ll 1$, $\log(\lambda_n) = (2n-1)\log \mu_0 t + \log \mu_0 + \log n/2^{n+1} + O(\mu_0 t)^2$ so T_n satisfies

$$\log \phi = (2n-1)\log(\mu_0 T_n) + \log \mu_0 + \log \frac{n}{2^{n-1}} + O(\mu_0 T_n)^2,$$

that is,

$$T_n \simeq \frac{1}{\mu_0}\left\{\frac{\phi 2^{n-1}}{n\mu_0}\right\}^{1/(2n-1)} \quad \text{hours.}$$

Rate of usage of components, U_n say, is n/T_n, i.e.,

$$u_n \simeq \left\{\frac{(n\mu_0)^{2n}}{\phi 2^{n-1}}\right\}^{1/(2n-1)} \quad \text{units/hour.}$$

For numerical illustration take, again, a desired system reliability of 10^{-7} failures/hour and components with $\lambda(t) = \mu_0^2 t$, $\mu_0 = 10^{-4}$ say. Table 1 gives system life and system component usage rates for single, duplicate, triplicate, etc., systems satisfying the quoted system reliability. (Figures for 50 and 200 channel systems are included to show the trends. Such systems are highly unlikely to be considered in practical applications.) As with systems composed of nonageing components, the proportionate gain in system life decreases as the number of channels increases. Further the component usage rate actually reverses and increases for large numbers of channels. The proportionate gain in system life from adding extra channels is finally outweighted by the extra components used. This ignores any other disadvantages of using high multiplicity systems such as extra weight, volume, etc. All these factors together combine to set an upper limit on the number of channels it is worthwhile using in practical design. For aircraft control systems, particularly in "fly by wire" applications, where system failure would bring a high probability of loss of the aircraft, triplicate and quadruplicate systems are currently most favored.

TABLE 1

n	T_n	U_n
1	1.00×10^1	1.00×10^{-1}
2	1.00×10^3	2.00×10^{-3}
3	2.66×10^3	1.13×10^{-3}
4	4.12×10^3	9.72×10^{-4}
5	5.28×10^3	9.47×10^{-4}
6	6.21×10^3	9.66×10^{-4}
7	6.97×10^3	1.00×10^{-3}
8	7.59×10^3	1.05×10^{-3}
9	8.11×10^3	1.11×10^{-3}
10	8.55×10^3	1.17×10^{-3}
50	1.26×10^4	3.96×10^{-3}
200	1.37×10^4	1.46×10^{-2}

Acknowledgments. The author would like to thank those colleagues whose discussions have contributed to the development of this material.

IONIZATION IN DIPROTIC ACIDS*

HOWARD REINHARDT†

In this note we present a combinatorial approach to the ionization of diprotic (DI − + PROT(ON) + − IC) acids, such as those in Fig. 1. In solution, three species exist: A molecule can be fully ionized, half-ionized or un-ionized. Expressions for the distribution of the species in a solution are commonly derived [2], [4] by assuming sequential ionization with different ionization constants for the two ionizations. However, the diprotic acid is sometimes considered as two independently ionizing monoprotic acids [1] and in such cases the intuitive argument involving sequential ionization seems contradictory.

Motivated by the suggestion of two independently ionizing monoprotic acids, we propose the following question as a model for the ionization: A room contains m men and

*Received by the editors September 20, 1980, and in revised form April 23, 1981.
†Department of Mathematics, University of Montana, Missoula, Montana 59812.

$w \leq 2m$ women. Each woman chooses at random and holds the hand of a man (i.e., each of the $\binom{2m}{w}$ configurations of women holding men's hands is equally likely). Borrowing from the language of urn models we say that a man is doubly-occupied, singly-occupied or unoccupied if his hands are held, respectively, by two, one or no women. What is the probability distribution of the numbers of doubly, singly and unoccupied men?

In this model, men correspond to acid ions and women to protons, so that doubly-occupied, singly-occupied and unoccupied men correspond, respectively to un-ionized, singly ionized and doubly ionized acid molecules.

The language of the problem is borrowed from Johnson and Kotz [3, p. 121] who suggest the following as arising in a chemical industry inquiry:

> There are n men in a room; n women enter the room, and each selects a man at random and holds one of his hands if not already held by another woman (so that no man can be selected by more than two women).
>
> Find the joint distribution of the number of men with (i) both hands, (ii) one hand, (iii) neither hand held by a woman.

Johnson and Kotz do not give specifics of the chemistry problem. Their model is not the appropriate model for diprotic acids; a chemist friend says that protons are interested in hands, not men.

Our problem is easily solved with techniques accessible to beginning students of probability. Let X_0, X_1, X_2 be respectively the number of un-, singly- and doubly-occupied men. We note that $X_0 + X_1 + X_2 = m$ and $X_1 + 2X_2 = w$. Hence $P(X_2 = j)$ =

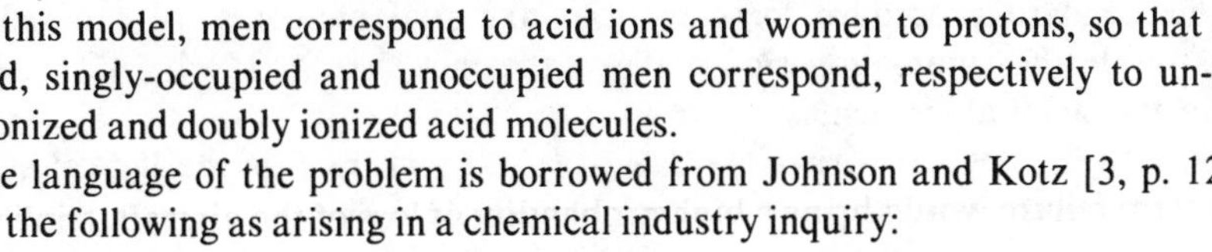

$$
\text{H—O—}\underset{\underset{\text{O}}{\parallel}}{\text{C}}\text{—CH}_2\text{—CH}_2\text{—CH}_2\text{—}\underset{\underset{\text{O}}{\parallel}}{\text{C}}\text{—O—H} \qquad \text{H—O—}\underset{\underset{\text{O}}{\parallel}}{\overset{\overset{\text{O}}{\parallel}}{\text{S}}}\text{—O—H}
$$

FIG. 1. *Two diprotic acids.*

$P(X_1 = w - 2j) = P(X_0 = m - w + j)$ and we need only compute $P(X_2 = j)$. There are $\binom{2m}{w}$ possible configurations. We arrive at a configuration with $X_2 = j$ by selecting the j doubly-occupied men ($\binom{m}{j}$ choices), selecting the singly occupied men from the remaining men ($\binom{m-j}{w-2j}$ choices) and finally selecting for the singly-occupied men the hand which is to be held (2^{w-2j} choices). Hence

$$
P(X_2 = j) = \binom{m}{j}\binom{m-j}{w-2j}2^{w-2j} \bigg/ \binom{2m}{w}.
$$

These numbers, summed over all possible values of j, must add to 1; that they do can be determined independently by examining the coefficient of x^w in the identity $(1 + x)^{2m} = (1 + 2x + x^2)^m$. It follows that, in particular,

$$
\sum \binom{m}{j}\binom{m-j}{w-2j}2^{w-2j} = \binom{2m}{w},
$$

an identity we shall use below.

One can find the mean and variance of X_2 by first finding the factorial moments

$$
E(X_2) = \sum_j j \binom{m}{j}\binom{m-j}{w-2j}2^{w-2j} \bigg/ \binom{2m}{w}
$$

and

$$E(X_2(X_2 - 1)) = \sum_j j(j - 1) \binom{m}{j}\binom{m - j}{w - 2j} 2^{w-2j}\bigg/\binom{2m}{w}.$$

Using the standard procedure for computing factorial moments we can reduce the fractions inside the summation sign and use the fact that $\sum_j \binom{m}{j}\binom{m-j}{w-2j}2^{w-2j} = \binom{2m}{w}$ to determine the factorial moments.

Alternatively, one can write $X_2 = I_1 + \cdots + I_m$ where I_r is the indicator random variable for the event "the rth male has both hands held." Then

$$E(I_r) = P(I_r = 1) = \binom{2m - 2}{w - 2}\bigg/\binom{2m}{w}$$

and, with $r \neq s$,

$$E(I_r I_s) = P(I_r = 1 \text{ and } I_s = 1) = \binom{2m - 4}{w - 4}\bigg/\binom{2m}{w}.$$

Linearity of expected value then allows computation of $E(X_2)$ and $E(X_2^2)$. Finally, with either starting place, one finds

$$E(X_2) = \frac{w(w - 1)}{4m - 2},$$

$$\text{var}(X_2) = \frac{w(w - 1)}{2(2m - 1)^2(2m - 3)} \cdot (4m^2 - 2m - 4mw + w^2 + w).$$

For the chemical application, m and w are both large. The results are conveniently expressed in terms of $\lambda = w/m$, the average number of protons bound to an acid molecule. If we let $Y_2 = X_2/m$, the fraction of unionized molecules, we find $E(Y_2) \doteq \lambda^2/4$ and var $Y_2 \doteq (4\lambda^2 - 4\lambda^3 + \lambda^4)/16m$. Since m is large, Chebyshev's inequality shows that, in any reaction, the fraction of unionized molecules will be essentially $\lambda^2/4$. The fractions of fully-ionized and half-ionized molecules will be $1 - \lambda + \lambda^2/4$ and $\lambda - \lambda^2/2$ respectively. Since λ depends only on a single ionization constant and the relative abundance of acid molecules and hydrogen ions, this result is satisfactory for the chemists' purposes. (A more sensitive result could be obtained from an appropriate central limit theorem.)

From a model builder's point of view, it is reassuring to learn that the results agree with experimental evidence and the traditional chemist's argument involving successive ionizations.

REFERENCES

[1] E. Q. ADAMS, *Relations between the constants of dibasic acids and of amphoteric electrolytes*, J. Amer. Chem. Soc., 38 (1916), 1503 pp.

[2] E. S. GOULD, *Mechanism and Structure in Organic Chemistry*, Holt, Rinehart and Winston, New York, 1959, pp. 202, 242.

[3] NORMAN J. JOHNSON AND SAMUEL KOTZ, *Urn Models and Their Applications*, John Wiley, New York, 1977, p. 121.

[4] G. W. WHELAND, *Advanced Organic Chemistry*, 3rd Edition, John Wiley, New York, 1960, p. 519, 523.

A PROBABILISTIC LOOK AT MIXING AND COOLING PROCESSES*

E. WAYMIRE†

1. Introduction. Since applications are receiving more emphasis in the core curriculum for mathematics majors, it is desirable to have examples which possess a depth of quality equal to that of the mathematical topics being illustrated. In the case of applications of logarithms and exponentials in the calculus it is becoming popular to include problems on mixing and cooling processes in addition to the standard growth and decay processes (see [3] and [4], for example). The purpose of this article is to illustrate that just as with the growth and decay process, the mixing and cooling processes also provide good examples to build on in probability courses. This is achieved by adapting the point of view of classical statistical mechanics. Aside from providing us with an opportunity for reinforcement of the basic understanding of these processes, consideration of these two problems also enables us to introduce fundamentally important modes of thought in classical physics within a context already familiar to most students.

2. Mixing processes. In a typical problem on mixing we are given a vessel S which holds a volume V of liquid. It is supposed that initially there is a concentration C_0 of solute (salt, for example) dissolved in the liquid. By the concentration of solute is meant the mass of solute per unit volume of liquid. Starting at time $t = 0$, a liquid-solute solution of concentration C_1 runs into the vessel at a constant rate r. It is assumed that the two solutions mix instantly, being kept uniform by stirring, and the excess drains off. One then asks for the concentration $C(t)$ of solution at time t as well as for $\lim_{t \to \infty} C(t)$. Of course it is found that

$$(2.1) \qquad C(t) = C_1 + (C_0 - C_1) e^{-rt/V}, \qquad t \geq 0,$$

$$(2.2) \qquad \lim_{t \to \infty} C(t) = C_1.$$

For our probabilistic model of a mixing process we shall adapt the point of view. used by Smoluchowski in his elegant analysis of Einstein's theory of concentration fluctuations under diffusion equilibrium (see [1] for an excellent review). It is assumed that the solute particles are identical in size, so that the total mass at any given time is proportional to the number of particles present. During each (small) unit of time Δt a Poisson-distributed number of solute particles with mean $\lambda = C_1 r \Delta t$ enter the vessel. In addition, provided that the solute is sufficiently dilute and well stirred, each solute particle present in the vessel independently of the others has probability $p = r\Delta t / V$ of being displaced from the vessel in each interval of time. Let X_n denote the number of solute particles in S after the nth time period. Then the flow equation is given by

$$(2.3) \qquad X_n = X_{n-1} - L_n + I_n, \qquad n = 1, 2, \cdots,$$

where L_n is the number of particles which left the vessel during the nth time period and I_n is the number of particles which entered the vessel. Observe that, given X_{n-1}, $X_{n-1} - L_n$ is (conditionally) binomially distributed with parameters X_{n-1} and $q = 1 - p$, and is independent of I_n. Let $\psi_n(z) = E(z^{X_n} | X_0 = N)$ denote the probability generating

* Received by the editors July 15, 1980, and in revised form April 3, 1981.

† Department of Mathematics, University of Mississippi, University, Mississippi, 38677.

function of X_n. Then

(2.4)
$$\psi_n(z) = \psi_{n-1}(p + qz)\, e^{-\lambda + \lambda z}, \qquad n = 1, 2, \cdots.$$

Iterating (2.4) we obtain, as probabilistic analogues to (2.1) and (2.2),

(2.5)
$$\psi_n(z) = [1 - q^n + q^n \cdot z]^N \cdot \exp\left\{-\frac{\lambda}{p}(1 - q^n) + \frac{\lambda}{p}(1 - q^n)z\right\},$$

(2.6)
$$\lim_{n \to \infty} \psi(z) = e^{-\lambda/p + \lambda z/p}.$$

In particular, the steady state distribution is Poisson with parameter λ/p.

In order to see that the stochastic model agrees with the deterministic model on a "macroscopic" level, let $m_n = E(X_n | X_0 = m_0)$. Then from the flow equation we obtain

(2.7)
$$m_n = q \cdot m_{n-1} + \lambda, \qquad n = 1, 2, \cdots.$$

So, again after an iteration, we have

(2.8)
$$m_n = \frac{\lambda}{p} + \left(m_0 - \frac{\lambda}{p}\right)q^n.$$

For a comparison with (2.1), write $t = n \cdot \Delta t$. Then,

$$\frac{\lambda}{p} = C_1 V, \quad m_0 = C_0 V \quad \text{and} \quad q^n = \left(1 - \frac{rt/V}{n}\right)^n \cong e^{-rt/V} \quad \text{for } N \text{ large } (\Delta t \text{ small}).$$

3. Cooling processes. Another important example from the physical sciences which is familiar to most students with only a calculus background is Newton's law of cooling (see [3] and [4]). Here the classical Ehrenfest model for heat exchange (see [2]) is a natural candidate for the probabilistic counterpart. Although the Ehrenfest model is traditionally used as an aid to understanding the irreversibility paradox in thermodynamics and statistical mechanics, a discussion based on Newton's law of cooling seems to be more readily accessible to students freshly out of calculus.

Newton's law of cooling states that the time rate of change of the temperature of a cooling substance is proportional to the difference between the temperature of the substance and that of the surrounding medium. Letting T_t denote the temperature of the substance at time t and letting E_0 denote the (equilibrium) temperature of the surrounding medium, we obtain

(3.1)
$$T_t = E_0 + (T_0 - E_0)\, e^{-\alpha t},$$

(3.2)
$$\lim_{t \to \infty} T_t = E_0,$$

where $\alpha > 0$ is the constant of proportionality.

The Ehrenfest model consists of two boxes I and II, together with $2N$ balls labelled $1, 2, \cdots, 2N$. Initially some of the balls are in box I and the remainder are in box II. At each instant of time a ball is selected at random from the balls numbered $1, 2, \cdots, 2N$ and is moved from the box in which it is contained to the other box. The number X_n of balls in box I at the nth period of time is regarded as a measurement of the temperature of the substance at time n. The probabilistic counterparts to (2.1) and (2.2) are obtained by calculating the one-step transition probabilities and the invariant initial distribution π_0 respectively. π_0 is binomial with parameters $2N$ and $\frac{1}{2}$ ([2]). At this point an analysis of the precise nature of the probabilistic analogue to

(3.2) offers an excellent opportunity for a discussion of the role of periodicity in Markov chains.

To see that the macroscopic law of Newton holds within the context of the present model, we let $T_n = E[X_n | X_0 = i]$ denote the average temperature of the substance when the initial temperature is $T_0 = i$. Writing $X_n = X_{n-1} + (X_n - X_{n-1})$ and conditioning this equation on X_{n-1} and X_0, we obtain the macroscopic evolution equation

$$(3.3) \qquad\qquad T_n = 1 + \left(1 - \frac{1}{N}\right) T_{n-1}, \qquad n = 1, 2, \cdots.$$

Upon iteration of (3.3) it then follows that

$$(3.4) \qquad\qquad T_n = N + (i - N)\left(1 - \frac{1}{N}\right)^n, \qquad n = 0, 1, \cdots,$$

$$(3.5) \qquad\qquad \lim_{n \to \infty} T_n = N.$$

Observe that $E_0 = N$ is the mean equilibrium temperature, since $E_{\pi_0}(X_n) = \frac{1}{2} \cdot (2N) = N$ is the mean temperature when the process is in the steady state π_0. For a further comparison with (3.1), note that $i = T_0$ is the initial temperature of the substance. Moreover, if τ denotes the time between transitions (*relaxation time*) then in time $t > 0$ there are $n = t/\tau$ transitions and

$$\left(1 - \frac{1}{N}\right)^n = \left(1 - \frac{t/N\tau}{n}\right)^n = e^{-t/N\tau}$$

for large n (small τ).

REFERENCES

[1] S. CHANDRASEKHAR, *Stochastic problems in physics and astronomy*, Rev. Modern Physics, 15 (1943), pp. 1–89.
[2] W. FELLER, *An Introduction to Probability Theory and its Applications*, vol. 1, John Wiley, New York, 1968.
[3] M. A. MUNEM AND D. J. FOULIS, *Calculus with Analytic Geometry*, Worth, New York, 1978.
[4] A. SHENK, *Calculus and Analytic Geometry*, Goodyear, Santa Monica, CA, 1977.

POISSON PROCESSES AND A BESSEL FUNCTION INTEGRAL*

F. W. STEUTEL[†]

Abstract. The probability of winning a simple game of competing Poisson processes turns out to be equal to the well-known Bessel function integral $J(x, y)$ (cf. Y. L. Luke, *Integrals of Bessel Functions*, McGraw-Hill, New York, 1962). Several properties of J, some of which seem to be new, follow quite easily from this probabilistic interpretation. The results are applied to the random telegraph process as considered by Kac [Rocky Mountain J. Math., 4 (1974), pp. 497–509].

Key words. Poisson process, Bessel function, random telegraph

*Received by the editors April 20, 1984, and in revised form August 29, 1984.
† Department of Mathematics and Computing Science, Eindhoven University of Technology, Eindhoven, the Netherlands.

1. Competing Poisson processes. Several problems can be described as follows: An object has to travel a distance x; it does so at unit speed, but it is obstructed at random moments and then held for a random period of time before it is allowed to continue. The object may be a particle moving between two electrodes, a person walking to a bus stop, or, as in [5, Problem 147], a book being read with random interruptions. The question is: What is the probability that the object reaches its destination at a moment not exceeding $x+y$? The situation may be modelled as a game of two competing (Poisson) renewal processes in the following way (see Fig. 1):

Let $X_1, Y_1, X_2, Y_2, \cdots$ be independent, exponentially distributed random variables with expectation one. Two persons, X and Y, take turns drawing lengths X_j and Y_j. Person X starts, and wins if the sum of his X_j exceeds x before the sum of Y's Y_j exceeds y.

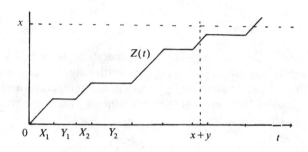

FIG 1. $N_x = 5$, $N_y = 3$; X loses.

More formally, if N_x and N_y are random variables defined by

$$N_x = \min\{n; X_1 + \cdots + X_n > x\},$$
$$N_y = \min\{n; Y_1 + \cdots + Y_n > y\},$$

then (remember that X starts)

(1) $$X \text{ wins} \Leftrightarrow N_x \leq N_y \Leftrightarrow X_1 + \cdots + X_{N_y} > x.$$

Remark. For our purposes the assumption that $EX_j = EY_j = 1$ for $j = 1, 2, \cdots$, is no restriction: replacing X_j and Y_j by X_j/λ and Y_j/μ, respectively, is equivalent to replacing x and y by λx and μy, respectively. The *process* $Z(t)$ depicted in Fig. 1, representing the distance travelled by the object at time t, would, of course, be changed by a transformation of the X_j and Y_j.

We shall use the following two well-known facts: $N_y - 1$ has a Poisson distribution with mean y, i.e.,

(2) $$P(N_y = n) = e^{-y} \frac{y^{n-1}}{(n-1)!} \qquad (n = 1, 2, \cdots),$$

and $X_1 + \cdots + X_n$ has a gamma distribution with density

(3) $$\frac{d}{dx} P(X_1 + \cdots + X_n \leq x) = e^{-x} \frac{x^{n-1}}{(n-1)!} \qquad (x > 0).$$

Now, let $J(x,y)$ be defined by (cf. Luke [4, p. 271])

(4)
$$J(x,y)=1-e^{-y}\int_0^x I_0(2\sqrt{yt})e^{-t}dt,$$

where I_0 is the modified Bessel function of order zero:

(5)
$$I_0(z)=\sum_0^\infty \frac{(z/2)^{2n}}{(n!)^2}.$$

Then we easily obtain
PROPOSITION 1.

(6)
$$P(N_x\leq N_y)=J(x,y).$$

Proof. By (1)–(5) we have

$$P(N_x\leq N_y)=1-P(N_x>N_y)=1-P(X_1+\cdots+X_{N_y}\leq x)$$

$$=1-\sum_{n=1}^\infty P(N_y=n,X_1+\cdots+X_n\leq x)$$

$$=1-\sum_{n=1}^\infty e^{-y}\frac{y^{n-1}}{(n-1)!}\int_0^x e^{-t}\frac{t^{n-1}}{(n-1)!}dt$$

$$=1-e^{-y}\int_0^x I_0(2\sqrt{yt})e^{-t}dt=J(x,y).$$

Remark. Srivastava and Kashyap [6, pp. 77, 78] consider an equivalent interpretation, in the context of a randomized random walk; there the interpretation remains implicit and is not pursued.

2. Properties of $J(x,y)$. Several properties of $J(x,y)$ follow immediately from (6). We list the following six together with their simple proofs.
 (i) $J(0,y)=P(X_1>0)=1$,
 (ii) $J(x,0)=P(X_1>x)=e^{-x}$.

From (2) and its counterpart for N_x (independent of N_y) it follows that

$$P(N_x=N_y)=\sum_1^\infty P(N_x=n,N_y=n)$$

$$=\sum_1^\infty e^{-x}\frac{x^{n-1}}{(n-1)!}e^{-y}\frac{y^{n-1}}{(n-1)!}=e^{-x-y}I_0(2\sqrt{xy}).$$

From this we conclude using (6) that
 (iii) $J(x,y)+J(y,x)=1+P(N_x=N_y)=1+e^{-x-y}I_0(2\sqrt{xy})$,
and especially
 (iv) $J(x,x)=\frac{1}{2}+\frac{1}{2}e^{-2x}I_0(2x)$.
Conditioning on $X_1=u$, with density e^{-u}, we have

$$P(N_x\leq N_y)=\int_0^x \left(1-P(N_y\leq N_{x-u})\right)e^{-u}du+\int_x^\infty e^{-u}du,$$

or in view of (5)

$$\text{(v) } J(x,y)=1-\int_0^x J(y,x-u)e^{-u}\,du,$$

which seems to be new. Rewriting (v) as

$$e^x J(x,y)=e^x-\int_0^x J(y,v)e^v\,dv,$$

and differentiating with respect to x, using (4) we recover (iii):

$$\text{(vi) } \frac{\partial}{\partial x}J(x,y)=1-J(x,y)-J(y,x)=-e^{-x-y}I_0(2\sqrt{xy}).$$

Several other relations given in [4] are easily obtained from (i)–(vi). In §3 we collect some asymptotic results.

3. Asymptotics. From the probabilistic interpretation the following limit relations are quite obvious (it is easy to give estimates; also compare (v)):

$$\lim_{x\to\infty} J(x,y)= \lim_{x\to\infty} P(N_x\le N_y)=0,$$
$$\lim_{y\to\infty} J(x,y)= \lim_{y\to\infty} P(N_x\le N_y)=1.$$

For both x and y large we have the following very simple relation, which seems related to expansions in [2] involving the error function, but which seems to be new in this form. Its proof is a simple consequence of the asymptotic normality of Poisson random variables with large means.

PROPOSITION 2. *For $x\to\infty$ and $y\to\infty$*

$$\text{(7)} \qquad J(x,y)=\Phi\left(\frac{y-x+1/2}{\sqrt{x+y}}\right)+O\left(\frac{1}{\sqrt{x}}+\frac{1}{\sqrt{y}}\right),$$

where Φ is the standard normal distribution function defined as

$$\Phi(u)=(2\pi)^{-1/2}\int_{-\infty}^u e^{-v^2/2}\,dv.$$

Proof.

$$J(x,y)=P(N_x-N_y\le 0)=P(N_x-N_y<\tfrac{1}{2}),$$

where the $\tfrac{1}{2}$ is the usual "continuity correction". As N_x-N_y is asymptotically normal with mean $x-y$ and variance $x+y$, it follows that

$$\text{(8)} \qquad J(x,y)=P\left(\frac{N_x-N_y-x+y}{\sqrt{x+y}}\le\frac{y-x+1/2}{\sqrt{x+y}}\right)\approx\Phi\left(\frac{y-x+1/2}{\sqrt{x+y}}\right).$$

That $J(x,y)$ actually satisfies (7) follows easily from the Berry–Esseen version of the central limit theorem (Feller [1, p. 542]).

Remark. Relation (7), of course, also holds without the term $\tfrac{1}{2}$. In practice the approximation (8) is much better than is suggested by (7). For values of x and y of 10 and higher it yields a result correct to about three decimal places. Two examples: $x = 10$ and $y = 20$ yields $J(10, 20) = 0.974206$ and $\phi(10.5\sqrt{30}) = \Phi(1.917) = 0.972$. For $x = y = 50$ we find $J(50, 50) = 0.519972$ and $\Phi(0.5/10) = \Phi(0.05) = 0.5199$. The abundance of tables of Φ makes the approximation (8) quite practical. To obtain good (proven) bounds is not so easy.

4. Relation with Kac's random telegraph model. In [3] Kac considers an (integrated) telegraph process $X(t)$ (in his formula (25) denoted by $x(t)$) that is closely related to the process $Z(t)$ of Fig. 1. The process $X(t)$ is constructed from the same X_j and Y_j as $Z(t)$; its graph is sketched in Fig. 2. Evidently, the processes $Z(t)$ and $X(t)$ are related by

$$(9) \qquad\qquad Z(t) = \tfrac{1}{2}(X(t) + t).$$

From Fig. 1 we immediately see that

$$Z(x + y) > x \Leftrightarrow N_x \leqq N_y,$$

and therefore by Proposition 1 we have, in view of (9),

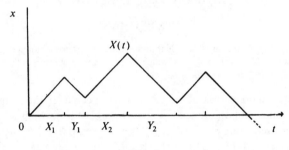

FIG. 2

PROPOSITION 3. *Let $F(x, t) = P(X(t) \leqq x)$ be the distribution function of $X(t)$. Then for $0 \leqq x \leqq t$*

$$(10) \qquad\qquad F(x, t) = 1 - J\left(\frac{t + x}{2}, \frac{t - x}{2}\right).$$

From Proposition 2 we then obtain, not very surprisingly,
COROLLARY.

$$F(x, t) \sim \Phi\left(\frac{x - 1/2}{\sqrt{t}}\right) \qquad (t \to \infty),$$

i.e., $X(t)$ is asymptotically normal with mean $\tfrac{1}{2}$ and variance t.

Remark 1. Of course, $X(t)$ is also asymptotically normal with mean zero and variance t; the $\tfrac{1}{2}$ will improve the approximation, though.

Remark 2. Since by (vi) (see also [4, p. 272]) J satisfies $J_{xy} + J_x + J_y = 0$, from (10) it follows that F satisfies the "telegrapher's" equation: $F_{tt} = F_{xx} - 2F_t$ as is proved in [3] for a more general F.

Acknowledgments. This note started as a simplified model for a problem in conductivity communicated to me by P. C. T. van der Laan. I am indebted to J. Boersma for identifying a more complicated expression for $P(N_x \leq N_y)$—involving an integral of I_1—as $J(x,y)$, and for references [3] and [6]. My thanks are due to W. K. M. Keulemans for calculating values of $J(x,y)$ on a computer.

REFERENCES

[1]   W. FELLER, *An Introduction to Probability Theory and Its Applications*, Vol. 2, 2nd ed., John Wiley, New York, 1971.

[2]   S. GOLDSTEIN, *On the mathematics of exchange processes in fixed columns*, I, *Mathematical solutions and asymptotic expansions*, Proc. Roy. Soc. A, 219 (1953), pp. 151–171.

[3]   M. KAC, *A stochastic model related to the telegrapher's equation*, Rocky Mountain J. Math., 4 (1974), pp. 497–509.

[4]   Y. L. LUKE, *Integrals of Bessel Functions*, McGraw-Hill, New York, 1962.

[5]   _____, *Problem Section*, Statist. Neerl., 37, 3 (1983), p. 160.

[6]   H. M. SRIVASTAVA AND B. R. K. KASHYAP, *Special Functions in Queueing Theory*, Academic Press, New York, 1982.

IN A RANDOM WALK THE NUMBER OF "UNIQUE EXPERIENCES" IS TWO ON THE AVERAGE*

DAVID NEWMAN†

Abstract. For any simple random walk of length $n > 0$ we consider the number, U, of places which are reached once and only once. The remarkable fact, which we alluded to in our title, is that the expected value of U, call it E_n, is always equal to 2, *independent of n*.

For any simple random walk of length $n > 0$ we consider the number, U, of places which are reached once and only once. The remarkable fact, which we alluded to in our title, is that the expected value of U, call it E_n, is always equal to 2, *independent of n*.

As far as we know, all the proofs of this result require some heavy machinery such as generating functions or binomial coefficient identities, or calculations with the so-called Catalan numbers. Our purpose, in this note, is to give an elementary proof.

To begin with, we note that E_1 is obviously equal to 2 so that our result amounts to

(I) $E_{n+1} = E_n$ *for* $n > 0$.

Consider a walk, W, of length $n + 1$ and think of it as a first step followed by a walk, W', of length n. Except possibly the initial point of W, the uniquely visited points of W and W' are the same. If W never returns to its initial point then this point is a unique experience which is never encountered in W', and we have $U(W) = U(W') + 1$. If W returns precisely once, then its initial point is not a unique experience in W, but is one in W', and we have $U(W) = U(W') - 1$. Otherwise (if W returns more than once) the initial point is not unique in W or W' and we have $U(W) = U(W')$.

Thus (I) is an immediate consequence of the following result:

(II) *For* $n > 1$ *the number of walks which never "return" is equal to the number of walks which "return" precisely once.*

Proof. No formulas! We set up a 1-1 correspondence between these two types of walks. Consider the operators A and B as follows:

For any walk W which never returns, locate the last time $k < n$ where we are 1 unit

---

*Received by the editors August 4, 1983, and in revised form November 25, 1983.
†American Telephone and Telegraph Corporation, 550 Madison Avenue, New York, New York.

from the origin (k exists since the first step takes us 1 unit from the origin and since $1 <$ n). At step $k + 1$, W is 2 units from the origin, since we know W never returns. If this is the *last time* walk W is 2 units from the origin, define $A(W)$ as the walk which takes the very same steps as W up to k and reverses all the steps of W past k.

Clearly $A(W)$ is a walk which returns precisely once.

If step $k + 1$ is not the *last time* walk W is 2 units from the origin, locate the *last time* $m, k + 1 < m \leq n$ where W is 2 units from the origin, and the *next to last time* $l, k + 1 \leq l$ $< m$ where W is 2 units from the origin. We now break W into 3 segments:

$S1$—step $l + 1$ through step m,
$S2$—step 1 through step l,
$S3$—step $m + 1$ through step n (if $m < n$, otherwise omit).

Define $A(W)$ as the walk composed of segment $S1$ followed by segment $S2$ followed by segment $S3$ (if $S3$ exists).

$A(W)$ returns precisely once, at the end of segment $S1$, since segments $S2$ and $S3$ never return to their respective initial points, and all 3 segments set off in the same direction.

Next, let W be a walk which returns precisely once, say at the jth step. If $j = n - 1$ or $j = n$ or if the $(j + 1)$st step in W is in the *same direction* as the jth step, define $B(W)$ as the walk which takes the same steps as W up to $j - 1$ and reverses all the steps after that.

If, alternatively, W takes 2 or more steps after the jth step in the *opposite direction* of the jth step, locate the *last time* $k, j + 2 \leq k \leq n$ where W is 2 units from the origin.

As before, define 3 segments:

$S1$—step 1 through step j,
$S2$—step $j + 1$ through step k,
$S3$—step $k + 1$ through step n (if $k < n$, otherwise omit).

Define $B(W)$ as the walk composed of segment $S2$ followed by segment $S1$ followed by segment $S3$ (if $S3$ exists).

$B(W)$ is a walk which never returns. (Observe that in the second case, segments $S1$, $S2$, and $S3$ all stay on the same side of their respective initial points, and segment $S2$, the first piece, terminates 2 units away from the origin.)

To conclude the proof, note that A and B are inverse operations, so we have the desired 1-1 correspondence.

A Coin Tossing Problem

Problem 77-11, by DANNY NEWMAN (Stanford University).

If one tosses a fair coin until a head first appears, then the probability that this event occurs on an even numbered toss is exactly $\frac{1}{3}$. For this procedure, the expected number of tosses equals 2. Can one design a procedure, using a fair coin, to give a success probability of $\frac{1}{3}$ but have the expected number of tosses less than 2?

Solution by R. D. FOLEY (Virginia Polytechnic Institute).

Let N be a random variable which represents the number of coin tosses. If there exists a finite number n such that

$$\Pr(N < n) = 1,$$

(i.e., we know for certain we make only a finite number of tosses), then the probability of success is $i2^{-n}$, where i is the number of successful sequences of n coin tosses. However $i2^{-n}$ does not equal $\frac{1}{3}$ for any integers i and n. Hence, we know that for any n there exists at least one sequence of n heads and tails in which we would not have stopped. Thus for all n,

$$\Pr(N > n) \geq 2^{-n}.$$

Now,

$$E(N) = \sum_{n=0}^{\infty} \Pr(N > n) \geq \sum_{n=0}^{\infty} 2^{-n} = 2.$$

Thus it is impossible to develop a procedure using a fair coin to give a success probability of $\frac{1}{3}$ but have the expected number of tosses less than 2.

Solution by J. C. BUTCHER (University of Auckland, Auckland, New Zealand).

Let p be given in $[0, 1]$; then amongst all procedures based on tossing of fair coins which result in a probability of success equal to p, we define $N(p)$ as the infimum of the expected number of tosses. We will show that $N(p) = 2$ unless $p = 0$ or 1, in which case $N(p) = 0$, or $p = n/2^m$ for m and n positive integers with n odd. In this last case we will show that $N(p) = 2(1 - 2^{-m})$. Furthermore, we show that an expected number of tosses equal to $N(p)$ can be achieved. From our result, it follows that a success probability of $\frac{1}{3}$ cannot be achieved with a lower expected number of tosses than 2.

We first outline a procedure that gives the expected number we have quoted and then prove it is optimal. The procedure is to construct a binary fraction whose rth digit is 0 for a tail and 1 for a head on the rth toss. As soon as a digit differs from the binary representation of p we terminate the procedure and deem it a success if the binary fraction formed by the toss is less than the binary representation of p. In the case when $p = n/2^m$, we also terminate if the fraction found after m tosses equals p. Clearly, the expected number of tosses required by this procedure is 2 in the general case or $2(1 - 2^{-m})$ in the case $p = n/2^m$.

To prove that a smaller expected number is not possible, we use the fact that $N(p) \leq 2$ in all cases, and consider the subprocedures after a first toss when $p \neq 0, \frac{1}{2}$ or 1. If we were to require further tosses if either a head or a tail was recorded in the first toss, then the overall procedure would require at least two tosses. Hence, in the case of one of the outcomes we must terminate the experiment with a success (if $p > \frac{1}{2}$) or a failure (if $p < \frac{1}{2}$) and for the other outcome the subprocedure must have a probability of success $2p - 1$ (if $p > \frac{1}{2}$) or $2p$ (if $p < \frac{1}{2}$). If this is carried out in the optimal way, we see that

$$N(p) = \begin{cases} 1 + \frac{1}{2}N(2p) & (0 < p < \frac{1}{2}), \\ 1 + \frac{1}{2}N(2p - 1) & (\frac{1}{2} < p < 1), \end{cases}$$

and iterated use of this gives the required result.

Comment by M. R. BROWN (Yale University).

There is no better procedure, in a suitable "decision tree" model. This is an immediate consequence of Theorem 2.1 of D. E. Knuth and A. C. Yao, *The complexity of nonuniform random number generation*, Algorithms and Complexity, J. F. Traub, ed., Academic Press, New York, 1976, pp. 357–428.

Expected Number of Stops for an Elevator

Problem 72-20, by D. J. NEWMAN (Temple University).

P persons enter an elevator at the ground floor. If there are N floors above the ground floor and if the probability of each person getting out on any floor is the same, determine the expected number of stops until the elevator is emptied.

Solution by PETER BRYANT and PATRICK E. O'NEIL (IBM Cambridge Scientific Center).

Define the random variable x_i to have the value 1 if the ith floor is a stop and 0 otherwise. Note that $E(x_i) = P(x_i = 1)$. The probability that no one stops at floor i, $P(x_i = 0)$, is the probability that P persons acting independently make a choice other than floor i, each with probability $(1 - 1/N)$. Thus

$$P(x_i = 0) = (1 - 1/N)^P.$$

Obviously,

$$P(x_i = 1) = 1 - P(x_i = 0) = 1 - (1 - 1/N)^P.$$

Now the expected number of stops is given by

$$E\left(\sum_{i=1}^{N} x_i\right) = \sum_{i=1}^{N} E(x_i) = N[1 - (1 - 1/N)^P].$$

A few remarks are in order. Note that the solution $N[1 - (1 - 1/N)^P]$ is asymptotic to $N[1 - \exp(-A)]$ if P/N tends to a limit A as N tends to infinity. Second, since the argument depends only on the independence of each individual in the elevator, the result may be generalized. Assume that the probability of person k stopping at floor i is p_{ik}, $i = 1, \cdots, N$, $k = 1, \cdots, P$. Then the expected number of stops is given by

$$\sum_{i=1}^{N} \left[1 - \prod_{k=1}^{P} (1 - p_{ik})\right].$$

If $p_{ik} = p_i$, $i = 1, \cdots, N$, $k = 1, \cdots, P$, then the expected number of elevator stops is

$$\sum_{i=1}^{N} [1 - (1 - p_i)^P].$$

Using Jensen's inequality, it is easy to show that this expression is maximized when $p_i = 1/N$, $i = 1, \cdots, N$. For the case where there is no restriction on p_{ik}, a simple argument shows that the maximum expected number of stops is equal to $\min(N, P)$.

Solution by F. W. STEUTEL (Technische Hogeschool Twente, Enschede, the Netherlands).

The description allows for several models. We assume the following: the P supposedly indistinguishable persons entering the elevator at the ground floor are labeled $x_1, x_2, \cdots, x_P$, indicating the number of the floor at which they will get out. All possible combinations are equally likely. This model is equivalent to the distribution of P indistinguishable balls over N cells (cf. W. Feller, *An Intro-duction to Probability Theory and its Applications*, vol. 1, John Wiley, New York, 1968, p. 38). The number of distributions equals $\binom{N + P - 1}{P}$.

Denoting by K the number of floors where people leave the elevator, we have $P(K = k) = $ prob (all balls are distributed over k cells, none of which are empty). Hence

$$P(K = k) = \binom{N}{k}\binom{P - 1}{P - k} \div \binom{N + P - 1}{P} \qquad k = 1, \cdots, M,$$

where $M = \min(N, P)$. We therefore have

$$\binom{N + P - 1}{P} E(K) = \sum_{k=1}^{M} k \binom{N}{k}\binom{P - 1}{P - k} = N \sum_{j=0}^{M-1} \binom{N - 1}{j}\binom{P - 1}{P - j - 1}$$

$$= N \binom{N + P - 2}{P - 1}.$$

It follows that

$$E(K) = \frac{NP}{N + P - 1}.$$

Higher moments can be obtained in the same way. Other occupancy models can be treated similarly.

Editorial note. As several of the solvers pointed out, the combinatorial part of this problem is simply a version of the classical occupancy problem. The elements of the sample space are distributions of P elevator passengers among N exit floors. If the P passengers are considered to be distinguishable, the sample space contains N^P points, and in the equally likely model, the expected number of elevator stops is $N[1 - (1 - 1/N)^P]$. If the passengers are considered to be in-distinguishable, the sample space contains $\binom{N + P - 1}{P}$ points, and in the equally likely model, the expected number of elevator stops is $NP/(N + P - 1)$. One solver, S. H. Saperstone, considered both cases. Several solvers discussed the probability distribution for the number of stops and/or higher moments of the distribution. Other solvers pointed out that these results are available in stan-dard references [1, p. 101], [2, Chap. 5]. Several solvers obtained the expected value for the floor on which the elevator is emptied. For the distinguishable case, the expected value of the terminal floor is $N - \sum_{k=1}^{N-1} (k/N)^P$, and for the indis-tinguishable case, the result is $(NP + 1)/(P + 1)$.

B. A. Powell has considered more realistic models of the elevator problem in [3]. In a forthcoming book, *An Elementary Description of the Combinatorial Basis of Thermodynamics*, T. A. Ledwell uses the elevator problem as an example to develop the techniques needed in statistical thermodynamics. In particular, he discusses in detail the "thermodynamic limit" of large P and N.

F. C. Roesler (Imperial Chemical Industries Limited) points out that Schrödinger considered the classical occupancy problem (distinguishable case) in the analysis of experimental data from cosmic ray counters [4]. Schrödinger asks the following question. If P cosmic rays bombard an assembly of N cosmic ray counters, what is the probability that exactly k of the counters go off? The problem considered by Schrödinger is especially interesting, since in this context special importance is given to the inverse problem of estimating P given the observed value of k. Roesler phrases the inverse problem for the elevator in the following way. "Having observed from the indicator lights that the elevator has stopped k times, what is the best guess for the number of passengers who entered on the ground floor?" [C.C.R.]

REFERENCES

[1] W. FELLER, *Introduction to Probability Theory and Its Applications*, vol. 1, John Wiley, New York, 1968.
[2] J. RIORDAN, *An Introduction to Combinatorial Analysis*, John Wiley, New York, 1958.
[3] D. P. GAVER AND B. A. POWELL, *Veriability in Round-Trip Times for an Elevator Car During Up-Peak*, Transportation Sci., 5 (1971), p. 169.
[4] E. SCHRÖDINGER, *A Combinatorial Problem in Counting Cosmic Rays*, Proc. Phys. Soc., A, LXVII (1951), p. 1040.

A Parking Problem

Problem 60-11, by M. S. KLAMKIN (University of Alberta), D. J. NEWMAN (Temple University) AND L. SHEPP (University of California, Berkeley).

Let $E(x)$ denote the expected number of cars of length 1 which can be parked on a block of length x if cars park randomly (with a uniform distribution in the available space).

Show that $E(x) \sim cx$ and determine the constant c.

Solution by ALAN G. KONHEIM (I.B.M.) and LEOPOLD FLATTO (Reeves Instrumentation Corp.).

We first show that $E(x)$ satisfies the integral equation

$$(1) \qquad E(x) = 1 + \frac{2}{x-1} \int_0^{x-1} E(t)\, dt \qquad\qquad (x > 1).$$

The probability that the rear end of the first car lies in the interval $(t, t + dt)$ is $1/(x-1)\, dt$ (for $0 \leq t \leq x - 1$). The interval $[0, x]$ is then decomposed into the two intervals $[0, t]$ and $[t + 1, x]$ in which the expected numbers of cars are $E(t)$ and $E(x - 1 - t)$, respectively. Integrating t over $[0, x - 1]$, we obtain (1)

Multiplying (1) by $(x - 1)$ and differentiating, we obtain

$$(2) \qquad (x-1)E'(x) + E(x) = 1 + 2E(x - 1), \qquad (x > 1),$$

If

$$(3) \qquad F(s) = \int_0^\infty e^{-sx} E(x)\, dx = \int_1^\infty e^{-sx} E(x)\, dx, \qquad R(s) > 0,$$

then the Laplace transform satisfies the differential equation

$$(4) \qquad F'(s) + (1 + 2e^{-s}/s)F(s) = -e^{-s}/s^2, \qquad R(s) > 0,$$

which has as general solution

$$F(s) = K \exp\left\{ -\int_1^s (1 + 2e^{-t}/t)\, dt \right\}$$

(5)

$$+ \int_s^\infty \frac{e^{-t}}{t^2} \exp\left\{ \int_s^t (1 + 2e^{-u}/u)\, du \right\} dt.$$

Since $E(t) = 0, 0 \leq t < 1$ and $E(t) \leq t$ $(1 \leq t < \infty)$ it follows that $F(s) < (2/s)e^{-s}$ (s real and greater than 1), so that $K = 0$. Therefore,

(6) $$F(s) = \int_s^\infty \frac{e^{-t}}{t^2} \exp\left\{ \int_s^t (1 + 2e^{-u}/u)\, du \right\} dt,$$

which gives upon several changes of variables

(7) $$F(s) = \frac{e^{-s}}{s^2} \int_s^\infty \exp\left\{ - 2\int_s^t \frac{1 - e^{-u}}{u}\, du \right\} dt.$$

From (7), we see that

$$F(s) \sim \frac{c}{s^2} \quad \text{as} \quad s \to 0^+$$

where

$$c = \int_0^\infty \exp\left\{ - 2\int_0^t \frac{1 - e^{-u}}{u}\, du \right\} dt.$$

Using a well known Tauberian theorem (Widder, *The Laplace Transform*, p. 192), we have

$$\int_1^x E(t)\, dt \sim cx^2/2 \quad \text{as} \quad x \to \infty,$$

and a standard argument shows

$$E(x) \sim cx \quad \text{as} \quad x \to \infty.$$

Also solved by the proposers.

Editorial Note:

   This problem was obtained third-hand by the proposers and attempts were made to track down the origin of the problem. These efforts were unsuccessful until after the problem was published. Subsequently, H. Robbins, Stanford University, has informed me that he had gotten the problem from C. Derman and M. Klein of Columbia University in 1957 and that in 1958 he had proven jointly with A. Dvoretzky that

(8) $$E(x) = cx - (1 - c) + O(x^{-n}), \qquad n \geq 1,$$

plus other results like asymptotic normality of x, etc. They had intended to publish their results but did not when they found that A. Renyi had published a paper proving (8) in 1958, i.e., "*On a One-Dimensional Problem Concerning Random Place Filling,*" Mag. Tud. Akad. Kut. Mat. Intézet. Közleményei, pp. 109–127. Also, (8) is proven by P. Ney in his Ph.D. thesis at Columbia.

   A reference to the Renyi paper was also sent in by T. Dalenius (University of California, Berkeley).

   An abstract of the Renyi paper was sent in anonymously from the National

Bureau of Standards. The abstract appears in the International Journal of Abstracts: Statistical Theory and Method, Vol. I, No. 1, July 1959, Abstract No. 18. According to the abstract, there is a remark due to N. G. DeBruijn in the Renyi paper stating that a practical application of the Renyi result is in the parking problem that was proposed here. In addition, the constant c has been evaluated to be 0.748.

An Unfriendly Seating Arrangement

Problem 62-3, by DAVE FREEDMAN (University of California, Berkeley) AND LARRY SHEPP (Bell Telephone Laboratories).

There are n seats in a row at a luncheonette and people sit down one at a time at random. They are unfriendly and so *never* sit next to one another (no moving over). What is the expected number of persons to sit down?

Solution by HENRY D. FRIEDMAN (Sylvania Applied Research Laboratory).

Let E_n be the desired expected number and number the seats consecutively from the left. The first person sits down in the ith seat, $i = 1, 2, \cdots, n$, leaving $i - 2$ and $n - i - 1$ seats (interpret as zero if negative) to his left and right, respectively, available to the unfriendly second person. This permits us to write recursively, for fixed i, $E_n = 1 + E_{i-2} + E_{n-i-1}$. Taking an average over all i, we obtain

$$(1) \qquad E_n = 1 + \frac{1}{n} \sum_{i=1}^{n} (E_{i-2} + E_{n-i-1}) = 1 + \frac{2}{n} \sum_{i=1}^{n} E_{i-2}$$

with $E_{-1} = E_0 = 0$.

It now follows from (1) that the generating function for E_n, $F(x) = \sum_{n=1}^{\infty} E_n x^n$, satisfies the differential equation

$$F'(x) - 2x(1 - x)^{-1}F(x) = (1 - x)^{-2}$$

with the initial condition $F(0) = 0$. Whence,

$$(2) \qquad F(x) = (1 - e^{-2x})(1 - x)^{-2}/2.$$

By expanding (2) into a power series in x, we find that

$$(3) \qquad E_n = \sum_{i=0}^{n-1} (n - i)(-2)^i/(i + 1)!$$

For numerical calculations, we rewrite (3) into

$$E_n = \sum_{i=0}^{\infty} S_i - \sum_{i=n}^{\infty} S_i$$

where S_i is the summand of (3). It now follows that $-\sum_{i=n}^{\infty} S_i$ lies between 0 and $(-2)^{n+1}/(n + 2)!$ and may be taken as an error term which is negligible for large n. Then

$$E_n \sim \sum_{i=0}^{\infty} S_i = n(1 - e^{-2})/2 + (1 - 3e^{-2})/2.$$

DAVID ROTHMAN (Harvard University) considered the more general problem where m seats are vacated on each side of a seated person and establishes a recurrence relation for the expected number $E_{2m}(n)$ of persons seated. He obtains the same results as before for $m = 1$ and gives the asymptotic expression

$$E_{2m}(n) \sim (n + 2m + 1)A_m - 1$$

where

$$A_m = \int_0^1 \exp\left\{2\left[\sum_{i=1}^m \frac{(t^i - 1)}{i}\right]\right\} dt.$$

For $m = 0$ and $m = 1$,

$$A_m = \frac{(1 - e^{-2})^{2m/(m+1)}}{(m + 1)}$$

and this is also a good approximation for $m \geq 2$. By letting $m \to \infty$ and normalizing one obtains the solution to the Parking Problem (Problem 60–11, July 1962, pp. 257–258). The latter problem which is the continuous version of the problem here is to find the expected number, $E(x)$, of cars of length 1 which can be parked on a block of length x if cars park randomly (with a uniform distribution in the available space). Asymptotically, $E(x) \sim cx$ where

$$c = \int_0^\infty \exp\left\{-2\int_0^t \frac{1 - e^{-u}}{u}\, du\right\} dt \cong 0.748.$$

The approximation for c here is

$$\lim_{m \to \infty} (m + 1)A_m \sim (1 - e^{-2})^2 = 0.747646.$$

J. K. MacKenzie (RIAS) has solved this problem and the generalization given by Rothman above in his paper *Sequential Filling of a Line by Intervals Placed at Random and its Application to Linear Adsorption*, Jour. of Chem. Physics, Vol. 37, Aug., 1962, pp. 723–728. In this paper he notes that the case $m = 1$ has been considered by E. S. Page, J. Roy. Statist. Soc., B21, 364 (1959) and F. Downton, J. Roy Statist. Soc., B23, 207 (1961). The related problem where unit intervals are placed not discretely but continuously on a line has been considered by A. Rényi, Magyar Tudományos Akad. Mat. Kutató Int. Közleményei 3, 109 (1958) (see Math. Rev. 21, 577 (1960). The related problem where all possible nonoverlapping configurations of intervals are assumed equally likely has been discussed by J. L. Jackson and E. W. Montroll, J. Chem. Phys. 28, 1101 (1958) for the case $m = 1$ and by H. S. Green and R. Leipnik, Revs. Modern Phys. 32, 129 (1960) for the cases $m = 1$ and $m = \frac{3}{2}$. Jackson and Montroll also give some approximations to the two-dimensional problem.

On a Switching Circuit

Problem 60-8, by Herbert A. Cohen (Space Instrumentation Division, Acton Laboratories).

Each of n switches connected in series are activated by clocks set to go off at a fixed time. The clocks, however, are imperfect in that they are in error given by a normal distribution with mean 0 and standard deviation 1. What is the

standard deviation for the distribution of all switches being activated?
In particular, how does it behave asymptotically for large n?

Solution by A. J. Bosch (Technological University, Eindhoven, the Netherlands).

The time that clock i goes off (later than the fixed time) is t_i $(i = 1, \cdots, n)$ with $f(t) \sim N(0, 1)$. All switches have been activated if and only if the "latest" has been activated. Let $t = \max(t_1, \cdots, t_n)$, $G(t_0) = P(t < t_0) = F^n(t_0)$ (all clocks are independent), hence $g(t) = nF^{n-1}(t)f(t)$.

The variance $\sigma_n^2 = n \int_{-\infty}^{\infty} (t - \mu_n)^2 F^{(n-1)}(t)f(t)\, dt$. There is no explicit formula for it, but there are many tables: D. Teichroew, *Tables of expected values of order statistics and products of order statistics for samples of size twenty and less from the normal distribution*, Ann. Math. Statist., 27 (1956), pp. 410–426.

For the asymptotic distribution, see references in H. A. David, *Order Statistics*, John Wiley, New York, 1970, and in E. J. Gumbel, *Statistics of Extremes*, Columbia Univ. Press, New York, 1960.

Supplementary References
Stochastic Models

[1] J. R. Bailey, *"Estimation from first principles,"* Math. Gaz. (1973) pp. 169–174.
[2] R. E. Barlow and F. Proschan, *Statistical Theory of Reliability and Life Testing,* Holt, Rinehart & Winston, N.Y., 1975.
[3] G. A. Barnard, *"Two aspects of statistical estimation,"* Math. Gaz. (1974) pp. 116–124.
[4] M. N. Barber and B. W. Ninham, *Random and Restricted Walks: Theory and Applications,* Gordon and Breach, N.Y., 1970.
[5] D. R. Barr, *"When will the next record rainfall occur?"* MM (1972) pp. 15–19.
[6] M. S. Bartlett, *An Introduction to Stochastic Processes with Special References to Methods and Applications,* Cambridge University Press, Cambridge, 1980.
[7] V. E. Benes, *Mathematical Theory of Connecting Networks and Telephone Traffic,* Academic, N.Y., 1965.
[8] A. T. Bharucha-Reid, *Elements of the Theory of Markov Processes and their Applications,* McGraw-Hill, N.Y., 1960.
[9] D. Blackwell and M. A. Girschick, *The Theory of Games and Statistical Decisions,* Wiley, N.Y., 1954.
[10] L. Breiman, *Probability and Stochastic Processes,* Houghton Mifflin, Boston, 1969.
[11] A. Charnes and W. W. Cooper, *"Deterministic equivalents for optimizing and satisfying under chance constraints,"* Oper. Res. (1963) pp. 18–39.
[12] L. E. Clarke, *"How long is a piece of string?"* Math. Gaz. (1971) pp. 404–407.
[13] A. C. Cole, *"The development of some aspects of teletraffic theory,"* BIMA (1975) pp. 85–93.
[14] E. P. Coleman, *"Statistical decision procedures in industry I: Control charts by variables,"* MM (1962) pp. 129–143.
[15] R. B. Cooper, *Introduction to Queueing Theory,* Collier-Macmillan, N.Y., 1972.
[16] D. Cox, *Renewal Theory,* Methuen, London, 1962.
[17] D. R. Cox and H. D. Miller, *The Theory of Stochastic Processes,* Chapman and Hall, London, 1977.
[18] H. Cramer, *Mathematical Methods of Statistics,* Princeton University Press, Princeton, 1946.
[19] H. A. David, *Order Statistics,* Wiley, N.Y., 1970.
[20] M. DeGroot, *Optimal Statistical Decisions,* McGraw-Hill, N.Y., 1970.
[21] J. Derderian, *Maximin hedges,"* MM (1978) 188–192.
[22] P. Diaconis and B. Efron, *"Computer-intensive methods in statistics,"* Sci. Amer. (1983) pp. 116–130, 170.
[23] J. M. Dobbie, *"Search theory: A sequential approach,"* Naval Res. Logist. Quart. (1963) pp. 323–334.
[24] S. F. Ebey and J. J. Beauchamp, *"Larval fish, power plants, and Buffon's needle problem,"* AMM (1977) pp. 534–541.
[25] B. Efron and C. Morris, *"Steins' paradox in statistics,"* Sci. Amer. (1977) pp. 119–127.
[26] R. A. Epstein, *The Theory of Gambling and Statistical Logic,* Academic, N.Y., 1967.
[27] W. B. Fairley and F. Mosteller, eds., *Statistics and Public Policy,* Addison-Wesley, Reading, 1977.
[28] W. F. Feller, *An Introduction to Probability Theory and its Applications I, II,* Wiley, N.Y., 1961, 1971.

[29] J. Ford, *"How random is a coin toss?"* Phys. Today (1983) pp. 40–47.

[30] D. P. Gaver and G. L. Thompson, *Programming and Probability Models in Operations Research*, Brooks/Cole, Monterey, 1973.

[31] R. Geist and K. Trivedi, *"Queueing network models in computer system design,"* MM (1982) pp. 67–80.

[32] N. Glick, *"Breaking records and breaking boards,"* AMM (1978) pp. 2–26.

[33] B. K. Gold, *"Statistical decision procedures in industry II: Control charts by attributes,"* MM (1962) pp. 195–210.

[34] V. L. Graham and C. I. Tulcea, *Casino Gambling*, Van Nostrand Reinhold, 1978.

[35] J. S. Growney, *"Planning for interruptions,"* MM (1982) pp. 213–219.

[36] E. J. Gumbel, *Statistics of Extremes*, Columbia University Press, New York, 1958.

[37] D. Haghighi-Talab and C. Wright, *"On the distributions of records in a finite sequence of observations with an application to a road traffic problem,"* J. Appl. Probability (1973) pp. 556–571.

[38] A. Hald, *Statistical Theory with Engineering Applications*, Wiley, N.Y., 1962.

[39] N. A. J. Hastings and J. B. Peacock, *Statistical Distributions: A Handbook for Students and Practitioners*, Halsted Press, N.Y. 1975.

[40] J. M. Howell, *"Statistical decision procedures in industry III: Acceptance sampling by attributes,"* MM (1962) pp. 259–268.

[41] R. Isaacs, *"Optimal horse race bets,"* AMM (1953) pp. 310–315.

[42] R. Jagannathan, *"Chance-constrained programming with joint constraints,"* Oper. Res. (1974) pp. 358–372.

[43] N. L. Johnson and S. Kotz, *Urn Models and their Applications*, Wiley, N.Y., 1977.

[44] K. Jordan, *Chapters on the Classical Calculus of Probability*, Akademiai Kiado, Budapest, 1972.

[45] F. P. Kelly, *Reversibility and Stochastic Networks*, Wiley, N.Y., 1979.

[46] M. G. Kendall and P. A. P. Moran, *Geometrical Probability*, Griffin, London, 1963.

[47] M. S. Klamkin, *"On the uniqueness of the distribution function for the Buffon needle problem,"* AMM (1953) pp. 677–680.

[48] M. S. Klamkin and D. J. Newman, *"Extensions of the birthday surprise,"* J. Comb. Th. (1967) pp. 279–282.

[49] M. S. Klamkin and J. H. Van Lint, *"An asymptotic problem in renewal theory,"* Stat. Neerlandica (1972) pp. 191–196.

[50] L. Kleinrock, *Communication Nets: Stochastic Message Flow and Delay*, Dover, New York, 1972.

[51] D. E. Knuth, *"The toilet paper problem,"* AMM (1984) pp. 465–470.

[52] B. O. Koopman, *"Search and its optimization,"* AMM (1979) pp. 527–540.

[53] L. H. Liyange, C. M. Gulati and J. M. Hill, *A Bibliography on Random Walks*, Math. Dept., University of Wollongong, Australia.

[54] C. L. Mallows and N. J. A. Sloane, *"Designing an auditing procedure, or how to keep bank managers on their toes,"* MM (1984) pp. 142–151.

[55] N. R. Mann, R. E. Scafer and N. D. Singpurwalla, *Methods for Statistical Analysis of Reliability and Life Data*, John Wiley, Chichester, 1974.

[56] B. L. Miller and H. M. Wagner, *"Chance constrained programming with joint constraints,"* Oper. Res. (1965) pp. 930–945.

[57] R. G. Miller, et al, *Biostatistics Casebook*, Wiley, N.Y., 1980.

[58] M. H. Millman, *"A statistical analysis of casino black jack,"* AMM (1983) pp. 431–436.

[59] O. B. Moan, *"Statistical decision, procedures in industry IV: Acceptance sampling by variables,"* MM (1963) pp. 1–10.

[60] S. G. Mohanty, *Lattice Path Counting and Applications*, Academic Press, N.Y., 1979.

[61] F. Mosteller, et al, eds., *Statistics by Example. Vol. I, Exploring Data; Vol. II, Weighing Changes; Vol. III, Detecting Patterns; Vol. IV, Finding Models;* Addison-Wesley, Reading, 1973.

[62] G. F. Newell, *Applications of Queueing Theory*, Chapman and Hall, London, 1971.

[63] D. J. Newman and L. Shepp, *"The double dixie cup problem,"* AMM (1960) pp. 58–61.

[64] D. J. Newman and W. E. Weissblum, *"Expectation in certain reliability problems,"* SIAM Rev. (1967) pp. 744–747.

[65] G. C. Papanicolaou, *"Stochastic equations and their applications,"* AMM (1973) pp. 526–544.

[66] A. Renyi, *Probability Theory*, North-Holland, Amsterdam, 1970.

[67] H. Robbins, *"Optimal stopping,"* AMM (1970) pp. 333–343.

[68] M. Rosenblatt, ed., *Studies in Probability Theory*, MAA, 1978.

[69] H. Solomon, *Geometric Probability*, SIAM, 1978.

[70] F. Spitzer, *Principles of Random Walks*, Springer-Verlag, N.Y., 1976.

[71] L. D. Stone, *"Search theory: A mathematical theory for finding lost objects,"* MM (1977) pp. 248–256.

[72] ———, *Theory of Optimal Search*, Academic, N.Y., 1975.

[73] J. L. Synge, *"The problem of the thrown string,"* Math. Gaz. (1970) pp. 250–260.

[74] R. Syski, *Introduction to Congestion Theory in Telephone Systems*, Oliver and Boyd, London, 1960.

[75] L. Takars, *Combinatorial Methods in the Theory of Stochastic Processes*, Krieger, N.Y., 1977.

[76] K. Trivedi, *Probability and Statistics with Reliability*, Queueing, and Computer Science Applications, Prentice-Hall, N.J., 1982.

[77] S. VAJDA, *Probabilistic Programming*, Academic, N.Y., 1972.

[78] G. H. WEISS AND R. J. RUBIN, *"Random walks: Theory and selective applications,"* Adv. in Chem. Phys. (1982) pp. 363–505.

[79] S. S. WILKS, *Mathematical Statistics*, Wiley, N.Y., 1962.

[80] A. WUFFLE, *"The pure theory of elevators,"* MM (1982) pp. 30–37.

[81] S. ZACKS, *The Theory of Statistical Inference*, Wiley, N.Y., 1971.

[82] N. ZADEH, *"Computation of optimal poker strategies,"* Oper. Res. (1977) pp. 541–562.

11.1 Sports

ON MAXIMIZING THE PROBABILITY OF JUMPING OVER A DITCH*

SHAUL P. LADANY† AND JAGBIR SINGH‡

Abstract. A model is considered for a man running towards a ditch and attempting to jump over it. We investigate the take-off point he should aim for in order to maximize his chances of jumping across the ditch. It is shown that aiming as close as possible to the edge is not necessarily in his best interest. The optimal policy for this objective is then compared with the case when the objective is to maximize the expected distance jumped from the edge of the ditch. Finally, conditions for which the two optimal policies coincide are pinpointed.

1. Introduction. A man running towards a ditch in order to jump over it has to decide on the point at which he will aim for his take-off. If he aims at the near edge of the obstacle, he can "fall short" of the edge or "overshoot" it. If he falls short, the effective length of the jump over the obstacle will be reduced by the amount of his "undershooting". Thus, he might fall into the ditch instead of landing on the far edge. However, if he overshoots the edge, he also falls into the ditch without accomplishing the crossing.

The man may also consider aiming for a take-off line short of the edge. His chances of falling into the ditch at the near edge will decrease; but on the other hand, the effective length of the obstacle is increased as well. Therefore, his chances of falling into the ditch before reaching the far edge will increase. Thus, the problem is to find the optimal aiming line in front of the edge in order to maximize the probability of jumping over an obstacle of a certain width. There are several practical applications of the problem; for example, in cross-country competitions, commando crossing of creeks, and in broad-jump competition where the incentive is to break a record (see Brearley [2]).

2. The model. Let us assume that the real jumping distance, X, which is the shortest horizontal distance between the tip of the forward shoe at take-off and the aftermost contour of the rear leg's footprint after landing, is a continuous random variable. Further, we express the accuracy of take-off as the distance, Y, between the tip of the forward shoe at take-off and the take-off line aimed for. Thus, Y is positive when "overshooting" and negative when "falling short" of the take-off mark. We assume that X and Y are independently and normally distributed with means μ_x and μ_y and variances σ_x^2 and σ_y^2, respectively.

When a jumper aims for a take-off point that is a distance, a, before the near edge of the ditch, which is of width L, the probability of successfully jumping over the ditch is

$$(1) \qquad F(a) = P(X + Y - a \geq L, Y \leq a).$$

The problem is to choose a, so that (1) is maximized.

* Received by the editors April 30, 1975, and in revised form April 22, 1977.

† Department of Industrial Engineering and Management, Ben-Gurion University of the Negev, Beer-Sheva, Israel.

‡ Department of Statistics, School of Business Administration, Temple University, Philadelphia, Pennsylvania 19122.

In order to evaluate (1) using charts of bivariate normal distributions, it is necessary to rewrite it in terms of standardized normal variables. To do this, let

$$U = \frac{(X+Y-L)-(\mu_x+\mu_y-L)}{\sqrt{\sigma_x^2+\sigma_y^2}}, \qquad V = \frac{Y-\mu_y}{\sigma_y},$$

$$h(a) = \frac{a-(\mu_x+\mu_y-L)}{\sqrt{\sigma_x^2+\sigma_y^2}}, \quad \text{and} \quad k(a) = \frac{a-\mu_y}{\sigma_y}.$$

Then,

$$F(a) = P(U \geq h(a), V \leq k(a)).$$

Notice that the correlation coefficient between the standardized normal variables U and V is $\rho = \sigma_y/\sqrt{\sigma_x^2+\sigma_y^2}$.

It is easy to see that $F(a)$ can be rewritten as:

$$F(a) = P(U \geq h(a)) - P(U > h(a), V \geq k(a))$$
$$= \Phi(-h) - L(h, k, \rho)$$

where $\Phi(h)$ is the standard normal cumulative distribution function and $L(h, k, \rho) = P(U \geq h, V \geq k)$ is the bivariate normal probability integral. It can be shown (see [1]) that $F(a)$ can be expressed as:

$$F(a) = \Phi(-h) - L\left(h, 0, \frac{(\rho h - k)(\mathrm{sgn}\, h)}{\sqrt{h^2 - 2\rho hk + k^2}}\right) - L\left(k, 0, \frac{(\rho k - h)(\mathrm{sgn}\, k)}{\sqrt{h^2 - 2\rho hk + k^2}}\right)$$

(2)
$$+\begin{cases} 0 & \text{if } hk > 0 \text{ or } hk = 0 \text{ and } (h+k) \geq 0, \\ \frac{1}{2} & \text{otherwise,} \end{cases}$$

where

$$\mathrm{sgn}\, x = \begin{cases} 1 & \text{if } x \geq 0, \\ -1 & \text{if } x < 0. \end{cases}$$

The values of $L(h, 0, \cdot)$ and $L(k, 0, \cdot)$ can be read from graphs of h versus ρ in [1] with constant contour lines such that $L(h, 0, \rho) = 0.01, \cdots, 0.50$. Thus, using the graphs and (2), $F(a)$ can be evaluated numerically for various values of a.

3. Numerical example. Experiments were performed with a cross-country runner over a period of ten training sessions. During this period, the athlete made no effort to improve his jumping or aiming ability. Between peak training sessions, his running quality was at a relatively low level and insufficient for improving his running or (indirectly) his jumping performance. He was requested to jump aiming for various take-off lines, and the values of X and Y were collected.

The athlete in the experiment was an ex-long jumper and therefore, his jumping performances were of high standards. The normal distribution was fitted to the observed data for the distribution of X as well as for Y (both in

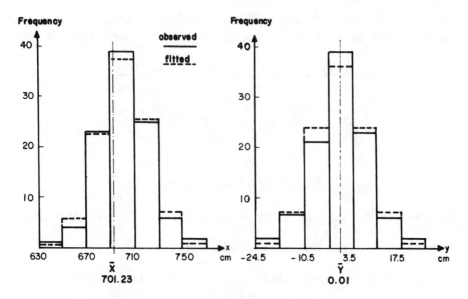

FIG. 1. *Fitted and observed frequency distributions of the length of jumps X and aiming accuracy Y.*

centimeters), and the results are:

$$X \sim N(701.23, 20.44^2), \qquad Y \sim N(0.01, 7.50^2).$$

The fitted and the observed frequency distributions are shown in Fig. 1. In both cases, the chi-square test accepts the null hypothesis at the 0.05 level of significance. The assumption of independence of X and Y is validated since the correlation coefficient between them is not significantly different from zero at the 0.05 level of significance.

The values of $F(a)$ in (2) were calculated for various values of a and L for the jumping and the aiming characteristics of the given athlete. The results are illustrated in Fig. 2. Several conclusions become evident. First, and not surprisingly, the probability of crossing the obstacle is a decreasing function of its length L. However, the probability of crossing is more sensitive to changes in the values of L around 700 centimeters (approximately the expected jumping distance) compared to those changes in L away from 700 centimeters. Second, the probability of crossing the obstacle is very sensitive to the value of a; the sensitivity is higher for the values of a below the optimal a than for values above it. Third, the absolute change in $F(a)$ due to one unit change in a from its optimal level increases as L decreases. Fourth, the optimal value of a, a_0, is very small for large values of L and it increases with decreasing L. This relationship is illustrated in Fig. 3. (Note that the curve is smoothed and hand-fitted and therefore it does not necessarily pass through the calculated points). For $L = 740$, a_0 is zero; i.e., the runner should aim at the edge of the ditch for his take-off. At the other extreme, for $L \leq 600$, a_0 is 30; and since it is about four standard deviations of the aiming ability from the edge of the ditch, it provides the probability of approximately one of not falling in the ditch at the near edge.

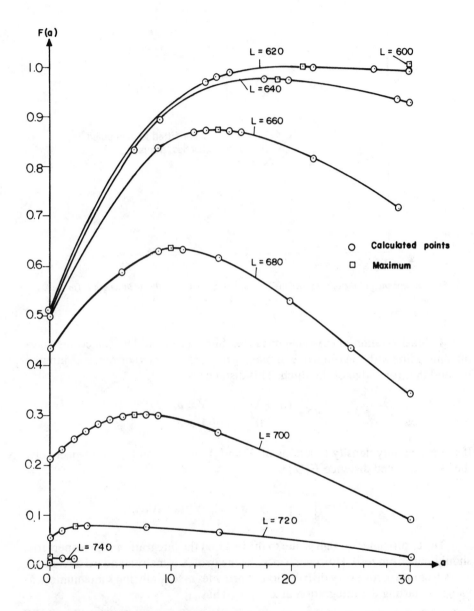

FIG. 2. *The probability of crossing a ditch, F(a), as a function of the location of the aiming line, a, for various ditch lengths L (a and L in cm).*

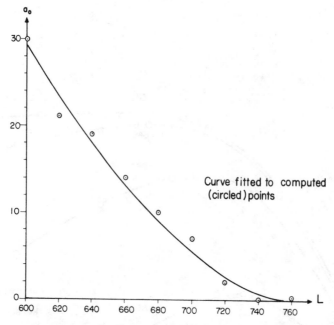

FIG. 3. *Location of the optimal aiming line, a_0, as a function of the ditch length L (in cm.).*

4. Maximization of expected distance. Suppose the athlete selects his take-off aiming line with the objective of maximizing the expected distance D jumped beyond the front edge of the ditch. This distance is

$$D = \begin{cases} Y+X-a & \text{if } Y \le a, \\ 0 & \text{if } Y > a. \end{cases}$$

If the probability density functions of X and Y are $f(x)$ and $g(y)$, respectively, then the expected distance $E(D)$ is:

$$E(D) \approx \int_{-\infty}^{a} \int_{-\infty}^{\infty} (y+x-a)g(y)f(x)\, dy\, dx.$$

The approximation sign is due to the fact that the integration with respect to x should start from zero. However, since the mean of X is many standard deviations away from zero, for every distribution, it provides essentially the same numerical value as starting the integration at $x = -\infty$. Thus,

$$E(D) = \int_{-\infty}^{a} yg(y)\, dy + (\mu_x - a) \int_{-\infty}^{a} g(y)\, dy$$

$$= \int_{-\infty}^{a'} (y\sigma_y + \mu_y)\phi(y)\, dy + (\mu_x - a) \int_{-\infty}^{a'} \phi(y)\, dy,$$

where $\phi(y)$ is the density of the standard normal distribution and $a' = (a - \mu_y)/\sigma_y$.

Hence,

$$E(D)=\sigma_y\int_{-\infty}^{a'} y\phi(y)\, dy+(\mu_y+\mu_x-a)\Phi(a').$$

The optimal value of a or equivalently of a' is obtained when $E(D)$ is maximized. To this end set

$$\frac{dE(D)}{da}=\sigma_y a'\phi(a')\frac{da'}{da}-\Phi(a')+(\mu_y+\mu_x-a)\phi(a')\frac{da'}{da}$$

equal to zero. Since $da'/da = 1/\sigma_y$, it follows that

(3) $$\phi(a')/\Phi(a')=\sigma_y/\mu_x.$$

Equation (3) provides a' maximizing $E(D)$, since it can be shown that $d^2E(D)/da^2<0$. The graphic solution of (3) for any value of σ_y/μ_x is shown in Fig. 4, providing a solution of $a'=2.69$ for the pertinent ratio of $\sigma_y/\mu_x = 7.50/701.23 = 0.01069$. Thus, the optimum policy, for the given athlete to maximize the expected length of jump, is to aim at a line which is given by

$$a = a'\sigma_y+\mu_y = 2.69(7.50)+0.01 = 20.18 \text{ cm}$$

in front of the edge. Comparison of this result, with the recommended values of a when the maximization of the probability to cross the ditch is the objective (see Fig. 2) illustrates clearly that in most cases to use the a which maximized $E(D)$ would reduce significantly the probability of crossing the ditch. Only when the

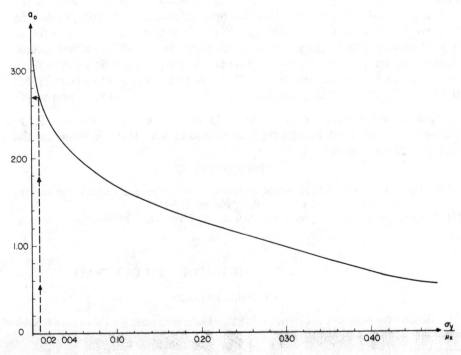

FIG. 4. *Optimal value of a_0 as a function of σ_y/μ_x.*

width of the ditch is significantly less than the average jumping ability of the athlete, 640, 620, or 600 centimeters (which are approximately 3, 4, or 5 standard deviations below the mean, respectively), do the practical implications of both policies coincide.

5. Concluding remarks. The optimal distance between the edge of a ditch and the arbitary take-off line that a person should select to aim for in order to maximize his probability of crossing it depends on the jumping and the aiming ability as well as on the width of the ditch. It is clear that for jumping over a ditch much narrower than his jumping ability, the athlete should aim much ahead of it; and that for crossing an obstacle considerably wider than his average jumping ability he should aim at the edge of the ditch. However, for jumping over a distance which is shorter than his average jumping ability by an amount of 1, 2, 3, 4, and 5 standard deviations of his jumping ability, the athlete is recommended to aim at a line which is in front of the edge by, respectively, 1.3, 1.9, 2.5, and 4 standard deviations of his aiming ability. On the other hand, for jumping over a distance which is exactly his average jumping ability, or 1, 2, or 3 standard deviations (of jumping ability) above it, it is suggested that he aim at a distance which is respectively 0.9, 0.3, 0 and 0 standard deviations of his aiming ability in front of the edge. It has been further shown that a ditch crossing policy motivated by the desire to leap the greatest expected distance is inconsistent with the desire to maximize the probability to succeed in the crossing (except in the trivial cases when the width of the ditch is considerably lower than the average jumping ability of the person).

A few words of caution are appropriate concerning the accuracy of the derived results. The usual error of reading the values from the graphs of the bivariate normal distributions [1] is of the order of magnitude of 0.01, so that the calculated values of the probabilities $F(a)$ are subject to such an error of approximation. This might not lead to the adoption of a different nonoptimal aiming line, except in the unimportant cases where $F(a)$ is practically insensitive to changes in a. For example, for $L = 700$, $F(a)$ was calculated to be 0.28555, 0.29812, 0.30036, 0.30280, and 0.29900 for $a = 5, 6, 7, 8$ and 9, respectively.

Acknowledgments. We are grateful to the referee and Professor R. P. Clickner for their many helpful suggestions and to Miss Ruth Williams for the typing of the manuscript.

REFERENCES

[1] WILLIAM H. BEYER, ed., *Handbook of Tables for Probability and Statistics*, 2nd ed., The Chemical Rubber Co., Cleveland, Ohio, 1968, pp. 147–150.
[2] M. N. BREARLEY, *The long jump miracle of Mexico City*, Math. Mag., 45 (1972), pp. 241–246.

PROBABILITY OF A SHUTOUT IN RACQUETBALL*

JOSEPH B. KELLER†

Abstract. The probability of a player winning a shutout in racquetball is calculated as a function of the probability that he wins any particular rally.

*Received by the editors October 7, 1983. This research was supported by the Office of Naval Research, the Air Force Office of Scientific Research, the Army Research Office, and the National Science Foundation.

†Departments of Mathematics and Mechanical Engineering, Stanford University, Stanford, California 94305.

To win a game of racquetball it is necessary to win 21 points, and to win a point it is necessary to have the serve. If the loser has zero points at the end of the game, he is said to be shut out. We seek the probability P that player A shuts out player B, assuming that A has probability p of winning any particular rally.

Let $P(n)$ be the probability that A wins n points while B wins zero points with A having the serve initially. Then $P(n)$ satisfies the recursion relation

$$(1) \qquad\qquad P(n) = P(n-1)P(1).$$

Repeated application of (1), or induction on n, yields

$$(2) \qquad\qquad P(n) = [P(1)]^n.$$

To find $P(1)$, we note that A can win one point while B wins zero points in either of two ways. Either A can win the first rally and thus win one point, or he can lose the first rally and thus lose the serve but then win the second rally to regain the serve. After that he is in the same condition he started from, and again he has the probability $P(1)$ of winning one point before B wins any. Thus $P(1)$ satisfies the equation

$$(3) \qquad\qquad P(1) = p + (1-p)\,pP(1).$$

Solving (3) yields

$$(4) \qquad\qquad P(1) = \frac{p}{1 - (1-p)p}.$$

Now (2) and (4) give

$$(5) \qquad\qquad P(n) = \left(\frac{p}{1 - p + p^2}\right)^n.$$

When A has the serve initially, then the probability that he shuts out B is just $P(21)$, where $P(n)$ is given by (5). However if the initial server is determined by the toss of a fair coin or racquet, then A has probability $\tfrac{1}{2}$ of having the serve initially. If B has the serve, A must win it and then win 21 points. Thus the probability P of a shutout is

$$(6) \qquad\qquad P = \tfrac{1}{2}P(21) + \tfrac{1}{2}pP(21).$$

Upon combining (5) and (6) we get the final result

$$(7) \qquad\qquad P = \frac{1+p}{2}\left(\frac{p}{1 - p + p^2}\right)^{21}.$$

The values of P for a few values of p are:

p	1.	.9	.85	.84	.842	.5
P	1.	.753	.534	.490	.500	.0001504

(8)

Thus the probability of a shutout is .5 when $p = .842$, which means that on the average A wins 5.33 rallies for each one that B wins. For evenly matched players $p = .5$ and the probability that A shuts out B is only .0001504 or one in 6649 games.

I want to thank Ralph Levine for having proposed this problem, and for his comments on the results.

PROBABILITY OF WINNING A GAME OF RACQUETBALL*

DAVID J. MARCUS[†]

Abstract. The probability of a player winning a game of racquetball is calculated as a function of the probability p of his winning a rally when he serves and the probability q of his winning a rally when his opponent serves.

In Classroom Notes in the April 1984 issue, Keller calculated the probability of a shutout in racquetball assuming that the probability of winning a rally is independent of who serves. For those of you who have enough trouble just winning, let alone getting a shutout, we calculate the probability that player A wins the game. Since a player usually wins more rallies on his own serve, we only assume A has probability p of winning a rally when he serves and probability q of winning a rally when B serves. To win a game A must win 21 points, and to win a point it is necessary to have the serve.

Let $P_A(n)$ be the probability A wins the next n points assuming A has the serve. The reasoning leading to equation (3) of Keller's note gives

$$(1) \qquad P_A(1) = p + (1-p)qP_A(1)$$

and

$$(2) \qquad P_A(1) = \frac{p}{1-(1-p)q}.$$

Also from Keller's note,

$$(3) \qquad P_A(n) = P_A(1)^n.$$

Similarly,

$$(4) \qquad P_B(1) = \frac{1-q}{1-q(1-p)}.$$

Assume A serves first. For $0 \leq j \leq 21$ and $b = (b_0, b_1, \cdots, b_{20})$, let $P(j|b)$ be the probability A wins j points with B winning b_i points between A's (i)th and $(i+1)$th points, $i = 0, 1, \cdots, j-1$. Then with the convention $P(0|b) = 1$, we have for $j \geq 1$ that

$$(5) \qquad P(j|b) = \begin{cases} P_A(1)P(j-1|b) & \text{if } b_{j-1} = 0, \\ P_A(1)qP_B(b_{j-1})(1-p)P(j-1|b) & \text{if } b_{j-1} \neq 0. \end{cases}$$

To see this suppose A has just won his $(j-1)$th point. For B to win $b_{j-1} \neq 0$ points he must win the rally (probability $= 1-p$), then win b_{j-1} points (probability $= P_B(b_{j-1})$), and then lose the next rally (probability $= q$). Then A can win his next point (probability $= P_A(1)$).

Let n equal the sum of the b_i (i.e. the number of points B wins) and let m equal the number of b_i which are not zero. Then

$$(6) \qquad P(21|b) = P_A(21)P_B(n)(q(1-p))^m.$$

---

*Received by the editors October 19, 1984, and in final form April 22, 1985.

[†] The Analytic Sciences Corporation, 1 Jacob Way, Reading, Massachusetts 01867.

There are 21 choose m ways of choosing which b_i are nonzero and for each choice there are $(n-1)$ choose $(m-1)$ ways of placing the n points in the m boxes so each has at least one point.

Hence the probability A wins is

(7)

$$\Pr(p,q) = P_A(21) + \sum_{n=1}^{20} \sum_{m=1}^{n} \binom{21}{m}\binom{n-1}{m-1} P_A(21) P_B(n)(q(1-p))^m$$

$$= \left(\frac{p}{1-(1-p)q}\right)^{21}\left\{1 + \sum_{n=1}^{20}\left(\frac{1-q}{1-q(1-p)}\right)^n \sum_{m=1}^{n}\binom{21}{m}\binom{n-1}{m-1}(q(1-p))^m\right\}.$$

The probability of A winning if B serves first is

(8)
$$1 - \Pr(1-q, 1-p).$$

Therefore, if they flip a coin to decide who serves first, the probability of A winning the game is

(9)
$$\Pr_{\text{coin}}(p,q) = \tfrac{1}{2}(1 + \Pr(p,q) - \Pr(1-q, 1-p)).$$

Table 1 gives some values. The top number is $\Pr$ and the bottom is $\Pr_{\text{coin}}$.

TABLE 1

p \ q	0.4	0.5	0.6
0.4	0.038	0.179	0.516
	0.034	0.167	0.5
0.5	0.209	0.522	0.845
	0.192	0.5	0.833
0.6	0.530	0.826	0.970
	0.5	0.808	0.966

So for evenly matched players the serve is worth approximately two games out of one hundred.

REFERENCE

J. B. KELLER, *Probability of a shutout in racquetball*, this Review, 26(1984), pp. 267–268.

POSTSCRIPT

L. C. Ford and C. L. Winter; and also J. Goldstein, independently, extended the Keller note in the same way.

Supplementary References
Sports

[1] R. M. ALEXANDER, *"Optimum walking techniques for quadrupeds and bipeds,"* J. Zool. (London) (1980) pp. 97–117.
[2] ———, *"Walking and running,"* Amer. Sci. (1984) pp. 348–354.
[3] R. M. ALEXANDER AND A. S. JAYES, *"Fourier Analysis of Forces exerted in walking and running,"* J. Biomechanics (1980) pp. 383–390.

[4] I. ALEXANDROV AND P. LUCHT, "Physics of sprinting," Amer. J. Phys. (1981) pp. 254–257.

[5] Animal Locomotion, Halsted, N.Y., 1977.

[6] S. J. BRAMS AND P. D. STRAFFIN, JR., "Prisoner's dilemma and the professional sports draft," AMM (1979) pp. 80–88.

[7] P. J. BRANCAZIO, Sport Science, Physical Laws and Optimum Performance, Simon and Schuster, N.Y., 1984.

[8] M. N. BREARLY, "The long jump miracle of Mexico City, MM (1972), 241–246.

[9] H. BRODY, "Physics of the tennis racket," Amer. J. Phys. (1979) pp. 482–487.

[10] ———, "Physics of the tennis racket II: The 'sweet spot,'" Amer. J. Phys. (1981) pp. 816–819.

[11] D. BURGHES AND M. O'CARROLL, "Mathematical models for weightlifting," BIMA (1980) pp. 155–159.

[12] S. CARLSON, How Man Moves, Heinemann, London, 1972.

[13] S. CHAPMAN, "Catching a baseball," Amer. J. Phys. (1968) pp. 868–870.

[14] E. COOK, Percentage Baseball, M.I.T. Press, Cambridge, 1966.

[15] E. COOK AND D. L. FINK, Percentage Baseball and the Computer, Waverly, N.Y., 1972.

[16] J. C. COOKE, "The boundary layer and 'seam' bowling," Math. Gaz. (1955) pp. 196–199.

[17] J. E. COUNSILMAN, The Science of Swimming, Prentice-Hall, N.J., 1968.

[18] C. B. DAISH, The Physics of Ball Games, English Universities Press, 1972.

[19] P. DAVIDOVITS, Physics in Biology and Medicine, Prentice-Hall, N.J., 1975.

[20] J. M. DAVIES, "Aerodynamics of golf balls," J. Appl. Phys. (1949) pp. 821–828.

[21] J. E. DRUMMOND, "Why does a high jump cross bar fall off?," Math. Gaz. (1981) pp. 182–185.

[22] G. DYSON, The Mechanics of Athletics, University of London Press, 1977.

[23] H. Y. ELDER AND E. R. TRUEMAN, eds., Aspects of Animal Movement, Cambridge University Press, Cambridge, 1980.

[24] R. A. FREEZE, "An analysis of baseball batting order by Monte Carlo simulation," Operations Research (1974) pp. 728–735.

[25] C. FROHLICH, "Aerodynamic drag crisis and its possible effect on the flight of baseballs," Amer. J. Phys. (1984) pp. 325–334.

[26] ———, "Aerodynamic effects on discus flight," Amer. J. Phys. (1981) pp. 1125–1132.

[27] ———, "Do springboard divers violate angular momentum conservation?," Amer. J. Phys. (1979) pp. 583–592.

[28] ———, "Effect of wind and altitude on record performances in foot races, pole vault, and long jump," Amer. J. Phys. (1985) pp. 726–730.

[29] ———, "The physics of somersaulting and twisting," Sci. Amer. (1980) pp. 154–164.

[30] D. GALE, "Optimal strategy for serving in tennis," MM (1971) pp. 197–199.

[31] D. F. GRIFFING, The Dynamics of Sports, Mohican Publ., Ohio, 1982.

[32] R. A. GROENEVELD AND G. MEEDEN, "Seven game series in sports," MM (1975) pp. 187–192.

[33] C. R. HAINES, "Old curves in a new setting," Math. Gaz. (1977) pp. 262–266.

[34] C. HALL AND C. SWARTZ, "The effect of handicap stroke location on golf matches," Math. Modelling (1981) pp. 153–159.

[35] ———, "The effect of handicap stroke location on best-ball golf scores," Math. Modelling (1981) pp. 161–167.

[36] J. G. HAY, The Biomechanics of Sports Techniques, Prentice-Hall, N.J., 1985.

[37] J. KARNEHM, Understanding Billiards and Snooker, Pelham, 1976.

[38] P. KIRKPATRICK, "Batting the ball," Amer. J. Phys. (1963) pp. 606–613.

[39] J. B. KELLER, "Mechanical aspects of athletics," Proc. 7th U.S. Nat. Cong. Appl. Mech., Boulder, Co. (1975) pp. 22–26.

[40] ———, "Optimal velocity in a race," AMM (1974) pp. 474–480.

[41] ———, "A theory of competitive running," Phys. Today (1973) pp. 42–47.

[42] S. P. LADANY, "Optimal starting height for pole-vaulting," Oper. Res. (1975) pp. 968–978.

[43] S. P. LADANY AND R. E. MACHOL, Optimal Strategies in Sports, North-Holland, Amsterdam, 1977.

[44] S. P. LADANY, J. W. HUMES AND G. P. SPHICAS, "The optimal aiming line," Oper. Res. (1975) pp. 495–506.

[45] H. LIN, "Newtonian mechanics and the human body: Some estimates of performance," Amer. J. Phys. (1978) pp. 15–18.

[46] G. R. LINDSAY, "Strategies in baseball," Oper. Res. (1963) pp. 477–501.

[47] B. B. LLOYD, "The energetics of running: An analysis of world records," Advancement. Sci. (1966) pp. 515–530.

[48] R. E. MACHOL AND S. P. LADANY, eds., Management Science in Sports, North-Holland, Amsterdam, 1976.

[49] A. G. MACKIE, "A difficulty factor for pots at snooker," BIMA (1984) pp. 66–69.

[50] ———, "Mathematics in sport," BIMA (1980) pp. 2–6.

[51] ———, "The mathematics of snooker," BIMA (1982) pp. 82–88.

[52] F. J. MAILLARDET, "The swing phase of locomotion," Eng. Medicine (1977) pp. 67–75.

[53] J. K. R. MANNING, "Mathematics of duplicate bridge tournaments," BIMA (1979) pp. 201–206.

[54] R. MARGARIA, Biomechanics and Energetics of Muscular Exercise, Oxford University Press, London, 1976.

[55] E. A. MARSHALL, "A dynamic model for the stride in human walking," Math. Modelling (1983) pp. 391–415.

[56] T. A. McMAHON, Muscles, Reflexes, and Locomotion, Princeton University Press, Princeton, 1984.

[57] S. MOCHON AND T. A. McMAHON, "Ballistic walking: An improved model," Math. Biosci. (1980) pp. 241–260.

[58] F. MOSTELLER, *"The world series competition,"* J. Amer. Statist. Assoc. (1952) pp. 355–380.

[59] J. M. NEVIN AND P. J. JACKSON, *"An interesting property of tennis rackets and dynamically similar rigid bodies,"* BIMA (1977) pp. 154–156.

[60] D. J. NEWMAN, *"How to play baseball,"* AMM (1960) pp. 865–868.

[61] W. PAISH, *Discus Throwing,* British Amateur Athletic Board, London, 1976.

[62] T. J. PEDLEY, ed., *Scale Effects in Animal Locomotion,* Academic, London, 1977.

[63] S. M. POLLAK, *"A model for evaluating golf handicapping,"* Oper. Res. (1974) pp. 1040–1050.

[64] J. G. PURDY, *"Computer analysis of champion athletic performance,"* Res. Quart. (1974) pp. 391–397.

[65] ———, *"Least squares model for the running curve,"* Res. Quart. (1974) pp. 224–238.

[66] P. S. RIEGEL, *"Athletic records and human endurance,"* Amer. Sci. (1981) pp. 285–290.

[67] H. W. RYDER, H. J. CARR AND P. HERGET, *"Future performance in footracing,"* Sci. Amer. (1976) pp. 109–118.

[68] F. J. SCHEID, *"A least squares family of cubic curves with an application to golf handicapping,"* SIAM J. Appl. Math. (1972) pp. 77–83.

[69] E. SCHRIER AND W. ALLMAN, eds., *Newton at the Bat,* Scribner, N.Y., 1984.

[70] M. STOB, *"A supplement to 'A mathematician's guide to popular sports',"* AMM (1984) pp. 277–282.

[71] J. STRAND, *"Physics of long-distance running,"* Amer. J. Phys. (1985) pp. 371–373.

[72] T. TECH, *"The ranking of incomplete tournaments: A mathematician's guide to popular sports,"* AMM (1983) pp. 246–266.

[73] M. S. TOWNSEND, *Mathematics in Sport,* Horwood, Chichester, 1984.

[74] E. A. TROWBRIDGE AND W. PAISH, *"Mechanics of athletics,"* BIMA (1981) pp. 144–146.

[75] K. WELLS AND K. LUTTGEN, *Kinesiology,* Saunders, Philadelphia, 1976.

[76] F. R. WHITT AND D. G. WILSON, *Bicycling Science,* M.I.T. Press, Cambridge, 1982.

[77] M. WILLIAMS AND H. R. LISSNER, *Biomechanics of Human Motion,* Saunders, Philadelphia, 1962.

[78] S. T. WILLIAMS AND J. UNDERWOOD, *Science of Hitting,* Simon and Schuster, N.Y., 1982.

12. Miscellaneous

A LATTICE SUMMATION USING THE MEAN VALUE THEOREM FOR HARMONIC FUNCTIONS*

PAUL K. MAZAIKA†

The equations for finding bulk properties of crystals are commonly solved by Fourier series; however, the resulting solutions can be multiple sums with slow convergence. Convergence may be improved using one of the theta function transformations on the sum, but the analysis is detailed, and the resulting functions may not be readily available on a computer. In this note, we give an example where a transformation based on the mean value theorem for harmonic functions can be used to form a rapidly converging sum of elementary functions.

Consider the following multiple sum arising in the solution for viscous flow past an infinite three-dimensional periodic array of spheres (Hasimoto [1]):

$$(1) \qquad S_1(\mathbf{r}) = \frac{1}{\pi} \sum_{\mathbf{n} \neq 0} \frac{e^{-2\pi i(\mathbf{n} \cdot \mathbf{r})}}{n^2}$$

where

$$\mathbf{n} = n_1 \mathbf{i} + n_2 \mathbf{j} + n_3 \mathbf{k}, \qquad n_1, n_2, n_3 = 0, \pm 1, \pm 2, \cdots,$$

are points in a three-dimensional unit lattice, $n = |\mathbf{n}|$, and the summation is an infinite triple sum over all points in the lattice (except $\mathbf{n} = 0$). We wish to find the constant term in the expansion of $S_1(\mathbf{r})$ about $\mathbf{r} = 0$, however, the series is singular at $\mathbf{r} = \mathbf{0}$, and only conditionally convergent for $\mathbf{r} \neq \mathbf{0}$.

The Laplacian of $S_1(\mathbf{r})$ is given by

$$(2) \qquad \nabla^2 S_1(\mathbf{r}) = -4\pi \sum_{\mathbf{n}} \delta(\mathbf{r} - \mathbf{n}) + 4\pi$$

where $\delta(\mathbf{r}) = \delta(x)\,\delta(y)\,\delta(z)$ is the Dirac delta function in three dimensions, and we have used Poisson's summation formula

$$(3) \qquad \sum_{\mathbf{n}} e^{-2\pi i(\mathbf{n} \cdot \mathbf{r})} = \sum_{\mathbf{n}} \delta(\mathbf{r} - \mathbf{n}).$$

Removal of the singularity at $\mathbf{r} = \mathbf{0}$ and the constant term from the right-hand side makes equation (2) homogeneous for $|\mathbf{r}| < 1$. Hence, a harmonic function, $T_0(\mathbf{r})$, can be defined by

$$(4) \qquad T_0(\mathbf{r}) = S_1(\mathbf{r}) - \frac{1}{r} - \frac{2}{3}\pi r^2$$

since

$$(5) \qquad \nabla^2 \frac{1}{r} = -4\pi\delta(\mathbf{r}).$$

---

*Received by the editors December 31, 1980, and in revised form May 1, 1983. This work was part of the author's Ph.D. thesis in Applied Mathematics at the California Institute of Technology in 1974.

†The Aerospace Corporation, Los Angeles, California 90009.

Note that $T_0(\mathbf{0})$ is the regular part of $S_1(\mathbf{r})$ at $\mathbf{r} = \mathbf{0}$.

Harmonic functions satisfy the mean value theorem (Courant and Hilbert [2]):

$$(6) \qquad T(\mathbf{r}) = \frac{1}{4\pi R^2} \iint_{|s| = R} T(\mathbf{r} + \mathbf{s})\, ds.$$

Successive applications of this theorem yield new, faster converging, representations for $T_0(\mathbf{r})$, and we will denote them successively by $T_1(\mathbf{r})$, $T_2(\mathbf{r})$, etc. Thus,

$$(7) \qquad T_1(\mathbf{r}) = \frac{1}{4\pi R_1^2} \iint_{|s| = R_1} \left[S_1(\mathbf{r} + \mathbf{s}) - \frac{1}{|\mathbf{r} + \mathbf{s}|} - \frac{2}{3}\pi(\mathbf{r} + \mathbf{s})^2 \right] ds$$

for points $\mathbf{r}$ such that a sphere of radius R_1 centered at $\mathbf{r}$ does not enclose a nonzero lattice point.

The series for S_1 must be uniformly convergent in order to be integrated term by term, so we require that $|\mathbf{r}| \neq R_1$ to avoid the singularity at $\mathbf{r} + \mathbf{s} = \mathbf{0}$. Then, for the integral involving S_1,

$$(8) \qquad \frac{1}{4\pi R_1^2} \iint_{|s| = R_1} S_1(\mathbf{r} + \mathbf{s})\, ds = \frac{1}{4\pi R_1^2} \sum_{n \neq 0} \frac{1}{\pi n^2} \iint_{|s| = R_1} e^{-2\pi i(\mathbf{n} \cdot (\mathbf{r} + \mathbf{s}))}\, ds$$

$$= \frac{1}{2\pi R_1} \frac{1}{\pi} \sum_{n \neq 0} \frac{e^{-2\pi i(\mathbf{n} \cdot \mathbf{r})} \sin 2\pi R_1 n}{n^3}$$

and the sum now converges like n^{-3} instead of n^{-2}. Applying the theorem to the other terms of $T_0(\mathbf{r})$, we obtain the new representation

$$(9) \qquad T_1(\mathbf{r}) = \frac{1}{\pi} \frac{1}{2\pi R_1} \sum_{n \neq 0} \frac{e^{-2\pi i(\mathbf{n} \cdot \mathbf{r})} \sin 2\pi R_1 n}{n^3} - \frac{2}{3}\pi(r^2 + R_1^2) - \frac{1}{R_1}.$$

We have assumed that $R_1 > |\mathbf{r}|$ in the integral of the $|\mathbf{r} + \mathbf{s}|^{-1}$ term, otherwise the $1/R_1$ term in (9) is replaced by $1/r$. In particular, if $R_1 = 1/2$, then (9) is valid for $|\mathbf{r}| < 1/2$.

After N iterations of the mean value theorem, the representation becomes

$$(10) \qquad \begin{aligned} T_N(\mathbf{r}) = & -\frac{1}{R_1} - \frac{2}{3}\pi(R_1^2 + R_2^2 + \cdots + R_N^2 + r^2) \\ & + \frac{1}{\pi} \sum_{n \neq 0} \left[\frac{e^{-2\pi i(\mathbf{n} \cdot \mathbf{r})}}{n^{N+2}} \prod_{j=1}^{N} \left(\frac{\sin 2\pi R_j n}{2\pi R_j} \right) \right] \end{aligned}$$

where there is now rapid algebraic convergence of the sum. Each successive representation at a point $\mathbf{r}$ uses the previous representation in a sphere about $\mathbf{r}$, so the region of validity of $T_N(\mathbf{r})$ decreases as N increases. For example, if $R_j = 2^{-j}$ for $j = 1, 2, \cdots, N$, then $T_N(\mathbf{r})$ is valid for $|\mathbf{r}| < 2^{-N}$.

To evaluate $T_N(\mathbf{0})$, we neglect the terms in the sum for which n is larger than an upper bound, u. The magnitude of the neglected sum is bounded by the volume integral

$$(11) \qquad \begin{aligned} E = & \frac{1}{\pi} \int_0^{2\pi} \int_0^{\pi} \int_u^{\infty} \frac{1}{n^{N+2}} \prod_{j=1}^{N} \left(\frac{1}{2\pi R_j} \right) n^2 \sin\varphi\, d\varphi\, d\theta\, dn \\ = & \frac{4}{N-1} \cdot \frac{1}{u^{N-1}} \cdot \prod_{j=1}^{N} \left(\frac{1}{2\pi R_j} \right). \end{aligned}$$

Choosing $R_j = 2^{-j}$ (for $j = 1, 2, \cdots, N$), $N = 4$, and $u = 20$, the maximum error is about 10^{-4}, and we numerically find that $T_N(\mathbf{0}) = -2.8373$.

The above calculation sums about 4200 terms after using the eightfold symmetry of the triple sum at $\mathbf{r} = \mathbf{0}$. This is beyond hand calculation, but easily evaluated with a small computer program involving only elementary functions. In contrast, Hasimoto uses a generalized theta transformation to convert $S_1(\mathbf{r})$ into a three-dimensional sum of incomplete Γ-functions, and expands each Γ-function in the neighborhood of zero. The resulting exponential convergence is very desirable for hand calculations, but the incomplete Γ-function is not a commonly stored computer function. Thus, the summation using the mean value theorem transformation is easier to implement when a computer is available.

For further information about lattice sums and other techniques for summation, we refer the reader to the articles by Zucker [3], and Chaba and Pathria [4], and the references therein.

REFERENCES

[1] H. HASIMOTO, *On the periodic fundamental solutions of the Stokes equations and their application to viscous flow past a cubic array of spheres*, J. Fluid Mech., 5 (1959), pp. 317–328.
[2] R. COURANT AND D. HILBERT, *Methods of Mathematical Physics, Vol.* II, Interscience, New York, 1962.
[3] I. J. ZUCKER, *Functional equations for poly-dimensional zeta functions and the evaluation of Madelung constants*, J. Phys. A:Math. Gen., 9 (1976), pp. 499–505.
[4] A. N. CHABA AND R. K. PATHRIA, *Evaluation of lattice sums using Poisson's summation formula* III, J. Phys. A:Math. Gen., 9 (1976), pp. 1801–1810.

SURFACE INTEGRAL OF ITS MEAN CURVATURE VECTOR*

DENIS BLACKMORE[†] AND LU TING[‡]

Abstract. We present two methods for proving that the integral of the mean curvature vector over a closed surface vanishes.

Key words. principal curvature, mean curvature, Laplace formula, diffeomorphism, differential forms, pullback, Stokes' theorem

AMS (MOS) 1980 subject classifications. 53A05, 53A07, 76V05

Introduction. In the study of two-phase problems, e.g., air bubbles in water, it is well known that the resultant force acting on a segment of the interface has to vanish since an interface has no inertia. Therefore, we have

$$(1) \qquad \iint_A (p^+ - p^-)\mathbf{N}\, dA + \sigma \oint_{\partial A} \mathbf{N} \times d\mathbf{X} = 0,$$

where σ_1, $p^+ - p^-$ and $\mathbf{N}$ denote respectively the surface tension coefficient, the pressure difference across the interface and its unit normal vector pointing towards the side with pressure p^+. For an infinitesimal surface element, (1) yields the Laplace's formula

$$(2) \qquad p^+ - p^- = \sigma(\kappa_1 + \kappa_2),$$

---
* Received by the editors November 13, 1984, and in revised form May 2, 1985.
† Department of Mathematics, New Jersey Institute of Technology, Newark, New Jersey 07102.
‡ Courant Institute of Mathematical Sciences, New York University, New York, New York 10012. The research of this author was partially supported by the Office of Naval Personnel under contract N000-14-81-K-0002.

where κ_1 and κ_2 are the principal curvatures of the interface. For a closed interface S, (1) leads to

(3)
$$\iint_S (p^+ - p^-)\mathbf{N}\, dA = 0$$

for any imposed $p^+ - p^-$ distribution. This is consistent with physics in that the resultant force acting on a closed interface of zero inertia has to vanish.

Using (2), we then get

(4)
$$\iint_S M\mathbf{N}\, dA = 0$$

where M denotes the mean curvature. The obvious question is whether (4) is valid only for a special class of surfaces? In that case, it would imply that a closed interface has to belong to that class. Instead, we find that (4) is valid for every closed surface for which the coefficients of the first and second fundamental forms are continuous.

THEOREM. *Let S be a closed ($=$ compact and boundaryless), orientable surface of class C^2 with mean curvature M and unit outer normal $\mathbf{N}$. Then (4) holds.*

We shall prove this theorem by two methods. In the first method, we shall prove that for a surface area element, the surface integral of $2M\mathbf{N}$ is equal to the contour integral $-\oint \mathbf{N} \times d\mathbf{X}$. In essence, we are rederiving the Laplace's formula while demonstrating the difference between the curvature vector of a surface curve and its normal curvature vector. In the second method, we use the general formulas for the mean curvature and surface normal and then show that $M\mathbf{N}$ can be put in a divergence form to which Stokes' theorem is applied.

Method I. Let us consider the area element of the surface $\mathbf{X}(u,v)$ bounded by constant u, $u + du$, v and $v + du$ lines. If the parametric lines are lines of curvature [1], the corresponding first and second fundamental forms are

(5) $\mathrm{I} = d\mathbf{X} \cdot d\mathbf{X} = E\, du^2 + G\, du^2,$
(6) $\mathrm{II} = d\mathbf{X}^2 \cdot \mathbf{N} = e\, du^2 + g\, dv^2.$

Let t_1 and t_2 denote respectively the unit tangent vectors along the lines of constant v and u, with $\mathbf{t}_1 \times \mathbf{t}_2 = \mathbf{N}$. Let s_1 and s_2 be the arc length along the constant v and u lines, we have $ds_1 = \sqrt{E}\, du$ and $ds_2 = \sqrt{G}\, dv$. The line integral in (1) becomes

(7)
$$\oint \mathbf{N} \times d\mathbf{X} = \left\{ \frac{\partial}{\partial u}\left(t_1 \sqrt{G}\right) + \frac{\partial}{\partial v}\left(t_2 \sqrt{E}\right) \right\} du\, dv.$$

We note that along a constant v line,

$$\frac{1}{\sqrt{E}} \frac{\partial}{\partial u} t_1 = \frac{\partial}{\partial s_1} t_1 = \mathbf{K}_1 = \eta_1 \mathbf{N} + (\kappa_g)_1 t_2,$$

where $\mathbf{K}_1$ is the curvature vector to the curve and has two components normal and tangential to the surface, respectively. The normal and tangent curvatures are

$$\kappa_1 = \frac{e}{E} \quad \text{and} \quad (\kappa_g)_1 = -\frac{1}{\sqrt{EG}} \frac{\partial}{\partial v} \sqrt{E}.$$

Similarly we have

$$\frac{1}{\sqrt{G}} \frac{\partial}{\partial v} \mathbf{t}_2 = \frac{\partial}{\partial s_2} \mathbf{t}_2 = \mathbf{K}_2 = \kappa_2 \mathbf{N} + (\kappa_g)_2 t_1,$$

with

$$\kappa_2 = \frac{g}{G} \quad \text{and} \quad (\kappa_g)_2 = \frac{1}{\sqrt{EG}} \frac{\partial}{\partial u} \sqrt{E} \,.$$

The first term on the right side of (7) can be written as

$$-\frac{\partial}{\partial u}\left(t_1\sqrt{G}\right) du\, dv = -\sqrt{EG}\, du\, dv \left[\kappa_1 N + (\kappa_g)_1 t_2 + t_1\left(\frac{\partial}{\partial u}\sqrt{G}\right)\!/\sqrt{EG}\right].$$

The last term in the bracket represents the stretching of the arc length from constant u to $u + du$ and is equal to $-(\kappa_g)_2 t_1$. This is consistent with the well-known result that κ_g is the curvature of the projection of the curve on the tangent plane. It is clear that the contribution of the stretching of the arc length in the first term on the right side of (7) cancels the contribution of the tangent curvature vector in the second term and vice versa. Consequently, (7) becomes

$$\oint N \times d\mathbf{X} = -(\kappa_1 + \kappa_2)N\sqrt{EG}\, du\, dv = -(\kappa_1 + \kappa_2)N\, dA.$$

We then recover the Laplace formula from (1). If the surface A is covered by two families of curvature lines, we have

$$(8) \qquad\qquad \iint_A 2MN\, dA = -\oint_{\partial A} N \times d\mathbf{X}.$$

If A has a finite number of umbilical points, (8) remains valid since their measure is zero. Equation (8) is true also when on A there is a line segment L of umbilical points. Since the line is of measure zero, there is no contribution to the surface integral on the left side of (8). On the other hand, the line integral around a cut on L vanishes due to mutual cancellation and therefore does not contribute to the right side of (8) either. If a portion of the surface lies in a plane, (8) is still valid since the contributions of the planar portion to both sides of (8) are zero. Equation (8) holds if the surface has a spherical portion, as the vector identity

$$2 \iint_\Sigma N\, dA = \oint_{\partial\Sigma} \mathbf{X} \times d\mathbf{X}$$

for a C^2 surface Σ can be used to show that the spherical contributions to both sides of (8) are identical. When A is a closed surface S, we have (4). If in addition the surface is C^3, the preceding line of argument can actually be generalized to include every closed surface.

Using a slight modification of this approach, we can show the validity of (8) employing orthogonal lines instead of curvature lines. For a closed surface S we have again (4), since it can be shown that every closed C^2 surface can be covered by an orthogonal net if finitely many exceptional points are excluded.

Method II. It is convenient to introduce some standard notation. Let $\mathbf{X} = (x, y, z)$ be the position vector in $\mathbb{R}^3$. Given any point $p \in S$ there exist an open neighborhood V of p in $\mathbb{R}^3$, an open neighborhood U of the origin in the uv plane and a C^2 diffeomorphism $\psi: U \to S \cap V$ such that $\psi(0,0) = p$. We define $\mathbf{X}(u,v) = \psi(u,v)$ and call $\mathbf{X} = \mathbf{X}(u,v)$ a local parametrization of S. Since ψ is a diffeomorphism,

$$\mathbf{X}_u \times \mathbf{X}_v \neq 0 \quad \text{in } U,$$

and by selecting u and v corresponding to the appropriate orientation of S we obtain

$$\mathbf{N} = \mathbf{X}_u \times \mathbf{X}_v / \| \mathbf{X}_u \times \mathbf{X}_v \|$$

in local u,v-coordinates. With this notation, the components of the vector-valued form $M\,N\,dA$ (with respect to the standard orthonormal basis $\mathbf{i}, \mathbf{j}, \mathbf{k}$ in $\mathbb{R}^3$) are

$$M \frac{\partial(y,z)}{\partial(u,v)} du \wedge dv, \quad M \frac{\partial(z,x)}{\partial(u,v)} du \wedge dv \quad \text{and} \quad M \frac{\partial(x,y)}{\partial(u,v)} du \wedge dv.$$

Let these 2-forms be denoted by ω_1, ω_2 and ω_3, respectively. The proof of (4) follows if we can show each of these 2-forms is exact: i.e., there are 1-forms θ_1, θ_2 and θ_3 defined on S such that

$$\omega_l = d\theta_l, \qquad l=1,2,3.$$

For then by Stokes' theorem (see [2])

$$\iint_S \omega_l = \iint_S d\theta_l = \oint_{\partial S} \theta_l = \oint_\phi \theta_l = 0, \qquad l=1,2,3.$$

We claim that the required 1-forms are as follows: $\theta_1 = \langle \mathbf{N}, \mathbf{k} \rangle dy - \langle \mathbf{N}, \mathbf{j} \rangle dz$, $\theta_2 = -\langle \mathbf{N}, \mathbf{k} \rangle dx + \langle \mathbf{N}, \mathbf{i} \rangle dz$ and $\theta_3 = \langle \mathbf{N}, \mathbf{j} \rangle dx - \langle \mathbf{N}, \mathbf{i} \rangle dy$, where $\langle \cdot, \cdot \rangle$ denotes the standard inner (dot) product in $\mathbb{R}^3$. To be more precise, we should actually use the pullback to S of each of these forms, but our abuse of notation is harmless. It is clear that these forms are globally C^1 on S. As all the verifications are essentially the same, it will suffice to prove $d\theta_3 = \omega_3$. This is accomplished most easily by working locally as follows: We may assume (after an isomorphic transformation of coordinates if necessary) that

$$(9) \qquad\qquad \frac{\partial(x,y)}{\partial(u,v)} > 0$$

on U. Hence

$$\omega_3 = M \frac{\partial(x,y)}{\partial(u,v)} du \wedge dv = M\,dx \wedge dy.$$

It is easily verified (see [1, 3]) that M has the following form in local coordinates:

$$M = \frac{1}{2} \frac{\langle \mathbf{N}, \mathbf{X}_{uu} \rangle \| \mathbf{X}_v \|^2 - 2 \langle \mathbf{N}, \mathbf{X}_{uv} \rangle \langle \mathbf{X}_u, \mathbf{X}_v \rangle + \langle \mathbf{N}, \mathbf{X}_{vv} \rangle \| \mathbf{X}_u \|^2}{(\| \mathbf{X}_u \| \| \mathbf{X}_v \|)^2 - \langle \mathbf{X}_u, \mathbf{X}_v \rangle^2}.$$

By virtue of (9), we may assume that the local parametrization is

$$\mathbf{X}(x,y) = (x,y,h(x,y)),$$

where h is of class C^2. In terms of these coordinates it is easy (see [3]) to show that

$$\mathbf{N} = \frac{(-h_x, -h_y, 1)}{\left(1 + h_x^2 + h_y^2\right)^{1/2}}$$

and

$$M = \frac{(1 + h^2 x) h_{yy} - 2 h_y h_y h_{xy} + \left(1 + h_y^2\right) h_{xx}}{\left(1 + h_x^2 + h_y^2\right)^{3/2}}.$$

We observe now that in xy-coordinates

$$d\theta_3 = \frac{\left(1+h_x^2\right)h_{yy} - 2h_x h_y h_{xy} + \left(1+h_y^2\right)h_{xx}}{\left(1+h_x^2+h_y^2\right)^{3/2}} dx \wedge dy$$

$$= M\, dx \wedge dy = \omega_3.$$

This completes the proof.

In conclusion, we note that this result can be generalized as follows:

Let S be a closed C^2 hypersurface of Euclidean n-space with normal vector **N** and curvature 2-form Ω. Then

$$\int_S (\text{trace}\,\Omega)\mathbf{N}\, dA = 0.$$

REFERENCES

[1] D. J. STRUIK, *Lectures on Classical Differential Geometry*, Addison-Wesley, Reading, MA, 1961, pp. 73–82, 127–130.
[2] F. WARNER, *Foundations of Differential Manifolds and Lie Groups*, Scott, Foresman & Co., Glenview, IL, 1971, pp. 145–148.
[3] M. DoCARMO, *Differential Geometry of Curves and Surfaces*, Prentice-Hall, Englewood Cliffs, NJ, 1976, pp. 153–163.

A SIMPLE PROOF THAT THE WORLD IS THREE-DIMENSIONAL*

TOM MORLEY[†]

Abstract. The classical Huygens' principle implies that distortionless wave propagation is possible only in odd dimensions. A little known clarifications of this principle, due to Duffin and Courant, states that radially symmetric wave propagation is possible only in dimensions one and three. This paper presents an elementary proof of this result.

1. Introduction. The title is, of course, a fraud. We prove nothing of the sort. Instead we show that radially symmetric wave propagation is possible only in dimensions one and three.

In 1864 James Clark Maxwell discovered the fundamental laws of electromagnetism; see [5]. Maxwell's theory predicted the existence of electromagnetic radiation, i.e., electromagnetic waves. It was not until 1888 that Heinrich Rudolf Hertz discovered radio waves in the laboratory. (By the way, Hertz was a student of Helmholtz.) There can be little doubt that this discovery and subsequent technological advances have had a profound effect on modern life. What would the world be like without radio, television, and global instantaneous (or nearly so) communication? It is the purpose of this note to give a short elementary proof that this state of affairs can exist only in three dimensions. In particular:

THEOREM. *Radially symmetric distortionless wave propagation is possible only in*

*Received by the editors June 26, 1984, and in revised form August 30, 1984.
†School of Mathematics, Georgia Institute of Technology, Atlanta, Georgia 30332.

dimensions one and three. However, in one dimension there is no attenuation.

For precise definitions of these terms see §2.

This theorem was proved by R. J. Duffin in 1952 [3], and is mentioned by R. Courant in [2]. (It is not known whether Courant knew of Duffin's work.) Neither Courant nor Duffin ever published a proof. The present proof, however, is different and considerably more elementary than Duffin's original proof, and is suitable for presentation in the typical junior-senior level ODE–PDE course.

2. Radial wave propagation. Consider the n-dimensional wave equation

$$\text{(W)} \qquad \sum u_{x_i x_i} = \frac{1}{c^2} u_{tt}.$$

A radially symmetric wave is a solution of (W) that depends only on t and

$$r = \left(x_1^2 + x_2^2 + \cdots + x_n^2 \right)^{1/2}.$$

Setting $v(r,t) = u(x,t)$ we obtain, by the chain rule, the n-dimensional radially symmetric wave equation

$$\text{(RW)} \qquad v_{rr} + \frac{n-1}{r} v_r = \frac{1}{c^2} v_{tt}.$$

DEFINITION. Distortionless radially symmetric wave propagation is possible if there are functions $\alpha(r) > 0$, $\delta(r) > 0$, $\delta(0) = 0$, and $\alpha(1) = 1$ such that given any "reasonable" f, the function

$$\alpha(r) f(t - \delta(r))$$

is a solution of (RW). The function $\alpha(\cdot)$ is termed the attenuation, and the function $\delta(\cdot)$ is the delay. If α is identically 1 then there is no attenuation.

It should be noted that "reasonable" can be quite unrestrictive; the class of polynomials or trigonometric polynomials will suffice.

Proof of theorem. If distortionless radially symmetric wave propagation is possible, then given any reasonable f the function $v(r,t) = \alpha(r) f(t - \delta(r))$ is a solution of (RW). Computing partial derivatives:

$$v_{tt} = \alpha f'',$$
$$v_r = \alpha' f - \alpha \delta' f',$$
$$v_{rr} = \alpha'' f - \alpha' \delta' f' - (\alpha' \delta' + \alpha \delta'') f' + \alpha \delta'^2 f''.$$

Plugging these values into (RW), we obtain

$$\text{(*)} \qquad \alpha'' f - \alpha' \delta' f' - (\alpha' \delta' + \alpha \delta'') f' + \alpha \delta'^2 f'' + \frac{n-1}{r} (\alpha' f - \alpha \delta' f') = \frac{\alpha}{c^2} f''.$$

In the above computations, the arguments of the functions have been deleted for notational convenience. For instance, f is an abbreviation for $f(t - \delta(r))$.

The only possible way for (*) to hold for all reasonable f is for the coefficients of f'', f' and f to each be equal to zero. Equating the coefficient of f'' to zero, gives

$$\text{(1)} \qquad \alpha \delta'^2 = \frac{\alpha}{c^2}.$$

Together with $\delta > 0$ and $\delta(0) = 0$, we deduce that

$$(2) \qquad\qquad \delta = \frac{r}{c}, \quad \delta' = \frac{1}{c}, \quad \delta'' = 0.$$

Plugging this into (*) and then considering the coefficients of f gives

$$(3) \qquad\qquad \alpha'' + \frac{n-1}{r}\alpha' = 0.$$

Similarly, the f' terms give

$$(4) \qquad\qquad \frac{1}{c}\left(2\alpha' + \frac{n-1}{r}\alpha\right) = 0.$$

Solutions of (3) and (4) are of the form Kr^{β}, where K and β are constants. Plugging this guess for α into (3) and (4) gives:

$$(3') \qquad\qquad \beta(\beta-1) + (n-1)\beta = 0,$$

$$(4') \qquad\qquad 2\beta + (n-1) = 0.$$

Equations (3') and (4') only have a solution for β if $n=1$ or $n=3$. However, plugging in $n=1$ gives $\alpha(r) = 1$, and thus there is no attenuation. Of course, a world without attenuation would be unbearably noisy.

Acknowledgments. The author would like to thank R. J. Duffin for historical discussions. The author would also like to thank an anonymous referee for contributing the final sentence.

REFERENCES

[1] V. Burke, R. J. Duffin and D. Hazony, *Distortionless wave propagation in inhomogeneous media and transmission lines*, Quart. Appl. Math., 34 (1974), pp. 183–194.

[2] R. Courant, *Hyperbolic partial differential equations and application*, in Modern Mathematics for the Engineer, E. Beckenback, ed., McGraw-Hill, New York, 1956.

[3] R. J. Duffin, unpublished, 1952.

[4] H. R. Hertz, *Electric Waves*, Stafford, 1893.

[5] J. C. Maxwell, *A dynamical theory of the electromagnetic field*, Proc. Roy. Soc. London, 13 (1864), pp. 531–536.

[6] ———, *A Treatise on Electricity and Magnetism*, third ed. 1891, reprinted Dover, New York, 1954.

General Supplementary References
Physical and Mathematical Sciences

[1] M. J. Ablowitz and H. Segur, *Solitions and the Inverse Scattering Transform*, SIAM, 1981.

[2] H. Akashi and S. Levy, *"The motion of an electric bell,"* AMM (1953) pp. 255–259.

[3] T. B. Akrill and C. J. Millar, *Mechanics, Vibrations and Waves*, Murray, Great Britain, 1974.

[4] H. C. Andrews and C. L. Patterson, *"Outer product expansions and their uses in digital image processing,"* AMM (1975) pp. 1–13.

[5] K. G. Beauchamp, *Walsh Functions and their Applications*, Academic, London, 1975.

[6] H. Berliner, *"Computer backgammon,"* Sci. Amer. (1980) pp. 64–72.

[7] H. J. Berstein and A. V. Phillips, *"Fiber bundles and quantum theory,"* Sci. Amer. (1981) pp. 123–137.

[8] M. P. Blake and W. S. Mitchell, *Vibration and Acoustic Measurement Handbook*, Spartan Books, N.Y., 1972.

[9] W. E. BLEICK, *"Fourier analysis of engine unbalance by contour integration,"* AMM (1956) pp. 466–472.

[10] M. BRAUN, *"Mathematical remarks on the Van Allen radiation belt,"* SIAM Rev. (1981) pp. 61–93.

[11] K. E. BULLEN, *An Introduction to the Theory of Seismology,* Cambridge University Press, Cambridge, 1979.

[12] D. N. BURGHES, *"Mathematical modelling in geography,"* BIMA (1979) pp. 254–260.

[13] J. C. BURNS, *"Differential equations for the flow of a solution of varying concentration,"* AMM (1968) pp. 705–710.

[14] P. A. BURROUGHS, *"The application of fractal ideas to geophysical phenomena,"* BIMA (1984) pp. 36–42.

[15] P. CHADWICK, *"Mathematical aspects of smoking,"* Math. Spectrum (1969.70) pp. 14–21.

[16] S. CHILDRESS, *Mechanics of Swimming and Flying,* Cambridge University Press, N.Y., 1981.

[17] R. J. CHORLEY AND P. HAGGETT, eds., *Models in Geography,* Methuen, London, 1967.

[18] J. M. CRADDOCK AND M. G. COLGATE, *"The use of eigenvalues for smoothing and prediction,"* BIMA (1974) pp. 152–160.

[19] J. P. COLE AND C. A. M. KING, *Quantitative Geography,* Wiley, N.Y., 1968.

[20] R. COURANT AND D. HILBERT, *Methods of Mathematical Physics I, II,* Interscience, N.Y., 1953.

[21] C. R. DEETER AND A. A. J. HOFFMAN, *"Energy related mathematical models: Annotated bibliography,"* Energy Conversion (1978) pp. 189–227.

[22] R. K. DODD, J. C. EILBECK, J. D. GIBBON AND H. G. MORRIS, *Solitons and Nonlinear Wave Equations,* Academic, London, 1982.

[23] P. G. DRAZIN, *Solitons,* Cambridge University Press, N.Y., 1983.

[24] J. A. DUTTON, *The Ceaseless Wind: An Introduction to the Theory of Atmospheric Motion,* McGraw-Hill, N.Y., 1976.

[25] ———, *"Fundamental theorems of climate theory – some proved, some conjectured,"* SIAM Rev. (1982) pp. 1–33.

[26] P. P. G. DYKE, A. O. MOSCARDINI AND E. H. ROBSON, *Offshore and Coastal Modeling,* Springer-Verlag, N.Y., 1985.

[27] K. EISEMANN, *"Number-theoretic analysis and extensions of The most complicated and fantastic card trick ever intented,"* AMM (1984) pp. 284–289.

[28] R. E. EWING, *The Mathematics of Reservoir Simulation,* SIAM, 1984.

[29] W. E. FARRELL, D. P. MCKENZIE AND R. L. PARKER, *"On the note emitted from a mug while mixing instant coffee,"* Proc. Camb. Phil. Soc. (1969) pp. 365–367.

[30] R. P. FEYNMAN, R. B. LEIGHTON AND M. SANDS., *The Feynman Lectures on Physics: Vol. I, Mainly Mechanics, Radiation and Heat,* 1963; *Vol. II, Mainly Electromagnetism and Matter,* 1964; *Vol. III, Quantum Mechanics,* 1965; Addison-Wesley, Reading.

[31] M. E. FISHER AND M. F. SYKES, *"Excluded volume problem and the Ising model of ferromagnetism,"* Phys. Rev. (1959) pp. 45–58.

[32] B. F. FITZGERALD, *Development in Geographical Method,* Oxford University Press, Oxford, 1974.

[33] J. A. FOX AND D. A. HENSON, *"The aerodynamics of the channel tunnel,"* BIMA (1972) pp. 124–129.

[34] M. GITTERMAN AND V. HALPERN, *Qualitative Analysis of Physical Problems,* Academic, N.Y., 1981.

[35] J. A. HALTON, *"A retrospective and prospective survey of the Monte Carlo method,"* SIAM Rev. (1970) pp. 1–63.

[36] P. HARVEY, *"A problem of astronomical proportion,"* Math. Gaz. (1976) pp. 263–269.

[37] D. HILLEL, ed., *Applications of Soil Physics,* Academic, N.Y., 1980.

[38] G. J. HOLTINER, *Numerical Weather Prediction,* Wiley, N.Y., 1971.

[39] R. I. G. HUGHES, *"Quantum logic,"* Sci. Amer. (1981) pp. 202–213.

[40] J. A. JACOBS, *The Earth's Core,* Academic Press, N.Y., 1975.

[41] ———, *"The earth's core and geomagnetism,"* BIMA (1977) pp. 86–91.

[42] T. L. JAHN AND J. L. VOTTA, *"Locomotion of protozoa,"* Ann. Rev. Fluid Mech. (1972) pp. 93–116.

[43] M. KAC, *"Can one hear the shape of a drum?"* AMM, part II (1966) pp. 1–23.

[44] P. W. KASTELEYN, *"A soluble self-avoiding walk problem,"* Physica (1963) pp. 1329–1337.

[45] J. B. KELLER, *"Inverse problems,"* AMM (1976) pp. 107–118.

[46] ———, *"Some bubble and contact problems,"* SIAM Rev. (1980) pp. 442–458.

[47] J. B. KELLER AND S. R. RUBINOW, *"Swimming of flagellated microorganisms,"* Biophys. J. (1976) pp. 151–170.

[48] M. S. KLAMKIN, *"A moving boundary filtration problem or the cigarette problem,"* AMM (1958) pp. 680–684.

[49] S. J. KLINE, *Similitude and Approximation Theory,* McGraw-Hill, N.Y., 1965.

[50] D. E. KNUTH, *"Computer science and its relation to mathematics,"* AMM (1974) pp. 323–343.

[51] S. KULIK, *"A problem arising in yarn spinning,"* AMM (1958) pp. 680–684.

[52] E. H. LAND, *"The retinex theory of color vision,"* Sci. Amer. (1977) pp. 108–128.

[53] E. R. LAPWOOD AND T. USAMI, *Free Oscillations of the Earth,* Cambridge University Press, Cambridge, 1981.

[54] H. P. LAWTHER, JR., *"An application of number theory to the splicing of telephone cables,"* AMM (1935) pp. 81–91.

[55] P. D. LAX, *"The formation of shock waves,"* AMM (1972) pp. 227–241.

[56] N. N. LEBEDEV, I. P. SKALSKAYA AND Y. S. UFLYAND, *Worked Problems in Applied Mathematics,* Dover, N.Y., 1965.

[57] P. H. LEBLOND AND L. A. MYSAK, *"Ocean waves,"* SIAM Rev. (1979) pp. 289–328.

[58] S. E. LEVINSON AND M. Y. LIEBERMAN, *"Speech recognition by computer,"* Sci. Amer. (1981) pp. 64–76.

[59] J. LIGHTHILL, *"Aerodynamic aspects of animal flight,"* BIMA (1974) pp. 369–393.

[60] ——, *"Flagellar hydrodynamics,"* SIAM Rev. (1976) pp. 161–230.

[61] ——, *Mathematical Biofluiddynamics*, SIAM, 1975.

[62] M. S. LONGUET-HIGGINS, *"On slinky: The dynamics of a loose, heavy spring*, Proc. Camb. Phil. Soc. (1954) pp. 347–351.

[63] B. M. McCOY AND T. T. WU, *The Two-Dimensional Ising Model*, Harvard University Press, Cambridge, 1973.

[64] T. A. McMAHON, *Muscles, Reflexes, and Locomotion*, Princeton University Press, Princeton, 1984.

[65] R. M. MIURA, *"The Korteweg-DeVries equation: A survey of results,"* SIAM Rev. (1976) pp. 412–459.

[66] I. H. NICHOLSON, ed., *Modelling of Dynamical Systems I*, Peregrinus, 1980.

[67] A. S. ODEH, *"An overview of mathematical modeling of the behavior of hydrocarbon reservoirs,"* SIAM Rev. (1982) pp. 263–273.

[68] L. E. PAYNE, *"Isoperimetric in qualities and their application,"* SIAM Rev. (1967) pp. 453–488.

[69] C. J. PENNYCUICK, *Animal Flight*, Arnold, London, 1972.

[70] O. M. PHILLIPS, *The Heart of the Earth*, Freeman, San Francisco, 1968.

[71] C. POMERANCE, *"The search for prime numbers,"* Sci. Amer. (1982) pp. 136–147, 178.

[72] W. H. PRESS, *"Mathematical theory of the water bed,"* Amer. J. Phys. (1978) pp. 966–970.

[73] G. N. RAMACHANDRAN AND S. SRINIVASSAN, *Fourier Methods in Crystallography*, Wiley, N.Y., 1970.

[74] A. C. S. READHEAD, *"Radio astronomy by very-long-baseline interferometry,"* Sci. Amer. (1982) pp. 53–61, 154.

[75] C. REBBI, *"Solitons,"* Sci. Amer. (1979) pp. 92–116.

[76] W. H. REID, *Mathematical Problems in the Geophysical Sciences*, AMS, 1971.

[77] T. A. REINHOLD, *Wind Tunnel Modeling for Civil Engineering*, Cambridge University Press, N.Y., 1982.

[78] L. A. SEGAL, *Mathematics Applied to Cointinuum Mechanics*, Macmillan, N.Y., 1977.

[79] I. STACKGOLD, *Boundary Value Problems of Mathematical Physics I, II*, Macmillan, N.Y., 1968.

[80] K. STEWARTSON, *"D'Alembert's paradox,"* SIAM Rev. (1981) 308–343.

[81] G. R. STIBLITZ, *"An application of number theory to gear ratios,"* AMM (1938) pp. 22–31.

[82] SWAMP Group, *Ocean Wave Modeling*, Plenum, N.Y., 1985.

[83] W. TAPE, *"Analytic foundations of halo theory,"* J. Optical Soc. Amer. (1980) pp. 1175–1192.

[84] D. TATTERSFIELD, *Projects and Demonstrations in Astronomy*, Thornes, 1979.

[85] I. UGI, J. DUGUNDJI, R. KOPP AND D. MARQUARDING, *Perspectives in Stereochemistry*, Springer-Verlag, N.Y., 1984.

[86] K. WALTERS, *Rheometry*, Wiley, N.Y., 1975.

[87] T. W. WIETING, *The Mathematics of Chromatic Plane Ornaments*, Dekker, N.Y., 1982.

[88] D. C. WISPELAERE, ed., *Air Pollution Modelling and its Applications*, SRI, Menlo Park, 1983.

[89] P. YOUNG, *"Recursive approaches to time series analysis,"* BIMA (1974) pp. 209–224.

Part II Life Sciences

1. Population Models

POPULATIONS AND HARVESTING*

DAVID A. SÁNCHEZ†

Abstract. Both finite difference and ordinary differential equation models of population growth in the case in which the harvest rate is constant are discussed.

1. Introduction. In the study of equations describing population growth it is of interest to study the effects of harvesting (e.g., by hunting, fishing or disease) on the stability of the population. A simple case which is of qualitative and classroom interest is the case in which the harvest rate is constant, and we will discuss this below both for a finite difference equation model as well as an ordinary differential equation model of the population growth.

2. A finite difference model. Let x_n be the number of females capable of reproducing which exist at breeding period n, and let $R(x_n)$ be the average number of surviving females each produces. Assuming that each female reproduces only once, then the population growth can be described by the finite difference equation

$$(1) \qquad x_{n+1} = R(x_n)x_n, \qquad x_0 = A$$

where A is the initial number of females present. A straightforward stability analysis will show that those values $x = x_e$ for which $R(x_e) = 1$ will be points of equilibrium for the population; they will be stable if $R'(x_e) < 0$ and unstable if $R'(x_e) > 0$. A very good elementary discussion of finite difference models for populations can be found in the book by J. Maynard Smith [3]—see also his more recent advanced treatise [4].

If we suppose a certain fixed number $H > 0$ of females are harvested during each breeding period we obtain the equation

$$(2) \qquad x_{n+1} = R(x_n)x_n - H, \qquad x_0 = A$$

and one is interested in studying the effects of harvesting on an equilibrium state x_e of (1). One makes the usual linearization near x_e and thus $R(x) \cong 1 + k(x - x_e)$, and then letting $y_n = x_n - x_e$ (the deviation from equilibrium) leads to the equation

$$(3) \qquad y_{n+1} = (1 + kx_e)y_n - H, \qquad y_0 = A - x_e = \alpha,$$

where terms of second degree are ignored, i.e. this is a local analysis.

The equation (3) can be solved recursively to get

$$(4) \qquad y_{n+1} = \left(\alpha - \frac{H}{kx_e}\right)(1 + kx_e)^n + \frac{H}{kx_e}$$

and, since harvesting in the presence of instability can only make things worse, let us assume x_e is a stable equilibrium point. This implies in turn that $|1 + kx_e| < 1$ and

---

* Received by the editors May 1, 1976.

† University of California, Los Angeles, California 90024.

so $k < 0$ and from (4) we then obtain the relation

(5) $$x_{n+1} = (1 + kx_e)^n \left(\alpha + \frac{H}{|k|x_e} \right) + \left(x_e - \frac{H}{|k|x_e} \right).$$

Since $(1 + kx_e)^n \to 0$ as $n \to \infty$ then $x_{n+1} \to x_e - H/(|k|x_e)$ and the effect of harvesting is to reduce the size of the equilibrium population. For the population to survive we must have that $0 < H < |k|x_e^2 = H_c$ and we can summarize our analysis with the following plausible statement:

There is a critical level of harvesting H_c such that if $0 < H < H_c$ then the population survives with a lower equilibrium value, whereas if $H > H_c$ the population expires in finite time.

Note that if $H > H_c$ one can calculate from expression (5) how long it will take for the population to expire, i.e. when $x_{n+1} \leqq 0$.

2. A differential equation model. A well known differential equation for describing population growth is the logistic equation

(6) $$\frac{dx}{dt} = rx\left(1 - \frac{x}{k}\right), \qquad x(0) = x_0$$

where $x(t)$ is the population size at time t, and $r, k > 0$. Equilibrium values are $x_e = 0$ and $x_e = k$ (i.e. values where the right hand side vanishes), and the solutions are the well known logistic curves. Again we suppose a constant harvest rate $H > 0$ of the population and obtain

(7) $$\frac{dx}{dt} = rx\left(1 - \frac{x}{k}\right) - H, \qquad x(0) = x_0.$$

The equilibrium values are the zeros of the right hand side of (7) which is a quadratic and one finds that

(a) It has real positive roots if and only if $H \leqq rk/4 = H_c$. Otherwise the right-hand side of (7) is negative.

(b) For $H < rk/4$ the larger of the two roots is $x_e = \frac{1}{2}k[1 + (1 - H/H_c)^{1/2}]$.

Incidentally, one can solve (7) explicitly by a partial fractions expansion and quadrature, but that is not the matter of interest here.

Our conclusion from (a) and (b) is exactly the same as for the finite difference model, and a study of the direction field of (7) when $0 < H < H_c$ shows that x_e is a stable equilibrium point whereas the smaller root is unstable. If we note that $rx(1 - x/k)$ has a maximum $\text{Max} = rk/4$ we conclude that the critical harvest rate is the maximum possible growth rate of the population—a biologically plausible result.

If $H > H_c$ one can calculate the extinction time T_H by separation of variables. In particular if the initial population in (7) is $x_0 = k$, the equilibrium value for (6), then

$$T_H = \frac{4}{r}\left(\frac{H}{H_c} - 1\right)^{-1/2} \arctan\left(\frac{H}{H_c} - 1\right)^{-1/2}.$$

Plotting T_H against H for fixed H_c should fascinate those advocates of the doomsday syndrome.

3. Remarks. The effects of constant rate harvesting on one population or two populations in competition was discussed by the author and F. Brauer in [1].

The paper contains a comparison of the results above for the logistic equation with a simulated model by R. M. Miller and D. Botkin [2] of the effect of hunting on the Sandhill Crane (*Grus canadensis*) population. An excellent classroom research project might be to study the effects of harvesting or enrichment ($H < 0$) on some of the well known population models described in Smith's book [4]. More general harvesting rates, e.g. $H = 0$ if $x > A$, $H = H_0$ otherwise, could also be considered.

REFERENCES

[1] F. BRAUER AND D. A. SÁNCHEZ, *Constant rate population harvesting*: *Equilibrium ad stability*, Theoret. Population Biology, 8 (1975), pp. 12–30.
[2] R. MILLER AND D. BOTKIN, *Endangered species*: *Models and prediction*, Amer. Sci., 62 (1971), pp. 172–181.
[3] J. MAYNARD SMITH, *Mathematical Ideas in Biology*, Cambridge University Press, Cambridge, England 1968.
[4] ——, *Models in Ecology*, Cambridge University Press, Cambridge, England, 1974.

DETERMINISTIC POPULATION MODELS AND STABILITY*

H. BARCLAY AND P. VAN DEN DRIESSCHE†

Abstract. The stability of a linear differential equation system and of an analogous linear difference equation system is analyzed by Liapunov matrices. Explicit stability criteria are given for the two species Lotka–Volterra competition model and predator-prey model for population growth.

This note investigates some theoretical, deterministic models for population growth of interacting species. The n species Lotka–Volterra equations

$$(1) \qquad \frac{dN_i(t)}{dt} = r_i N_i(t)\left(1 - \sum_{j=1}^{n} \frac{\alpha_{ij}}{K_i} N_j(t)\right), \qquad \alpha_{ii} = 1, \quad i = 1, \cdots, n,$$

are commonly used as a model to describe the number of individuals N_i in species i. Here r_i (positive constant) is the intrinsic rate of increase, K_i (positive constant) is the carrying capacity of the ith species; and α_{ij} is the competition coefficient of the jth species on the ith, $\alpha_{ij} > 0$ when the jth species competes with the ith. Interest is in a state in which the n species coexist and for which the equilibrium point is asymptotically stable. For such an equilibrium to occur the system $dN_i/dt = 0$ must have a solution $N_i = N_i^* > 0$, $i = 1, \cdots, n$, and (1) leads to a linear system for the nontrivial equilibrium populations N_i^*. Strobeck [12] uses algebraic arguments to derive determinant conditions for these requirements (in the case of n species) to be satisfied. These results are well-known for two species; see Pielou [10] and references contained therein.

The above model treats time as a continuous variable, but some corresponding discrete systems have also been used to discuss population growth; work on these models is described in May [7] and Maynard Smith [9]. Discrete models are biologically more realistic in situations when reproduction is seasonal, and growth occurs at discrete intervals. Both continuous time variation, giving rise to a differential equation system;

* Received by the editors November 22, 1974, and in final revised form April 20, 1977.
† University of Victoria, Victoria, British Columbia, Canada V8W 2Y2.

and discrete time variation, giving rise to a difference equation system, are considered in this note. The stability of equilibrium points for both systems is analyzed by Liapunov matrix methods. Each system can also be solved by a more straightforward method (as indicated later in this note), but the Liapunov matrix method works for both problems and is a reasonable method for the discrete system. Conditions are given for the two species Lotka–Volterra competition model, and also for the two species predator-prey model, but the methods used can be extended to more species.

Consider first the continuous model given by (1) with $n = 2$, that is the two species competition equations. The nontrivial equilibrium populations are

(2) $N_1^* = (K_1 - \alpha_{12}K_2)/\alpha$, $N_2^* = (K_2 - \alpha_{21}K_1)/\alpha$,   where $\alpha = 1 - \alpha_{12}\alpha_{21}$.

Here it is assumed that the denominator is nonzero, and the equilibrium populations are required to be positive. In order to consider the stability of this equilibrium let $N_i = N_i^*(1 + n_i)$, $i = 1, 2$. On substituting in (1) with $n = 2$ and linearizing in n_i the following system results:

(3) $$\frac{d}{dt}\begin{bmatrix} n_1 \\ n_2 \end{bmatrix} = \begin{bmatrix} -x_1 & -x_1\alpha_{12}N_2^*/N_1^* \\ -x_2\alpha_{21}N_1^*/N_2^* & -x_2 \end{bmatrix}\begin{bmatrix} n_1 \\ n_2 \end{bmatrix},$$

where $x_1 = r_1 N_1^*/K_1$, $x_2 = r_2 N_2^*/K_2$.

For a nonderogatory coefficient matrix, that is a matrix for which the minimal and characteristic polynomials are identical, this system can be written in companion matrix form as

(4) $$\frac{dy}{dt} = Cy, \quad c_{11} = 0, \quad c_{12} = 1, \quad c_{21} = -x_1 x_2 \alpha, \quad c_{22} = -(x_1 + x_2).$$

This can be seen by forming a second order differential equation from system (3), or by using a linear transformation so that the coefficient matrix and C are similar; and is a convenient form for calculations.

Liapunov's theorem is used to determine when the eigenvalues of C have negative real parts, that is C is stable; for discussions of this theory see, for example, Barnett and Storey [1], Davies and James [3], Hahn [5]. Let Π be the class of positive definite symmetric matrices. Liapunov's theorem states that matrix C is stable iff there exists $G \in \Pi$ such that $C'G + GC = -I$, where C' is the transpose of C and I is the identity matrix. This is equivalent to choosing a positive definite Liapunov function $V = y'Gy$ such that for this system $dV/dt = -y'Iy$, a negative definite function. Carrying out the calculation for C given by (4) gives

$$G = -\frac{1}{2c_{21}c_{22}}\begin{bmatrix} (-c_{22}^2 + c_{21} - c_{21}^2) & c_{22} \\ c_{22} & (c_{21} - 1) \end{bmatrix}.$$

Sylvester's conditions give $G \in \Pi$ iff $c_{21} < 0$ and $c_{22} < 0$, which for (4) give $x_1 x_2 \alpha > 0$ and $x_1 + x_2 > 0$. Note that these can also be obtained straightforwardly from (3) by using the fact that the eigenvalues of the matrix all have negative real parts iff the trace of the matrix is negative and the determinant of the matrix is positive. These inequalities are satisfied for positive populations iff

(5) $\alpha > 0$,   that is, $\alpha_{12}\alpha_{21} < 1$,

the well-known condition for two species that interspecific competition is less than intraspecific. Species satisfying this inequality are called alpha-compatible (Riebesell [11]). This is a necessary but not sufficient condition for stability, as in addition the equilibrium values must be positive; thus (2) requires that $K_1 > \alpha_{12}K_2$ and $K_2 > \alpha_{21}K_1$.

A discrete analogue to the above linear continuous model, with the same equilibrium populations given by (2), is

(6)
$$\begin{bmatrix} n_1(t+1) \\ n_2(t+1) \end{bmatrix} = \begin{bmatrix} 1-x_1 & -x_1\alpha_{12}N_2^*/N_1^* \\ -x_2\alpha_{21}N_1^*/N_2^* & 1-x_2 \end{bmatrix} \begin{bmatrix} n_1(t) \\ n_2(t) \end{bmatrix}.$$

This system can also be written in companion matrix form as

$$y(t+1) = By(t), \quad b_{11} = 0, \quad b_{12} = 1,$$

(7)
$$b_{21} = -x_1x_2\alpha + x_1 + x_2 - 1,$$

$$b_{22} = 2 - (x_1 + x_2).$$

A Liapunov function can be chosen for this system (Hahn [5]) as $V(t) = y'Hy$, $H \in \Pi$, giving $V(t+1) - V(t) = -y'Iy$, where $B'HB - H = -I$. Carrying out this calculation for the matrix B in (7) gives

(8)
$$H = \frac{1}{\Delta} \begin{bmatrix} b_{22}^2(1+b_{21}) - (1-b_{21})(1+b_{21}^2) & -2b_{21}b_{22} \\ -2b_{21}b_{22} & 2(b_{21}-1) \end{bmatrix},$$

where $\Delta = (1+b_{21})(b_{22}+b_{21}-1)(b_{22}-b_{21}+1)$. Sylvester's conditions give $H \in \Pi$ iff $1+b_{21} > 0$, $1-b_{21}+b_{22} > 0$, $1-b_{21}-b_{22} > 0$. For (7), these give

(9.1)
$$\left(x_1 - \frac{1}{\alpha}\right)\left(x_2 - \frac{1}{\alpha}\right) < \frac{1}{\alpha^2},$$

(9.2)
$$\left(x_1 - \frac{2}{\alpha}\right)\left(x_2 - \frac{2}{\alpha}\right) > \frac{4}{\alpha^2} - \frac{4}{\alpha},$$

(9.3)
$$x_1x_2\alpha > 0.$$

For positive equilibrium populations this last inequality reduces to the alpha-compatibility condition for the continuous model. But in addition the discrete model requires for stability that the first two inequalities of (9) are satisfied, conditions which limit the magnitudes of the intrinsic growth rates. May [8] has recently derived similar inequalities for another difference equation analogue of the Lotka–Volterra competition model. As noted earlier (9.3) is the alpha-compatibility condition, so for the competition model under consideration $\alpha \in (0, 1)$. Taking these inequalities together the stability region reduces to $\{(x_1, x_2): 0 < x_1 < 2, 0 < x_2 < 2(2-x_1)/(2-\alpha x_1)\}$, which is shaded in Fig. 1(a).

Brauer [2] has analyzed a two species predator-prey model which is similar to the above continuous time model except that α_{21} is now negative as an increase in species one (the prey) produces an increase in species two (the predator). The method used above can be applied directly to this model and shows that if $K_1 > \alpha_{12}K_2$ then the equilibrium given by (2) with $\alpha_{21} < 0$ is stable. Here the system is trivially alpha-compatible.

The discrete analogue of this predator-prey model can be treated in the same way as the competition model above and inequalities (9) are unaltered except that $\alpha > 1$, and (9.3) is trivially satisfied. The other two inequalities place additional restrictions on the parameters for a stable equilibirium, again showing the same feature of continuous versus discrete models. Inequalities (9.1) and (9.2) now interact giving a stability region as sketched in Fig. 1(b), the stability region is shaded.

Maynard Smith [9] considers a discrete predator-prey model in which it is assumed that the number of offspring produced by each predator is proportional to the number of prey killed by that predator. This model can be obtained as a limiting case of that analyzed above, where in terms of Maynard Smith's parameters R and r, the maximum reproductive rate of prey and predator respectively,

$$x_1 x_2 \alpha_{12} \alpha_{21} = -(R-1)(r-1)/r, \qquad x_1 = (R-1)/r, \quad x_2 = 0,$$

Inequalities (9) then agree with Maynard Smith's results.

It should be noted that for one species obeying the discrete logistic equation, the inequalities (9) (with $\alpha = 1$) reduce to $N_1^* = K_1$, $r_1 \in (0, 2)$ as the well-known stability criterion. The discrete analogue (6) corresponds to replacing the derivative by the forward difference operator with unit step size. If arbitrary step size $h > 0$ is used, the result is modified to $r_1 \in (0, 2/h)$ (May [7]). The continuous logistic model is stable for $r_1 \in (0, \infty)$; thus the forward difference analogue requires a more stringent condition for stability. May bases his arguments on determinant relations for the differential and difference equation systems. For a unit step size he shows that the analogous forward difference system is stable when $|\lambda + 1| < 1$ where λ are the eigenvalues of the matrix of the differential system. Applying this result to system (3) or (4), and using Theorem 4.2 of Goldberg [4] or Lemma 1 of Levin [6], gives an alternative way to derive inequalities (9) for the difference system. These can also be derived by means of a bilinear mapping taking the left half plane onto the unit circle, see Taussky [13], van den Driessche [14].

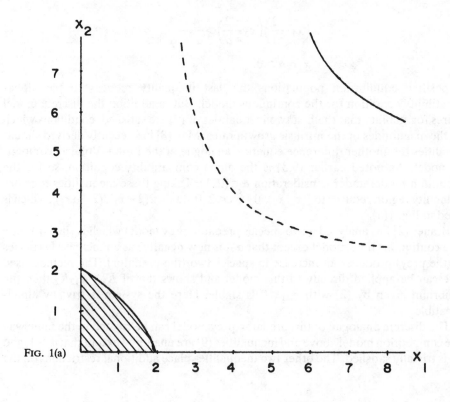

FIG. 1(a)

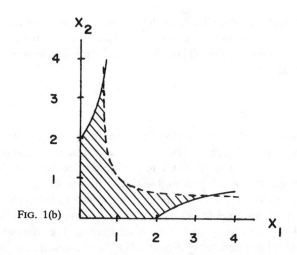

Fig. 1(b)

FIG. 1. *Stability regions in x_1, x_2 plane for discrete models, $x_1 = (r_1/K_1)N_1^*$, $x_2 = (r_2/K_2)N_2^*$, $\alpha = 1 - \alpha_{12}\alpha_{21}$. Broken curves are $x_2 = x_1/(\alpha x_1 - 1)$ for (9.1), solid curves are $x_2 = 2(x_1-2)/(\alpha x_1 - 2)$ for (9.2). Fig. 1(a) is competition model with $\alpha = \frac{1}{2} < 1$, Fig. 1(b) is predator-prey model with $\alpha = 2 > 1$.*

REFERENCES

[1] S. BARNETT AND C. STOREY, *Matrix Methods in Stability Theory*, Thomas Nelson, London, 1970.
[2] F. BRAUER, *A nonlinear predator-prey problem*, Ordinary Differential Equations, 1971 NRL-MRC Conference, L. Weiss, ed., Academic Press, New York, 1972, pp. 371–377.
[3] T. V. DAVIES AND E. M. JAMES, *Nonlinear Differential Equations*, Addison-Wesley, New York, 1966.
[4] S. GOLDBERG, *Introduction to Difference Equations*, John Wiley, New York, 1958.
[5] W. HAHN, *Stability of Motion*, Springer-Verlag, Berlin, 1967.
[6] S. A. LEVIN, *A mathematical analysis of the genetic feedback mechanism*, Amer. Natur., 106 (1972), pp. 145–164.
[7] R. M. MAY, *Stability and Complexity in Model Ecosystems*, Princeton University Press, Princeton, NJ, 1973.
[8] ———, *Biological populations with nonoverlapping generations: Stable points, stable cycles, and chaos*, Science, 186 (1974), pp. 645–647.
[9] J. MAYNARD SMITH, *Mathematical Ideas in Biology*, Cambridge University Press, London, 1968.
[10] E. C. PIELOU, *An Introduction to Mathematical Ecology*, John Wiley, New York, 1969.
[11] J. F. RIEBESELL, *Paradox of enrichment in competitive systems*, Ecology, 55 (1974), pp. 183–187.
[12] C. STROBECK, *N-species competition*, Ibid., 54 (1973), pp. 650–654.
[13] O. TAUSSKY, *Matrices C with $C^n \to 0$*, J. Algebra, 1 (1964), pp. 5–10.
[14] P. VAN DEN DRIESSCHE, *Stability of linear population models*, J. Theoret. Biol., 47 (1974), pp. 473–476.

POPULATION GROWTH IN A CLOSED SYSTEM*

ROBERT D. SMALL†

Abstract. Volterra's model for population growth in a closed system includes an integral term to indicate accumulated toxicity in addition to the usual terms of the logistic equation. Solutions to this equation are studied by means of singular perturbation techniques. Two major types of systems are discovered, those with populations immediately sensitive to toxins and those that succumb to toxins only after long times.

*Received by the editors March 5, 1981, and in revised form March 18, 1982.

†Department of Mathematics, University of New Brunswick, Box 4400, Fredericton, New Brunswick, E3B 5A3 Canada.

This note deals with a mathematical model of the accumulated effect of toxins on a population living in a closed system. Scudo [1] indicates in his review that Volterra proposed this model for a population $u(t)$ of identical individuals which exhibits crowding and sensitivity to "total metabolism":

$$(1) \qquad \frac{du}{dt} = au - bu^2 - cu \int_0^t u(\tau)d\tau.$$

If the integral term on the right is missing we have the well-known logistic equation with birth rate a and crowding coefficient b. The last term contains the integral that indicates the "total metabolism" or total amount of toxins produced since time zero. The individual death rate is proportional to this integral, and so the population death rate due to toxicity must include a factor u. The presence of the toxic term due to the system being closed always causes the population level to fall to zero in the long run, as will be seen shortly. The relative size of the sensitivity to toxins, c, determines the manner in which the population evolves before its fated decay. Equation (1) can be solved in quasi-closed form, the result of which is not very insightful. If we set $v = \int u(\tau)d\tau$, then we can eliminate t in favor of u and v and separate variables obtaining an inverse integral form for v.

A much more informative approach involves singular perturbation techniques which, in several cases, give good closed form approximations to the solutions. In order to establish the size of the solution and the time scale, we scale out the parameters of (1) as much as possible. There are four different ways to do this, each having its own importance. The scaled equations are

$$(2) \qquad \frac{du_1}{dt_1} = u_1 - u_1^2 - \frac{c}{ab} u_1 \int_0^{t_1} u_1(\tau_1)d\tau_1, \qquad \left[u_1(t_1) = \frac{b}{a} u\left(\frac{t_1}{a}\right) \right],$$

$$(3) \qquad \frac{c}{ab} \frac{du_2}{dt_2} = u_2 - u_2^2 - u_2 \int_0^{t_2} u_2(\tau_2)d\tau_2, \qquad \left[u_2(t_2) = \frac{b}{a} u\left(\frac{b}{c} t_2\right) \right],$$

$$(4) \qquad \frac{du_3}{dt_3} = u_3 - \frac{ab}{c} u_3^2 - u_3 \int_0^{t_3} u_3(\tau_3)d\tau_3, \qquad \left[u_3(t_3) = \frac{c}{a^2} u\left(\frac{t_3}{a}\right) \right],$$

$$(5) \qquad \frac{du_4}{dt_4} = \frac{ab}{c} u_4 - u_4^2 - u_4 \int_0^{t_4} u_4(\tau_4)d\tau_4, \qquad \left[u_4(t_4) = \frac{b^2}{c} u\left(\frac{b}{c} t_4\right) \right].$$

We consider the two cases c/ab small and c/ab large, with the case between being inferred from the former cases.

Case of c/ab small. This type of population is relatively insensitive to toxins. Initially, the population growth is limited primarily by crowding effects, the toxin build up dominating only in the long run. Equations (2) and (3) are the relevant ones here, the solutions u_1 and u_2 being identical but $t_2 = (c/ab)t_1$ being a slow time compared with t_1. If we neglect the small term of (2), the approximate solution is

$$(6) \qquad u_1 = \frac{1}{1 + (1/u_{10} - 1)e^{-t_1}},$$

where u_{10} is the value of u_1 at $t_1 = 0$. Neglecting the small term of (3), we obtain

$$\frac{du_2}{dt_2} = -u_2$$

and the initial condition $u_2(0) = 0$ or 1, the nontrivial solution being

(7) $$u_2 = e^{-t_2}.$$

Solution (6) provides the correct initial value but cannot remain valid over long times because for large t_1, du_1/dt_1 goes to zero and converts (2) into (3). We must start with (6) and then when the amplitude reaches a value close to 1, switch to (7). The amplitudes match because when t_1 is large relative to 1 so that $u_1 \cong 1$, t_2 is still of order c/ab and $u_2 \cong 1$. Thus we have a rapid rise along the logistic curve and then a slow exponential decay to zero. Back in the original variables we can obtain a uniformly valid solution by adding the two solutions and subtracting the common part. For c/ab small we obtain

(8) $$u(t) \cong \frac{a}{b}\left\{\left[1 + \left(\frac{a}{bu_0} - 1\right)e^{-at}\right]^{-1} + e^{-ct/b} - 1\right\}.$$

Case of c/ab large. Populations of this type are extremely sensitive to toxins. Their growth is immediately checked by toxic accumulations. Here the relevant equations are (4) and (5). Neglecting the small term in (4), we have the solution

(9) $$u_3 = \left(u_{30} + \frac{1}{2}\right)\text{sech}^2\left[\frac{\sqrt{2u_{30} + 1}}{2}t_3 - \tanh^{-1}((2u_{30} + 1)^{-1/2})\right]$$

found by trying solutions proportional to sech^2. If u_{30} is small compared to c/ab, this

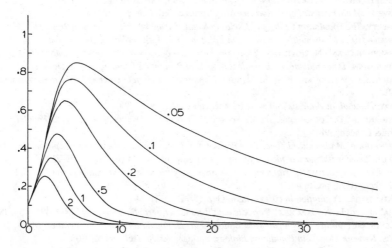

FIG. 1. *Numerical solution of $du/dt = u - u^2 - cu \int_0^t u(\tau)d\tau$ for $u(0) = 0.1$. The curves are demarked by their value of c.*

solution remains small and (4) remains valid. In the original variables we have the solution, valid for c/ab large and u_0 small compared to a/b,

(10) $$u(t) = \left(u_0 + \frac{a^2}{2c}\right)\text{sech}^2\left[\sqrt{2cu_0 + a^2}\,\frac{t}{2} - \tanh^{-1}\left(\frac{a}{\sqrt{2cu_0 + a^2}}\right)\right].$$

For u_0 large compared to a/b we have (5) which, while it cannot be solved in closed form, clearly exhibits decay of its solutions on a time scale, $t_4 = (c/ab)t_3$, fast compared to t_3.

Initially large amplitudes decay quickly until the solution joins solution (10).

Figure 1 illustrates the two major cases with the larger amplitude curves being the logistic curve with slow decay and the smaller amplitude ones being sech^2.

REFERENCE

[1] F. M. SCUDO, *Vito Volterra and theoretical ecology*, Theoret. Population Biol., 2 (1971), pp. 1–23.

Supplementary References
Population Models

[1] W. B. ARTHUR, *"Why a population converges to stability,"* AMM (1981) pp. 557–563.

[2] N. BAILEY, *The Mathematical Theory of Epidemics*, Hafner, New York, 1957.

[3] M. S. BARTLETT AND R. W. HIORNS, eds., *The Mathematical Theory of The Dynamics of Biological Populations*, Academic, London, 1973.

[4] M. BARTLETT, *Stochastic Population Models*, Methuen, London, 1960.

[5] A. T. BERGERUD, *"Prey switching in a simple ecosystem,"* Sci. Amer. (1983) pp. 130–141.

[6] L. L. CAVALLI-SFORZA AND W. A. BODMER, *The Genetics of Human Populations*, Freeman, San Francisco, 1971.

[7] A. J. COALE, *The Growth and Structure of Human Populations: A Mathematical Investigation*, Princeton University Press, Princeton, 1972.

[8] G. R. CONWAY, *"Mathematical models in applied ecology,"* Nature (1977) pp. 291–297.

[9] J. F. CROW AND M. KIMURA, *An Introduction to Population Genetics Theory*, Harper & Row, N.Y., 1970.

[10] G. D. ELSETH AND K. D. BAUMGARDNER, *Population Biology*, Van Nostrand, N.J., 1981.

[11] W. J. EWENS, *Mathematical Population Genetics*, Springer-Verlag, N.Y., 1979.

[12] J. P. FINERTY, *The Population Ecology of Cycles in Small Mammals*, Yale University Press, New Haven, 1980.

[13] J. C. FRAUENTHAL, *Mathematical Modeling in Epidemiology*, Springer-Verlag, N.Y., 1980.

[14] J. C. FRAUENTHAL AND N. GOLDMAN, *"Demographic dating of the Nukuaro society,"* AMM (1977) pp. 613–618.

[15] H. I. FREEDMAN, *Deterministic Mathematical Models in Population Ecology*, Marcel Dekker, N.Y., 1980.

[16] N. S. GOEL, S. C. MAITRA AND E. W. MONTROLL, *Nonlinear Models in Interacting Populations*, Academic, N.Y., 1971.

[17] L. HENRY, *Population: Analysis and Models*, Academic, N.Y., 1977.

[18] R. W. HIORNS AND D. COOKE, eds., *The Mathematical Theory of the Dynamics of Biological Populations II*, Academic, London, 1981.

[19] F. HOPPENSTEADT, *Mathematical Theories of Populations: Demographics, Genetics and Epidemics*, SIAM, 1975.

[20] S. KARLIN, *"Some mathematical models of population genetics,"* AMM (1972) pp. 699–739.

[21] S. KARLIN AND M. FELDMAN, *"Mathematical genetics: A hybrid seed for educators to sow,"* Int. J. Math. Educ. Sci. Tech. (1972) pp. 169–189.

[22] C. W. KILMISTER, *"Population in cities,"* Math. Gaz. (1976) pp. 11–24.

[23] A. C. LAZER AND D. A. SANCHEZ, *"Periodic equilibria under periodic harvesting,"* MM (1984) pp. 156–158.

[24] S. A. LEVIN, ed., *Studies in Mathematical Biology, Part II: Populations and Communities*, MAA, 1978.

[25] E. LEWIS, *Network Models in Population Biology*, Springer-Verlag, Heidelberg, 1977.

[26] D. LUDWIG, *Stochastic Population Theories*, Springer-Verlag, Heidelberg, 1974.

[27] R. MAY, *"Biological populations obeying difference equations: stable points, stable cycles and chaos,"* J. Theor. Biol. (1975) pp. 511–524.

[28] ———, *"Simple mathematical models with very complicated dynamics,"* Nature (1976) pp. 459–467.

[29] R. M. MAY AND G. F. OSTER, *"Bifurcations and dynamic complexity in simple ecological models,"* Amer. Naturalist (1976) pp. 573–599.

[30] R. M. MAY, et al., *"Management of multispecies fisheries,"* Science (1979) pp. 267–277.

[31] B. PARTLETT, *"Ergodic properties of populations I: The one-sex model,"* Theor. Population Bio. (1970) pp. 191–207.

[32] E. C. PIELOU, *Mathematical Ecology*, Wiley, N.Y., 1977.

[33] J. H. POLLARD, *Mathematical Models for the Growth of Human Populations*, Cambridge University Press, Cambridge, 1973.

[34] G. F. RAGGETT, *"Modelling the Eyam plague,"* BIMA (1982) pp. 221–226.

[35] D. A. SMITH, *"Human population growth: Stability or explosion,"* MM (1977) pp. 186–197.

[36] J. M. SMITH, *Models in Ecology*, Cambridge University Press, Cambridge, 1974.

[37] P. D. STRAFFIN, JR., *"Periodic points of continuous functions,"* MM (1978) pp. 99–105.

[38] M. WILLIAMSON, *The Analysis of Biological Populations*, Arnold, London, 1972.

2. Medical Models

MEDICAL APPLICATION OF FOURIER ANALYSIS*

D. E. RAESIDE,† W. K. CHU† AND P. A. N. CHANDRARATNA†

Abstract. An example of a successful application of Fourier analysis to the classification of medical echocardiograms is discussed and illustrated.

Ultrasound provides a means for the noninvasive examination of the interior of the human heart. In this technique a short burst of ultrasound is transmitted through the chest wall and "echoes" are received from underlying tissue interfaces which separate regions of differing acoustic impedance. By measuring the time which elapses between transmission of the ultrasound and reception of an echo, and assuming a particular value for the speed of sound in tissue, it is possible to locate the structures within the heart and, by many repetitions of the transmission-reception sequence, to quantify their temporal behavior. An echocardiogram is the graphical presentation of such data. Figure 1 shows a typical normal echocardiogram. Each of the several waveforms which appear in it is associated with the motion of a particular cardiac structure. During an echocardiographic examination, the patient is usually in the supine position, with the ultrasound transducer (acting both as the transmitter and the receiver) placed on the chest wall slightly lateral to the left sternal border and in the fourth intercostal space (the procedure is illustrated in Fig. 2). Figure 3 presents the heart cross section showing cardiac structures which the ultrasonic beam passes through.

This note describes an application of pattern recognition methodology to the following four-class problem: classification of the normal, mitral stenosis, mitral valve prolapse and idiopathic hypertrophic subaortic stenosis cardiac conditions using echocardiogram analysis. Since none of the presently utilized clinical criteria pos-

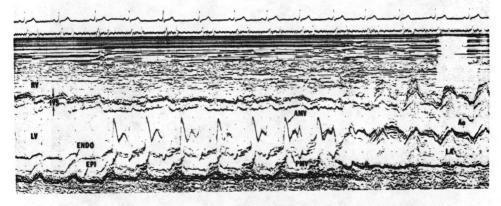

FIG. 1. *A typical, normal echocardiogram (AO = Aortic Root, LA = Left Atrium, AMV = Anterior Mitral Valve, PMV = Posterior Mitral Valve, IVS = Interventricular Septum, ENDO = Endocardium, EPI = Epicardium).*

* Received by the editors September 5, 1977.

† Department of Radiological Sciences and Division of Cardiology, University of Oklahoma Health Sciences Center, Oklahoma City, Oklahoma 73190.

sessed the capability of classifying waveforms for this four-class problem, it was decided to try Fourier analysis. The results for the classification of anterior mitral leaflet waveforms were excellent. In addition, Fourier analysis proved to be equally good for the classification of aortic root and left ventricular wall waveforms. The anterior mitral leaflet results are described below.

If $x(t)$ denotes the time-varying amplitude of an echocardiogram waveform, then the finite Fourier series representation of $x(t)$ can be expressed satisfactorily as

$$x(t) = \sqrt{C_0} + \sum_{i=1}^{N} \sqrt{C_n} \sin(2\pi n f_0 t + \theta_n) + e(t)$$

where f_0, the fundamental frequency of the series, is the reciprocal of the heart period.

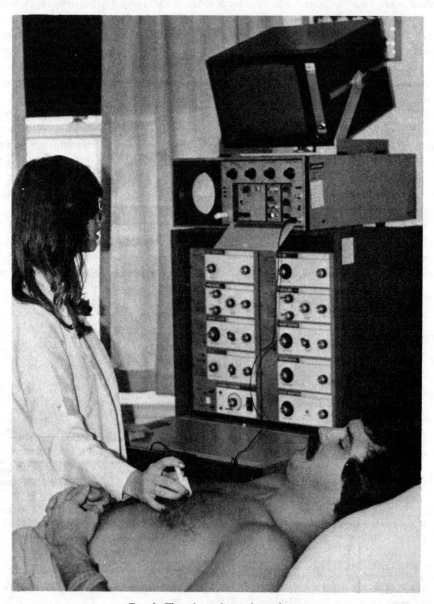

FIG. 2. *The echocardiographic technique.*

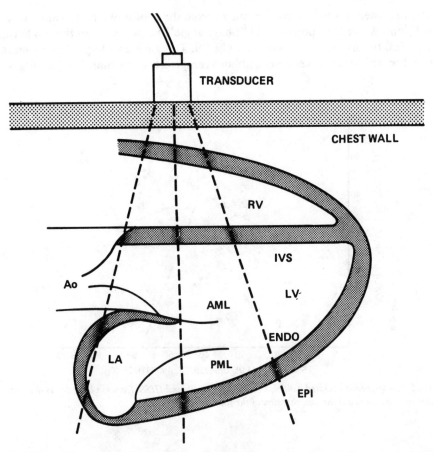

FIG. 3. *A cross section of the heart showing cardiac structures as the ultrasonic beam sweeps through the heart (RV = Right Ventricle, AO = Aortic Root, IVS = Interventricular Septum, LA = Left Atrium, AML = Anterior Mitral Leaflet, PML = Posterior Mitral Leaflet, LV = Left Ventricle, ENDO = Endocardium, EPI = Epicardium).*

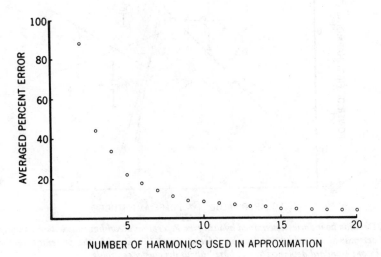

FIG. 4. *Approximation error versus the number of harmonics used in reconstruction.*

The term $e(t)$ denotes the "error" resulting from the choice of a finite value of N. The terms C_n and θ_n are the "power" and "phase angle", respectively, of the nth harmonic of the series. In this note N was taken to be 20. This gave an adequate representation of the echocardiogram waveforms with an average error less than 5% (see Fig. 4).

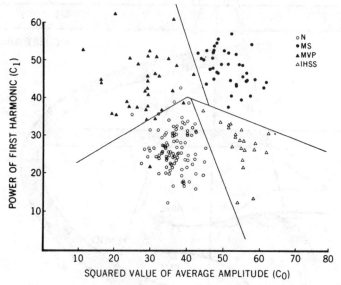

FIG. 5. *Scattergram associated with classes, N, MS, MVP, and IHSS. Decision boundaries determined by a human were drawn to aid the visualization of the four clusters.*

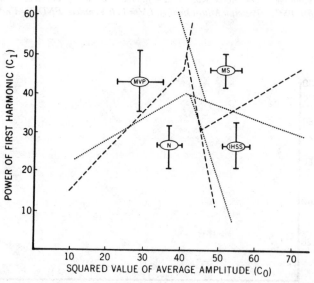

FIG. 6. *Decision boundaries determined by using the Perceptron algorithm (dashed lines) compared to the boundaries determined visually by a human (dotted lines). To simplify the presentation, the data are represented by one standard deviation "error bars" about the cluster centroids.*

To characterize the anterior mitral leaflet waveform, the coefficients $C_0, C_1, \cdots,$ C_{20} were utilized to form a 21-component feature vector for each of the 194 files in a data base consisting of 101 normal subjects (class N), 39 patients with mitral stenosis (class MS), 31 patients with mitral valve prolapse (class MVP) and 21 patients with idiopathic hypertrophic subaortic stenosis (class IHSS). Because of the complexity of dealing with vectors of high dimensionality, feature ranking was carried out to establish criteria which would allow the elimination of the less informative components of the feature vector (see any text on pattern recognition for a description of feature ranking, for example [1, Chap. 7]). This made it possible to truncate the feature vector summarizing the information contained in the anterior mitral leaflet waveform so as to include just the components C_0 and C_1. Figure 5 demonstrates the ability of the features C_0 and C_1 to discriminate between the classes N, MS, MVP, and IHSS, and Fig. 6 shows decision boundaries determined by the Perceptron algorithm [1, pp. 158–168] together with a comparison of these boundaries with those drawn by a human.

Acknowledgment. The authors would like to express their gratitude to Dr. R. E. D. Brown and Mr. H. Poehlmann for their interest in the study described here.

REFERENCE

[1] J. T. TOU AND R. C. GONZALEZ, *Pattern Recognition Principles*, Addison-Wesley, Reading, MA, 1974.

Estimation of the Median Effective Dose in Bioassay

Problem 81-12, by D. E. RAESIDE (University of Oklahoma, Health Sciences Center).

Consider a system subjected to a single stimulus intensity (dose level) x, and assume that the system responds in just one of two ways, "positively" or "negatively". Let the probability of a positive response be denoted by π, and assume that π is related to x through the quantal logistic dose response function

$$\pi = [1 + e^{-\beta(x-\gamma)}]^{-1}.$$

Use Bayesian estimation theory with a catenary loss function and a conjugate prior to determine an optimal estimate of the parameter γ (the parameter β is assumed known). Show that for certain parameter values and for large numbers of positive and negative responses this Bayes estimator is approximately the maximum likelihood estimator. See the paper *Optimal Bayesian sequential estimation of the median effective dose* by P. R. Freeman (Biometrika, 57 (1970), pp. 79–89) for a treatment of this problem using a quadratic loss function. See the paper *A class of loss function of catenary form* by D. E. Raeside and R. J. Owen (J. Statist. Phys., 7 (1973), pp. 189–195) for a summary of properties of the catenary loss function.

Solution by P. W. JONES *and* P. SMITH (Keele University, Staffordshire).

For a single dose level x let the number of responses in n' trials be r'. The likelihood is then

$$\frac{e^{r'\beta(x-\gamma)}}{[1 + e^{\beta(x-\gamma)}]^{n'}}, \qquad r' \leqq n'.$$

Suppose that γ is assigned a natural conjugate prior distribution with density

(1) $$f(\gamma) \alpha \frac{e^{r_0 \beta (x-\gamma)}}{[1 + e^{\beta(x-\gamma)}]^{n_0}}, \qquad 1 \le r_0 \le n_0, \quad n_0 \ge 2,$$

involving parameters r_0, n_0. Then by Bayes' theorem the posterior distribution of γ is (1) with (r_0, n_0) replaced by (r, n) where $r = r_0 + r'$ and $n = n_0 + n'$.

We require the minimum with respect to d of the integral

$$I = K \int_{-\infty}^{\infty} \frac{e^{r\beta(x-\gamma)} \cosh (a(\gamma - d)) \, d\gamma}{(1 + e^{\beta(x-\gamma)})^n} - 1,$$

where

$$\frac{1}{K} = \int_{-\infty}^{\infty} \frac{e^{r\beta(x-\gamma)} \, d\gamma}{(1 + e^{\beta(x-\gamma)})^n}.$$

This gives the Bayes estimator d^* of γ with respect to the catenary loss function $\cosh [a(\gamma - d)] - 1$, where a is a positive constant.

The integral can be written

(2) $$I = F_n(a, \beta, r, x)e^{-ad} + F_n(-a, \beta, r, x)e^{ad} - 1,$$

with

(3) $$F_n(a, \beta, r, x) = \frac{1}{2} K e^{r\beta x} \int_{-\infty}^{\infty} \frac{e^{\gamma(a - r\beta)} \, d\gamma}{(1 + e^{\beta(x-\gamma)})^n}.$$

The minimizer of (2) is

(4) $$d = d^* = \frac{1}{2a} \ln [F_n(a, \beta, r, x)/F_n(-a, \beta, r, x)].$$

The substitution $e^{-\beta\gamma} = u$ transforms (3) into

$$F_n(a, \beta, r, x) = \frac{K}{2\beta} e^{r\beta x} \int_0^{\infty} u^{r-1-a/\beta} (1 + e^{\beta x} u)^{-n} \, du$$

$$= \frac{K}{2\beta} e^{ax} B\left(r - \frac{a}{\beta}, n - r + \frac{a}{\beta}\right), \qquad r > \frac{a}{\beta},$$

from [1, p. 285]. Express the beta function $B(\cdot, \cdot)$ in terms of gamma functions ([1], p. 950) and substitute for F_n in (4):

(5) $$d^* = \frac{1}{2a} \ln \left[\frac{e^{2ax} \Gamma(r - a/\beta) \Gamma(n - r + a/\beta)}{\Gamma(r + a/\beta) \Gamma(n - r - a/\beta)}\right],$$

with (Bayes) risk given by substituting $d = d^*$ in (2). In order to obtain the asymptotic behavior of (5) for large n we require

$$\ln \Gamma\left(n - r + \frac{a}{\beta}\right) = n \ln n - n - \left(r - \frac{a}{\beta} + \frac{1}{2}\right) \ln n + \ln \sqrt{2\pi} + O\left(\frac{1}{n}\right),$$

which has been adapted from the asymptotic expansion of the logarithm of the gamma function given in [1, p. 940]. With a similar expression for $\ln \Gamma(n - r - a/\beta)$, the expansion for $d*$ becomes

$$d* = \frac{1}{\beta} \ln n + x + \frac{1}{2a} \ln \left[\frac{\Gamma(r - a/\beta)}{\Gamma(r + a/\beta)} \right] + O\left(\frac{1}{n}\right)$$

as $n \to \infty$ (n_0 finite). The parameter values must satisfy $r > a/\beta$.

The maximum likelihood estimator of γ is given by $x + \ln (n'/r' - 1)/\beta$, which has the same leading term as $d*$ for large n'.

REFERENCE

[1] I. S. GRADSHTEYN AND I. M. RYZHIK (1965), *Tables of Integrals, Series and Products*, Academic Press, New York.

A MODEL FOR DRUG CONCENTRATION*

L. E. THOMAS†

Abstract. A model, suitable for classroom use, is developed to describe the concentration of a drug in the body when the drug is administered repeatedly. The model shows that under certain conditions there are limiting values for the maximum and minimum concentrations of the drug in the system. Cases of random detoxification and enzymatic detoxification are discussed.

1. The model. In this note we show how a simple differential equations model can be used to predict the concentration of a drug in the body under certain circumstances. The results are generalizations of those presented by Rustagi [2]; the method of solution used was suggested in § 5 of [3].

We are interested in what happens to the concentration of a drug in the body when the drug is introduced into the body repeatedly. Will repeated application of the drug cause the concentration to become too large eventually? How should the time interval between applications be chosen so that the concentration does not become so large as to be dangerous or so small as to be ineffective?

We will answer these questions on the basis of the following assumptions: If $y(t)$ is the concentration of the drug at time t, we assume that (i) the rate of destruction of the drug depends only upon the amount of drug present, i.e., $y' = f(y)$. Appropriate forms for the function f will be considered later, but it makes sense to assume that (ii) f is a continuous, nonpositive, monotone decreasing function with $f(0) = 0$. At time $t = 0, T, 2T, \cdots$ a dose of the drug is administered in such a way that the concentration of the drug is increased by an amount d. We assume that (iii) the concentration is immediately increased throughout the body when the dose is administered.

Under these assumptions, there is a sequence of initial value problems which describes the situation. The important features of this sequence can be seen in the accompanying figure. Let $y_n(t)$ be the concentration of the drug on the interval

---

* Received by the editors October 20, 1975, and in revised form October 27, 1976.
† Department of Mathematics, Saint Peter's College, Jersey City, New Jersey 07306.

$(n-1)T < t < nT$. When the drug is first introduced into the body at time $t = 0$, the concentration of the drug rises from 0 to d, so $y_1(0) = d$. Thereafter the concentration decreases according to the equation $y_1' = f(y_1)$ to a value $y_1(T)$ when a booster dose is given raising the concentration to $y_1(T) + d$. This value becomes the initial value of y_2: $y_2(T) = y_1(T) + d$. After this jolt, the concentration of the drug again decreases, this time according to $y_2' = f(y_2)$ until it reaches $y_2(2T)$ when another dose is given, raising the concentration to $y_2(2T) + d$. This becomes the initial condition for y_3: $y_3(2T) = y_2(2T) + d$. This process continues, giving a sequence of initial value problems, the general case being

$$(1) \qquad y_{n+1}' = f(y_{n+1}), \qquad y_{n+1}(nT) = y_n(nT) + d$$

on the interval $nT < t < (n+1)T$.

The concentration of the drug immediately after the administration of the dose at time nT is $y_n(nT) + d$. The concentration then decreases until it reaches its minimum value $y_{n+1}((n+1)T)$ just before a new dose is given. Since $y_n(nT) + d$

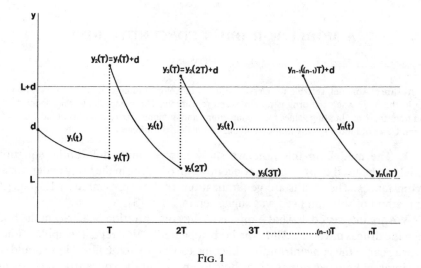

FIG. 1

and $y_{n+1}((n+1)T)$ give the maximum and minimum concentrations of the drug on the interval $nT < t < (n+1)T$, the questions raised at the beginning of this note can be answered by considering the sequence $\{y_n(nT)\}$.

In particular, if it could be shown that the sequence $\{y_n(nT)\}$ converges to a limit L, say, then we would know that the maximum concentration approaches $L + d$ and that the minimum approaches L. See Fig. 1.

2. Analysis. By separating the variables in (1) and integrating, we find

$$(2) \qquad \int_{d+x_n}^{x_{n+1}} \frac{dx}{f(x)} = \int_{nT}^{(n+1)T} dt = T,$$

where we have introduced x as a dummy variable of integration and set $x_n = y_n(nT)$ for convenience in notation.

Assume for the moment that $x_n = y_n(nT) \to L$. Then taking the limit $n \to \infty$ in

(2) we obtain

(3)
$$\int_{L+d}^{L} \frac{dx}{f(x)} = T.$$

It will be shown below (Theorem 1) that if there is a number L which satisfies equation (3), then $\{y_n(nT)\}$ does in fact converge to L. For the moment we use this result without proof to analyze two examples.

3. Examples. We will consider two explicit forms of the function f which may be appropriate under certain circumstances. The first case is that in which the concentration of the drug is reduced by random action of the system on the drug. (Such is the case with penicillin, for example.) Under random action, the rate at which the drug is destroyed is proportional to the amount of the drug present, so $f(x) = -kx, k > 0$. We can calculate the limiting maximum and minimum concentrations $L + d$ and L by using equation (3)

$$-\int_{L+d}^{L} \frac{dx}{kx} = T \quad \text{or} \quad \ln\left[\frac{L+d}{L}\right] = kT.$$

This last equation can be solved for L to obtain $L = d/(e^{kt} - 1)$. The limiting maximum concentration is $L + d = d/(1 - e^{-kT})$. This agrees with the result in [2] where the case of $f(x) = -kx$ was also discussed.

As another example, consider the case where the destruction of the drug is due to enzymatic action. In this case it is appropriate under some conditions to take $f(x) = -ax/(b + x)$, $a, b > 0$. This is the Michaelis–Menton equation which is derived in most texts on biological chemistry; see, for example, [1, pp. 225–229]. When the concentration of the drug is very small, the enzymatic case reduces to the case of random action, but when the concentration of the drug is large, as with repeated doses of an alcoholic beverage, the Michaelis–Menton equation provides a reasonable model. By putting $f(x) = -ax/(b + x)$ into (3), integrating, and solving for L, we find for the minimum and maximum concentrations

$$L = \frac{d}{e^{(aT-d)/b} - 1}, \qquad aT - d > 0,$$

and

$$d + L = \frac{d}{1 - e^{(d-aT)/b}}, \qquad aT - d > 0.$$

It is interesting to note that a solution exists in this case only if $d < aT$. If $d > aT$, then either the dosage is too large or the time interval too short (or both) for the concentration to be reduced sufficiently between applications of the drug for a limit to be established.

We note in passing that in both of these examples, the limiting concentration L increases if d increases and decreases if T increases. These results, which are not surprising, are consequences of Theorem 2 below.

4. Theorems.

THEOREM 1. *If f is a continuous, monotone decreasing, nonpositive function with $f(0) = 0$ which is such that the equation*

(4)
$$\int_{r+d}^{r} \frac{dx}{f(x)} = T$$

has solution r = L, then the sequence $\{x_n\}$ defined by (2) with $x_1 = y_1(T)$ converges to L.

Proof. Because of the monotonicity of f, the solution L of equation (4) is unique. We will show that $\{x_n\}$ is a monotone bounded sequence. Suppose that $x_1 < L$. Then (with $g(x) = (1/f(x))$),

$$\int_{x_2}^{L} g(x)\,dx = \int_{x_2}^{x_1+d} g(x)\,dx + \int_{x_1+d}^{L+d} g(x)\,dx$$

$$+ \int_{L+d}^{L} g(x)\,dx = \int_{x_1+d}^{L+d} g(x)\,dx < 0.$$

Thus, $x_2 < L$. Since

$$\int_{x_1+d}^{x_2} g(x)\,dx = \int_{L+d}^{L} g(x)\,dx$$

and $x_1 < L$, the monotonicity of f implies that $(x_1 + d) - x_2 < (L + d) - L$. Thus $x_1 < x_2$. The same reasoning, coupled with mathematical induction, will show that if $x_1 < L$, then $x_n < x_{n+1}$ for $n = 1, 2, 3, \cdots$. The sequence $\{x_n\}$ is therefore monotone increasing and bounded above by L. A similar argument will show that if $x_1 > L$, then $\{x_n\}$ is a monotone decreasing sequence bounded below by L. Should it happen that $x_1 = L$, then $x_n = L$, all n. In each case the sequence converges to L since the function

$$h(r) = \int_{r+d}^{r} g(x)\,dx$$

is continuous in r.

THEOREM 2. *For f as in Theorem 1, equation (4) defines r implicitly as a function of d and T. This function is monotone increasing in d (T fixed) and monotone decreasing in T (d fixed).*

Proof. The first statement of the theorem follows from the monotonicity of f. The remainder follows since $\partial r/\partial d$ is positive and $\partial r/\partial T$ is negative.

Project suggestions. Students may wish to formulate and solve other problems described by first order equations. Some which come to mind are a growing population periodically decimated by disease, a cooling body whose temperature is periodically changed drastically.

Acknowledgment. The author is indebted to the referees for their very helpful suggestions.

REFERENCES

[1] H. R. MAHLER AND E. H. CORDES, *Biological Chemistry*, Harper and Row, New York, 1966.
[2] J. S. RUSTAGI, *Mathematical models in medicine*, Internat. J. Math. Educ. Sci. Tech., 2 (1971), pp. 193–203.
[3] L. E. THOMAS AND W. E. BOYCE, *The behavior of a self-excited system acted upon by a sequence of random impulses*, J. Differential Equations, 12 (1972), pp. 438–454.

General Supplementary References
Life Sciences

[1] I. ADLER, "The consequences of contact pressure in phyllotaxis," J. Theor. Bio. (1977) pp. 29–77.

[2] H. R. ALKER, K. DEUTSCH AND A. H. STOETZEL, Mathematical Approaches to Politics, Jossey-Bass, San Francisco, 1973.

[3] D. C. ALTMAN, "Statistics in medical journals," Statistics in Medicine (1982) pp. 59–71.

[4] P. L. ANTONELLI, ed., Mathematical Essays on Growth and the Emergence of Form, University of Alberta Press, Edmonton, 1985.

[5] E. N. ATKINSON, R. BARTOSZYNSKI, B. W. BROWN AND J. R. THOMPSON, "On estimating the growth function of tumors," Math. BioSci. (1983) pp. 145–166.

[6] P. J. BACON AND D. W. MACDONALS, "To control rabies—vaccinate foxes," New Scientist (1980) pp. 640–645.

[7] N. T. J. BAILEY, The Mathematical Approach to Biology and Medicine, Wiley, N.Y., 1967.

[8] ———, Mathematics, Statistics and Systems for Health, Wiley, Chichester, 1977.

[9] ———, The Mathematical Theory of Infectious Diseases, Methuen, London, 1975.

[10] N. T. J. BAILEY, B. SENDOV AND R. TSANEV, eds., Mathematical Models in Biology and Medicine, North-Holland, Amsterdam, 1974.

[11] H. T. BANKS, Modeling and Control in the Biomedical Sciences, Springer-Verlag, N.Y., 1975.

[12] E. BASER, H. FLOHR, H. HAKEN AND A. J. MANDELL, Synergetics of the Brain, Springer-Verlag, Berlin, 1983.

[13] E. BATSCHELET, et al, "On the kinetics of lead in the body," J. Math. Biology (1979) pp. 15–23.

[14] A. C. BELL AND R. M. NEREM, eds., Advances in Bioengineering, Amer. Soc. Mech. Eng., N.Y., 1975.

[15] G. I. BELL, A. S. PERELSON AND G. H. PIMBLEY, JR., Theoretical Immunology, Dekker, N.Y., 1978.

[16] J. BERGER, W. BUHLER, R. REPGES AND P. TAUTU, eds., Mathematical Models in Medicine, Springer-Verlag, N.Y., 1976.

[17] M. BRAUN, "Probability survival distributions for cancerous tumors," Math. Modelling (1980) pp. 27–31.

[18] D. M. BURLEY, "Mathematical modelling in biology and medicine," BIMA (1979) pp. 261–266.

[19] V. CAPASSO, E. GROSSO AND S. L. PAVERI-FONTANA, Mathematics in Biology and Medicine, Springer-Verlag, N.Y., 1985.

[20] C. G. CARO, T. J. PEDLEY, R. C. SCHROTER, AND W. A. SEED, The Mechanics of Circulation, Oxford University Press, Oxford, 1978.

[21] E. R. CARSON, C. COBELLI AND L. FINKELSTEIN, The Mathematical Modeling of Metabolic and Endocrine Systems, Wiley, N.Y., 1983.

[22] D. O. CHANTER, "Mushroom initiation and Fredholm's equation," BIMA (1977) pp. 43–46.

[23] M. B. CHENOWETH, ed., Modern Inhalation Anesthetics, Springer-Verlag, Berlin, 1972.

[24] C. COBELLI AND R. N. BERGMAN, eds., Carbohydrate Metabolism: Quantitative Physiology and Mathematical Modelling, Wiley, Chichester, 1981.

[25] T. N. CORNSWEET, "Changes in the appearance of stimuli of very high luminance," Psych. Rev. (1962) pp. 257–273.

[26] J. D. COWAN, ed., Some Mathematical Questions in Biology IV, V, AMS, 1973 and 1974.

[27] J. CRONIN, Mathematics of Cell Electrophysiology, Dekker, New York, 1981.

[28] P. V. D. DRIESSCHE, Mathematical Problems in Biology, Springer-Verlag, N.Y., 1974.

[29] G. M. DUNN AND B. S. EVERITT, An Introduction to Mathematical Taxonomy, Cambridge University Press, Cambridge, 1982.

[30] M. EISEN, Mathematical Models in Cell Biology and Cancer Chemotherapy, Springer-Verlag, N.J., 1979.

[31] W. F. FORBES AND R. W. GIBBERD, "Mathematical models of carcinogenesis: A review," Math. Scientist (1984) pp. 95–110.

[32] J. C. FRAUENTHAL, Smallpox: When Should Routine Vaccination be Discontinued?, UMAP/Birkhauser, Boston, 1981.

[33] T. H. FROST, C. J. S. PETRIE AND T. A. M. SALEH, "Hydrodynamics of purging blood filled extra-corporeal circuits," BIMA (1978) pp. 205–211.

[34] Y. C. FUNG, Biomechanics: Mechanical Properties of Living Tissues, Springer-Verlag, N.Y., 1981.

[35] Y. C. FUNG, N. PERRONE AND M. ANLIKER, eds., Biomechanics: Its Foundations and Objectives, Prentice-Hall, N.J., 1972.

[36] S. A. GAUTHREAUX, ed., Animal Migration, Orientation and Navigation, Academic, N.Y., 1980.

[37] N. GOEL AND N. RICHTER-DYN, Stochastic Models in Biology, Academic Press, N.Y., 1974.

[38] H. J. GOLD, Mathematical Modeling of Biological Systems, Wiley, N.Y., 1977.

[39] B. GOODWIN, Temporal Organization in Cells, Academic Press, N.Y., 1963.

[40] J. S. GRIFFITH, Mathematical Neurobiology: An Introduction to the Mathematics of the Nervous System, Academic Press, N.Y., 1971.

[41] S. I. GROSSMAN AND J. E. TURNER, Mathematics for the Biological Sciences, Macmillan, N.Y., 1974.

[42] S. P. HASTINGS, "Some mathematical problems for neurobiology," AMM (1975) pp. 881–895.

[43] F. HEINMETS, ed., Concepts and Models in Biomathematics, Dekker, N.Y., 1969.

[44] G. T. HERMAN AND P. M. B. VITANYI, "Growth functions associated with biological development," AMM (1976) pp. 1–15.

[45] H. W. HETHCOTE, "Measles and rubella in the United States," Amer. J. Epidemiology (1983) pp. 2–13.

[46] E. C. HILDRETH, The Measurement of Visual Motion, M.I.T. Press, Cambridge, 1984.

[47] F. HOPPENSTEADT, Thresholds for Deterministic Epidemics, Springer-Verlag, N.Y., 1974.

[48] J. L. HOWLAND AND C. A. GRABE, JR., A Mathematical Approach to Biology, Heath, Lexington, 1972.

[49] M. IOSIFESCU AND P. TAUTU, Stochastic Processes and Applications in Biology and Medicine, Springer-Verlag, N.Y., 1973.

[50] N. N. JARDINE AND R. SIBSON, Mathematical Taxonomy, Wiley, N.Y., 1971.

[51] A. JOHNSSON, "Oscillatory water regulation in plants," BIMA (1976) pp. 22–26.

[52] D. S. JONES AND B. D. SLEEMAN, Differential Equations and Mathematical Biology, Allen & Unwin, London, 1983.

[53] J. N. KAUPUR, "Some problems in biomathematics," Int. J. Math. Educ. Sci. Tech. (1978) pp. 287–306.

[54] J. P. KERNEVEZ, Enzyme Mathematics, North-Holland, Amsterdam, 1980.

[55] N. KEYFITZ, Applied Mathematical Demography, Wiley, N.Y., 1977.

[56] N. KEYFITZ AND J. A. BEEKMAN, Demography Through Problems, Springer-Verlag, N.Y., 1984.

[57] J. F. C. KLINGMAN, Mathematics of Genetic Diversity, SIAM, 1980.

[58] T. KOHONEN, Self-Organization and Associative Memory, Springer-Verlag, N.Y., 1984.

[59] J. KRANZ, ed., Epidemics of Plant Diseases: Mathematical Analysis and Modeling, Springer-Verlag, N.Y., 1974.

[60] E. KUNS AND S. GRISOLIA, eds., Biochemical Regulatory Mechanisms in Eukaryotic Cells, Interscience, N.Y., 1971.

[61] S. A. LEVIN, ed., Some Mathematical Questions in Biology VI, AMS, 1974.

[62] ———, Studies in Mathematical Biology I, MAA, 1978.

[63] G. N. LEWIS, "A mathematical model for decompression," Math. Modelling (1983) pp. 489–500.

[64] H. H. LIEBERSTEIN, Mathematical Physiology, Elsevier, N.Y., 1973.

[65] D. A. McDONALD, Blood Flow in Arteries, Arnold, London, 1974.

[66] N. MacDONALD, Time Lags in Biological Models, Springer-Verlag, N.Y., 1978.

[67] D. MACHIN, "Statistical aspects of clinical trials," BIMA (1983) pp. 201–204.

[68] W. W. MAPLESON, "Diffusion in relation to the distribution of inhaled anaesthetics in the body," BIMA (1977) pp. 8–12.

[69] Mathematical Models for Environmental Problems, Proceedings of the International Conference at the University of Southampton, 1976.

[70] Mathematical Thinking in Behaviour Sciences, Readings from Scientific American, Freeman, San Francisco, 1968.

[71] D. E. MATTHEWS, ed., Mathematics and the Life Sciences, Springer-Verlag, N.Y., 1978.

[72] A. G. McKENDRICK, "Application of mathematics to medical problem," Proc. Edinburgh Math. Soc. (1926) pp. 98–130.

[73] H. MEINHARDT, Models of Biological Pattern Formation, Academic Press, N.Y., 1982.

[74] G. H. MEISTERS, "Tooth tables: Solution of a dental problem by vector algebra," MM (1982) pp. 274–280.

[75] B. G. MIRKIN AND S. N. RODIN, Graphs and Genes, Springer-Verlag, Berlin, 1984.

[76] ———, G. J. MITCHISON, "Phyllotaxis and the Fibonacci series," Science (1977) pp. 270–275.

[77] G. D. MOSTOE, ed., Mathematical Models of Cell Rearrangement, Yale University Press, New Haven, 1975.

[78] V. C. MOW AND W. M. LAI, Synovial Joint Biomechanics, SIAM Rev. (1980) pp. 275–317.

[79] J. D. MURRAY, Lectures in Nonlinear Differential Equation Models in Biology, Oxford University Press, Oxford, 1977.

[80] A. I. PACK, D. MURRAY-SMITH, R. J. MILLS, M. HOOPER AND J. TAYLOR, "Application of mathematical models in respiratory medicine," BIMA (1974) pp. 20–26.

[81] M. PAIVA, "Hypotheses underlying a continuous analysis of gas transport in a lung model," BIMA (1978) pp. 17–21.

[82] T. PAVLIDIS, Biological Oscillators: Their Mathematical Analysis, Academic Press, N.Y., 1973.

[83] T. J. PEDLEY, The Fluid Mechanics of Large Blood Vessels, Cambridge University Press, Cambridge, 1980.

[84] D. W. PFAFF, ed., Taste, Olfaction and the Central Nervous System, Rockefeller University Press, N.Y., 1985.

[85] E. C. PIELOU, Biogeography, Wiley, N.Y., 1981.

[86] R. L. PRENTICE AND A. S. WHITTEMORE, Environmental Epidemiology: Risk Assessment, SIAM, 1983.

[87] B. PYE, E. GHOSH AND B. HESS, eds., Biological and Biochemical Oscillators, Academic, N.Y., 1973.

[88] A. RAPOPORT, "Directions in mathematical psychology," AMM (1976) pp 85–106, 153–172.

[89] P. RAPP, "Mathematical techniques for the study of oscillations in biochemical control loops," BIMA (1976) pp. 11–21.

[90] N. RASHEVSKY, Mathematical Biophysics I, II, Dover, N.Y., 1960.

[91] D. M. RAUP, "Probabilistic models in evolutionary paleobiology," Amer. Scientist (1977) pp. 50–57.

[92] D. S. RIGGS, The Mathematical Approach to Physiological Problems, M.I.T. Press, Cambridge, 1970.

[93] R. ROSEN, ed., Foundations of Mathematical Biology I, II, III, Academic Press, N.Y., 1973.

[94] ———, *Optimality Principles in Biology,* Plenum Press, N.Y., 1967.

[95] S. I. RUBINOW, *Introduction to Mathematical Biology,* Wiley, N.Y., 1975.

[96] L. A. SEGAL, *Mathematical Models in Molecular and Cellular Biology,* Cambridge University Press, Cambridge, 1981.

[97] ———, *Modeling Dynamic Phenomena in Molecular and Cellular Biology,* Cambridge, University Press, N.Y., 1984.

[98] M. C. SHEPS AND J. A. MENKIN, *Mathematical Models of Conception and Birth,* University of Chicago Press, Chicago, 1973.

[99] D. P. SMITH AND N. KEYFITZ, *Mathematical Demography,* Springer-Verlag, Heidelberg, 1978.

[100] D. SMITH AND N. KEYFITZ, eds., *Mathematical Demography: Selected Papers,* Springer-Verlag, N.Y., 1977.

[101] J. M. SMITH, *Mathematical Ideas in Biology,* Cambridge University Press, Cambridge, 1971.

[102] G. G. STEEL, *Growth Kinetics of Tumors,* Oxford University Press, Oxford, 1977.

[103] R. THOM, *Mathematical Models of Morphogenesis,* Halsted, N.Y., 1983.

[104] D. W. THOMPSON, *On Growth and Form I, II,* Cambridge University Press, Cambridge, 1942.

[105] C. H. WADDINGTON, ed., *Toward a Theoretical Biology I, II, III, IV,* Edinburgh University Press, Edinburgh, 1972.

[106] T. H. WATERMAN AND H. J. MOROWITZ, eds., *Theoretical and Mathematical Biology,* Blaisdell, Waltham, 1965.

[107] A. WHITTEMORE AND J. B. KELLER, *"Quantitative theories of carcinogenesis,"* SIAM Rev. (1977) pp. 1–30.

[108] N. WIENER AND A. ROSENBLEUTH, *"The mathematical formulation of the problem of the conduction of impulses in a network of connected excitable elements,"* Arch. Inst. Cardiol. Mex. (1946) pp. 205–265.

[109] A. T. WINFREE, *The Geometry of Biological Time,* Springer-Verlag, N.Y., 1980.

[110] ———, *"Sudden cardiac death: A problem in topology,"* Sci. Amer. (1983) pp. 144–161, 170.

Appendix

1. An Unsolved Problem in the Behavioral Sciences

A Variant of Silverman's Board of Directors Problem*

Problem 78-9, by W. AIELLO AND T. V. NARAYANA (University of Alberta).

Suppose we assign positive integer weights to the vote of each member of a board of directors that consists of n members so that the following conditions apply:
   (1) Different subsets of the board always have different total weights so that there are no ties in voting (tie-avoiding).
   (2) Any subset of size k will always have more weight than any subset of size $k-1$ $(k = 1, \cdots, n)$ so that any majority carries the vote, abstentions allowed (nondistorting).

Kreweras, Wynne and Narayana [1] have given a solution (shown in table 1) for $n = 1, \cdots, 7$ that can be extended very easily from any n to $n+1$. It is conjectured that this is a *minimal dominance* solution. Here, an increasing sequence $(y_1, \cdots, y_n)$ is said to *dominate* another increasing sequence $(x_1, \cdots, x_n)$ if $y_i \geq x_i$ $(i = 1, \cdots, n)$. So a solution $(x_1, \cdots, x_n)$ is minimal dominant if no other solution $(y_1, \cdots, y_n)$ exists such that $x_i \geq y_i$ $(i = 1, \cdots, n)$.

TABLE 1
Nondistorting, tie-avoiding integer vote weights W_n.

Members, n	1	2	3	4	5	6	7
Totals, S_n	1	3	9	21	51	117	271
Column vectors of vote weights, W_n	<u>1</u>	2	4	7	13	24	46
		<u>1</u>	3	6	12	23	45
			<u>2</u>	5	11	22	44
				<u>3</u>	9	20	42
					<u>6</u>	17	39
						<u>11</u>	33
							<u>22</u>

The underlined values along the diagonal of vector elements are the I_n values, where:

$$I_1 = I_2 = 1 \quad \text{and} \quad I_{2n+1} = 2I_n,$$

$$I_{2n+2} = 2I_{2n+1} - I_n.$$

REFERENCE

[1] B. WYNNE AND T. V. NARAYANA, *Tournament configuration and weighted voting*, Cahiers du BURO, Paris, to appear.

*Although no Classroom Notes were submitted to *SIAM Review* in this field, because of its importance, I have included an available unsolved problem and a list of general references. *Editor.*

2. General Supplementary References in the Behavioral Sciences

[1] P. ABELL, *Model Building in Sociology*, Schocken, N.Y., 1971.

[2] R. D. ALBA, *"A graph-theoretic definition of a sociometric clique,"* J. Math. Sociol. (1973) pp. 113–126.

[3] K. J. ARROW, *Social Choice and Individual Values*, Yale University Press, New Haven, 1963.

[4] R. H. ATKIN, *Mathematical Structure in Human Affairs*, Crane and Russak, N.Y., 1974.

[5] ———, *Mathematical Structure in Human Affairs*, Heinemann, London, 1974.

[6] R. AXELROD, *The Evolution of Cooperation*, Basic Books, 1984.

[7] M. L. BALINSKI AND H. P. YOUNG, *"Apportionment schemes and the quota method,"* AMM (1977) pp. 450–455.

[8] ———, *Fair Representation: Meeting the Ideal of One Man, One Vote*, Yale University Press, New Haven, 1982.

[9] ———, *"The quota method of apportionment,"* AMM (1975) pp. 701–730.

[10] P. A. BALLONOFF, *Elementary Theory of Minimal Structures: Mathematical Foundations of Social Anthropology*, Mouton, The Hague, 1974.

[11] J. F. BANZHAF III, *"Multi-member electoral districts – do they violate the 'one-man-one vote' principle?"* Yale Law J. (1966) pp. 1309–1388.

[12] ———, *"Weighted voting doesn't work: A mathematical analysis,"* Rutgers Law Rev. (1965) pp. 317–343.

[13] D. BARTHOLOMEW, *Stochastic Models for Social Processes*, Wiley, N.Y., 1967.

[14] M. B. BECK, *Water Quality Management. A Review of the Development and Applications of Mathematical Models*, Springer-Verlag, N.Y., 1985.

[15] E. BELTRAMI, *The High Cost of Clean Water: Models for Water Quality Management*, UMAP/Birkhauser, Boston, 1982.

[16] ———, *Models for Public Systems Analysis*, Academic, N.Y., 1977.

[17] J. O. BERGER, *Statistical Decision Theory*, Springer-Verlag, N.Y., 1980.

[18] G. BIRKOFF, *"Mathematics and psychology,"* SIAM Rev. (1969) pp. 429–469.

[19] D. H. BLAIR AND R. A. POLLAK, *"Rational collective choice,"* Sci. Amer. (1983) pp. 88–95, 128.

[20] H. M. BLALOCK, JR., ed., *Quantitative Sociology: International Perspectives on Mathematical and Statistical Modelling*, Academic, N.Y., 1975.

[21] S. A. BOORMAN AND H. C. WHITE, *"Social structure from multiple networks: II. Role structures,"* Amer. J. Sociol. (1976) pp. 1384–1446.

[22] B. BRAINERD, *Introduction to the Mathematics of Language Study*, Elsevier, N.Y., 1971.

[23] S. J. BRAMS, *Game Theory and Politics*, Free Press, N.Y., 1975.

[24] ———, *The Presidential Election Game*, Yale University Press, New Haven, 1978.

[25] S. J. BRAMS AND P. C. FISHBURN, *Approval Voting*, Birkhauser, Boston, 1983.

[26] T. H. F. BRISSENDEN, *"Some derivations from the marriage bureau,"* Math. Gaz. (1974) pp. 250–257.

[27] J. BROCKNER AND J. Z. RUBIN, *Entrapment in Escalating Conflicts. A Social Psychological Analysis*, Springer-Verlag, N.Y., 1985.

[28] D. N. BURGHES AND A. D. WOOD, *Mathematical Models in the Social, Management and Life Sciences*, Horwood, Chichester, 1980.

[29] R. R. BUSH AND F. MOSTELLER, *Stochastic Models for Learning*, Wiley, N.Y., 1955.

[30] L. L. CAVALLI-SFORZA AND M. W. FELDMAN, *Cultural Transmission and Evolution: A Quantitative Approach*, Princeton University Press, Princeton, 1981.

[31] D. D. CHIRAS, *Environmental Science. A Framework for Decision Making*, Benjamin/Cummings, Menlo Park, 1985.

[32] J. S. COLEMAN, *The Mathematics of Collective Action*, Aldine, Chicago, 1973.

[33] S. COLE, *"Modelling the international order,"* Appl. Math. Modelling (1978) pp. 66–76.

[34] T. M. COOK AND B. S. ALPRIN, *"Snow and ice removal in an urban environment,"* Management Sci. (1976) pp. 227–234.

[35] L. COOLINS, ed., *Use of Models in the Social Sciences*, Tavistock, London, 1974.

[36] A. D. CULLISON, *"Identification by probabilities and trial by arithmetic (a lesson for beginners in how to be wrong with great precision),"* Houston Law Rev. (1969) pp. 471–518.

[37] D. DEUTSCH, *The Psychology of Music*, Academic, N.Y., 1982.

[38] L. E. DUBINS AND D. FREEDMAN, *"Machiavelli and the Gale-Shapely algorithm,"* AMM (1981) pp. 485–494.

[39] T. J. FARARO, *Mathematical Sociology: An Introduction to Fundamentals*, Wiley, N.Y., 1973.

[40] M. O. FINKELSTEIN, *Quantitative Methods of Law*, Free Press, N.Y., 1978.

[41] P. C. FISHBURN, *"Discrete mathematics in voting and group choice,"* SIAM J. Alg. Discrete Methods (1984) pp. 263–275.

[42] ———, *Mathematics of Decision Theory*, Mouton, The Hague, 1973.

[43] ———, *"Montonicity paradoxes in the theory of elections,"* Discrete Appl. Math. (1982) pp. 119–134.

[44] ———, *"Social choice functions,"* SIAM Rev. (1974) pp. 63–90.

[45] ——, *The Theory of Social Choice,* Princeton University Press, Princeton, 1973.

[46] P. C. FISHBURN AND S. J. BRAMS, "Paradoxes of preferential voting," MM (1983) pp. 207–214.

[47] D. S. FREEDMAN, R. FREEDMAN AND P. K. WHELPTON, "Size of family and preference for children of each sex," Amer. J. Sociology (1960) pp. 141–146.

[48] H. FREUDENTHAL, ed., *The Concept of the Model in Mathematics and the Social Sciences,* Riedel, Dordrecht, 1961.

[49] D. GALE AND L. S. SHAPELY, "College admissions and the stability of marriage," AMM (1962) pp. 9–14.

[50] D. GALE AND M. SOTOMAYOR, "Ms. Machiavelli and the stable matching problem," AMM (1985) pp. 261–268.

[51] M. W. GRAY, "Statistics and the law," MM (1983) pp. 67–81.

[52] D. F. GREENBERG, *Mathematical Criminology,* Rutgers University Press, New Brunswick, 1977.

[53] B. GROFMAN, "Fair apportionment and the Banzhaf index," AMM (1981) pp. 1–5.

[54] M. GROSS, *Mathematical Models in Linguistics,* Prentice-Hall, N.J., 1972.

[55] T. R. GURR, *Polimetrics: An Introduction to Macropolitics,* Prentice-Hall, N.J., 1972.

[56] P. HAGE AND F. HARARY, *Structural Models in Anthropology,* Cambridge University Press, Cambridge, 1983.

[57] R. L. HAMBLIN, R. B. JACOBSEN AND J. L. MILLER, *A Mathematical Theory of Social Change,* Wiley, N.Y., 1973.

[58] J. C. HARSANYI, *Rational Behavior and Bargaining Equilibrium in Games and Social Situations,* Cambridge University Press, Cambridge, 1977.

[59] P. HAYES, *Mathematical Models in the Social and Management Sciences,* Wiley, N.Y., 1975.

[60] M. HENDRIKSEN AND G. ORLAND, "On the juror utilization problem," Jurimetrics J. (1976) pp. 318–333.

[61] W. W. HILL, JR., "Prisoner's dilemma, a stochastic solution," MM (1975) pp. 103–105.

[62] J. HOLLAND AND M. D. STEUR, *Mathematical Sociology: A Selective Annotated Bibliography,* Socken, N.Y., 1979.

[63] A. J. JONES, *Game Theory: Mathematical Models of Conflict,* Wiley, N.Y., 1980.

[64] D. KAHNEMAN AND D. TVERSKY, "The psychology of preferences," Sci. Amer. (1982) pp. 160–173, 180.

[65] J. S. KELLY, *Arrow Impossibility Theorems,* Academic, N.Y., 1978.

[66] J. G. KEMENY AND J. L. SNELL, *Mathematical Models in the Social Sciences,* Blaisdell, Waltham, 1962.

[67] W. F. KEMP AND B. H. REPP, *Mathematical Models for Social Psychology,* Wiley, N.Y., 1977.

[68] C. W. KILMISTER, "Mathematics in the social sciences," in R. DUNCAN AND M. WESTON-SMITH, eds., *The Encyclopedia of Ignorance,* Pergamon, Oxford, 1977.

[69] A. KLEINER, "A weighted voting model," MM (1980) pp. 28–32.

[70] D. H. KRANTZ, R. D. LUCE, P. SUPPES AND A. TVERSKY, *Foundations of Measurement,* Academic, N.Y., 1971.

[71] R. H. KUPPERMAN AND H. A. SMITH, "The role of population defense in mutual deterrence," SIAM Rev. (1977) pp. 297–318.

[72] D. LAMING, *Mathematical Psychology,* Academic, N.Y., 1973.

[73] C. A. LAVE AND J. G. MARCH, *An Introduction to Models in the Social Sciences,* Harper and Row, N.Y., 1975.

[74] R. D. LUCE, "The mathematics used in psychology," AMM (1964) pp. 364–378.

[75] M. L. MANHEIM, *Hierarchical Structure: A Model of Planning and Design Processes,* M.I.T. Press, Cambridge, 1962.

[76] S. MARCUS, *Algebraic Linguistics: Analytical Models,* Academic, N.Y., 1967.

[77] G. McLACHLAN, ed., *Measuring for Management: Quantitative Methods in Health Services Management,* Oxford University Press, London, 1975.

[78] S. MERRILL, "Approval voting: A 'best buy' method for multi-candidate elections?" MM (1979) pp. 98–102.

[79] A. MIZRAHI AND M. SULLIVAN, *Finite Mathematics with Applications for Business and Social Sciences,* Wiley, N.Y., 1973.

[80] E. W. MONTROLL AND W. W. BADGER, *Introduction to Quantitative Aspects of Social Phenomena,* Gordon and Breach, N.Y., 1974.

[81] C. MUELLER, *Public Choice,* Cambridge University Press, Cambridge, 1979.

[82] M. OLINICK, *An Introduction to Mathematical Models in the Social and Life Sciences,* Addison-Wesley, Reading, 1978.

[83] H. H. PATTEE, ed., *Hierarchy Theory, The Challenge of Complex Systems,* Braziller, N.Y., 1973.

[84] M. H. PEARL AND A. J. GOLDMAN, "Policing the market place," MM (1977) pp. 179–185.

[85] D. W. RAE, *The Political Consequences of Election Laws,* Yale University Press, New Haven, 1971.

[86] A. RAPOPORT, *Mathematical Models in the Social and Behavioral Sciences,* Wiley, N.Y., 1983.

[87] ——, "Use of mathematics outside the physical sciences," SIAM Rev. Suppl. (1973) pp. 481–502.

[88] C. RENFREW AND K. L. COOKE, eds., *Transformations: Mathematical Approaches to Culture Change,* Academic, N.Y., 1979.

[89] F. RESTLE, *Mathematical Models in Psychology,* Penguin, England, 1971.

[90] F. RESTLE AND J. GREENE, *Introduction to Mathematical Psychology,* Addison-Wesley, Reading, 1970.

[91] F. S. ROBERTS, *Measurement Theory with Applications to Decision-Making, Utility, and the Social Science,* Addison-Wesley, Reading, 1979.

[92] ——, *Graph Theory and Its Applications to Problems of Society,* SIAM, Philadelphia, 1978.

[93] M. R. ROSENZWEIG AND M. R. LEIMAN, *Physiological Psychology*, Heath, Boston, 1982.

[94] D. G. SAARI, *"Apportionment methods and the house of representatives,"* AMM (1978) pp. 792–802.

[95] T. L. SAATY, *The Analytic Hierarchy Process*, McGraw-Hill, N.Y., 1980.

[96] S. B. SEIDMAN, *"Models for social networks: Mathematics in Anthropology and Sociology,"* UMAP J. (1985) pp. 19–36.

[97] H. A. SELBY, ed., *Notes of Lectures on Mathematics in the Behavioral Sciences*, MAA, 1973.

[98] G. A. SHAFER, *A Mathematical Theory of Evidence*, Princeton University Press, Princeton, 1976.

[99] R. N. SHEPARD AND L. A. COOPER, *Mental Images and their Transformations*, M.I.T. Press, Cambridge, 1982.

[100] H. A. SIMON, *"The uses of mathematics in the social sciences,"* Math. and Computers in Simulation (1978) 159–166.

[101] J. M. SMITH, *"The evolution of behaviour,"* Sci. Amer. (1978) pp. 176–194, 242.

[102] P. D. STRAFFIN, *Topics in the Theory of Voting*, Birkhauser, Boston, 1980.

[103] P. D. STRAFFIN, JR. AND B. GROFMAN, *"Parlimentary coalitions: A tour of models,"* MM (1984) pp. 259–274.

[104] R. M. STOGDILL, ed., *The Process of Model-Building in the Behavioral Sciences*, Ohio University Press, Columbus, 1970.

[105] L. H. TRIBE, *"Trial by mathematics, Precision and ritual in the legal process,"* Harvard Law Rev. (1971) pp. 1329–1391.

[106] E. R. TUFTE, ed., *The Quantitative Analysis of Social Problems*, Addison-Wesley, Reading, 1970.

[107] S. VAJDA, *Mathematics of Manpower*, Wiley, N.Y., 1978.

[108] H. M. WAGNER, *Principles of Management Science*, Prentice-Hall, N.J., 1975.

[109] B. J. WEST, ed., *Mathematical Models as a Tool for the Social Sciences*, Gordon and Breach, N.Y., 1980.

[110] H.C. WHITE, S. A. BOORMAN AND R. L. BREIGER, *"Social structures from multiple networks: I. Blockmodels for roles and positions,"* Amer. J. Sociol. (1976) pp. 730–779.

[111] A. WUFFLE, *"Mo Florina's advice to children and other subordinates,"* MM (1979) pp. 292–297.

[112] J. E. WARD II, *"The probability of election reversal,"* MM (1981) pp. 256–259.

[113] H. J. ZIMMERMAN, *"Testability and meaning of mathematical models in social sciences,"* Math. Modelling (1980) pp. 123–139.

[114] D. A. ZINNES AND J. V. GILLESPIE, *Mathematical Models in International Relations*, Praeger, N.Y., 1976.